NITROGEN TURNOVER IN THE SOIL-CROP SYSTEM

Developments in Plant and Soil Sciences

VOLUME 44

The titles published in this series are listed at the end of this volume.

Nitrogen Turnover
in the Soil-Crop System

Modelling of Biological Transformations,
Transport of Nitrogen and Nitrogen Use Efficiency.
Proceedings of a Workshop held at the Institute for Soil Fertility Research,
Haren, The Netherlands, 5–6 June 1990

Edited by
J.J.R. GROOT, P. DE WILLIGEN and E.L.J. VERBERNE
Institute for Soil Fertility Research
Oosterweg 92
9750 RA Haren, The Netherlands

Reprinted from *Fertilizer Research*, Volume 27 (2–3) 1991

Springer-Science+Business Media, B.V.

Library of Congress Cataloging-in-Publication Data

Nitrogen turnover in the soil-crop system : modelling of biological
  transformations, transport of nitrogen and nitrogen use efficiency /
  edited by J.J.R. Groot, P. de Willigen, E.L.J. Verberne.
       p.   cm. -- (Developments in plant and soil sciences ; v. 44)
     "Proceedings of a workshop held at the. Inst. for Soil Fertility
  Research, Haren, The Netherlands, 5-6 June 1990."
    ISBN 978-94-010-5518-5   ISBN 978-94-011-3434-7 (eBook)
    DOI 10.1007/978-94-011-3434-7
    1. Soils--Nitrogen content--Computer simulation--Congresses.
  2. Crops and soils--Computer simulation--Congresses. 3. Plants,
  Effect of nitrogen on--Computer simulation--Congresses. 4. Crops-
  -Nutrition--Computer simulation--Congresses. 5. Nitrogen cycle-
  -Computer simulation--Congresses.   I. Groot, J. J. R., 1958-   .
  II. Willigen, P. de. III. Verberne, E. L. J., 1963-   .
  IV. Series.
  S592.6.N5N58 1991
  631.8'47--dc20   90-26527

Contents

Financial support from the following Institutions is gratefully acknowledged:

Dutch Fertiliser Research Institute, The Hague, The Netherlands Fertiliser Manufacturers Association, Peterborough, England, Norsk Hydro Agrar, Agricultural Research Station Hanninghof, Dülmen, Germany, Commission of the European Communities, Directorate VI Agriculture.

Fertilizer Research **27**: vii, 1991.

Preface

In the Netherlands the Institute for Soil Fertility Research plays a major role in soil biological, soil physical and plant nutritional research on the availability of nitrogen to crops. Main subjects of research are nitrogen turnover in the crop-soil ecosystem through biological transformations, nitrogen transport through the soil and nitrogen losses by leaching, denitrification and volatilization, and nitrogen use efficiency of various crops and cropping systems. The current knowledge in the different fields of research is integrated in simulation models. Simulation models not only make it possible to summarize and structure knowledge, but also, after verification, to extrapolate the knowledge to situations different from the situations that have actually been studied.

Such research is also carried out in other European and non-European countries. To compare the various simulation models currently in use, a workshop was organized by the Institute for Soil Fertility Research on 5–6 June 1990 on the occasion of its centennial. The title of the workshop was 'Nitrogen turnover in the soil-crop ecosystem: modelling of biological transformations, transport of nitrogen and nitrogen use efficiency'. The 40 participants, who came from Canada and various European countries, were requested to run their model with data provided by the Institute prior to the workshop. Data from 18 cases were made available to the participants: three locations, three treatments, and two seasons. It was left to the participants to choose which cases to compute and in which form to present the results. The main purpose of the workshop was to bring together scientists working within the same field, and to provide them with an opportunity to present their work in detail and get acquainted with that of others.

The workshop was organized by J.J.R. Groot, P. de Willigen and E.L.J. Verberne, scientists at the Department of Fertilization and Plant Nutrition of the Institute for Soil Fertility Research. This volume contains the papers presented at the Workshop, as well as a paper on the data set supplied to the participants and a summarizing paper. Being a member of the editorial board of Fertilizer Research, I am deeply impressed by the way of this volume was produced by the editors. Publication of these very well edited proceedings so soon after the workshop is a major achievement. I also wish to compliment the authors of the papers on their outstanding contributions, which made this volume a 'must' for everyone working on modelling the nitrogen cycle in the soil-crop ecosystem.

J. J. Neeteson
Head, Department of Fertilization and Plant Nutrition
Institute for Soil Fertility Research,
Haren, The Netherlands
October 1990

Fertilizer Research **27**: 141–149, 1991.
© 1991 *Kluwer Academic Publishers.*

Nitrogen turnover in the soil-crop system; comparison of fourteen simulation models

P. de Willigen
Institute for Soil Fertility Research, P.O. Box 30003, 9750 RA Haren (Gn.), The Netherlands

Key words: Crop growth, leaching, nitrogen turnover, nitrogen uptake, simulation models

Abstract

Fourteen models of the nitrogen turnover in the soil-crop system were compared. The comparison comprised the processes included in the models, the methods of process description, as well as the results of simulations carried out with the same data set. It is concluded that simulation of the aboveground processes was less problematic than that of the belowground processes. None of the models could account for the loss of mineral nitrogen occurring shortly after application of fertilizer in late spring and early summer.

Introduction

It has long been known that nitrogen is one of the elements of which supply is most often limiting growth. Traditionally agricultural research has focussed on how to provide the crop with sufficient nitrogen to guarantee optimum yields. If efficiency was taken into account, the reason was usually to maximize financial returns rather than to minimize the danger of leaching. Nowadays agriculture is faced with a growing popular demand for a higher priority to prevention of minimization of pollution. The challenge for agricultural research is to devise production methods which hold no risk of losses causing pollution but which at the same time lead to economically acceptable yield levels. Simulation models, which describe in sufficient detail the turnover of nitrogen in the soil-crop system, can be of great assistance in understanding the interactions between the different processes involved. Also, they can point to gaps in our knowledge, and in doing so, can help in designing experiments that aim at clarification of poorly understood parts of the problem.

To celebrate its centennial the Institute for Soil Fertility Research organized a workshop on 5–6 June 1990. The purpose was to compare various simulation models of nitrogen turnover in the soil-crop system. The participants were requested to run their model with data provided by the Institute prior to the workshop. The main purpose of the workshop was to bring together scientists working in the same field, and to provide them with an opportunity to present their work in some detail and get acquainted with that of others. The data furnished were used to demonstrate the performance of the respective models.

This paper surveys the models presented, and summarizes the results. We will compare the models in terms of the processes they represent and of their description of these processes. Further we will point out for what purposes they seem best suited.

Description of models

In all, 18 papers were presented. Fourteen dealt with the data supplied. Table 1 gives the list of these 14 models, together with the names of the

Table 1. List of participating models

Code	Name of first author	Pages where model is discussed (Fert Res)	References for further information
[A]	Cabon	161–169	(1), (2), (3)
[B]	Ramos	171–180	(4)
[C]	Bergström	181–188	(5), (6)
[D]	Rijtema	189–198	(7)
[E]	Grant	199–213	(8), (9), (10), (11), (12)
[F]	Lafolie	215–231	—
[G]	Vereecken	233–243	(6), (13), (14)
[H]	Hansen	245–259	(15)
[I]	Groot	261–272/349–383	(16)
[J]	Kersebaum	273–281	(17)
[K]	Whitmore	283–291	(18), (19)
[L]	Mirschel	293–304	(20), (21)
[M]	Addiscott	305–312	(18)
[N]	Eckersten	313–329	(22), (15), (16)

References for further information, corresponding with Table 1.
 (1) Geng QZ (1988) Modélisation conjointe du cycle de l'eau et du transfert des nitrates dans un système hydrologique. Thèse de doctorat, Ecole Normale Supérieure des Mines de Paris, Paris
 (2) Girard G (1975) Modèle global ORSTROM 1974. Première application de modèle à discrétisation spatiale sur le bassin versant de la crique Grégoire en Guyanne. Atelier hydrologique sur les modèles mathématiques, ORSTROM
 (3) Ledoux E (1980) Modélisation intégrée des écoulements de surface et des écoulements souterrains sur un bassin hydrologique. Thèse de Docteur-Ingenieur, Ecole Nationale Supérieure des Mines de Paris, Paris
 (4) Wagenet RJ and Hutson JL (1989) LEACHM: Leaching Estimation and Chemistry Model: A process based model of water and solute movement transformations, plant uptake and chemical reactions in the unsaturated zone. Continuum Vol 2 Version 2 Water Resources Inst, Cornell Univ, Ithaca, NY
 (5) Jansson PE and Halldin S (1979) Model for annual water and energy flow in layered soil. In: Halldin S (ed) Comparison of Forest Water and Energy Exchange Models, pp 145–163. Copenhagen: International Society for Ecological Modelling
 (6) Johnsson H, Bergström L, Jansson PE and Paustian K (1987) Simulated nitrogen dynamics and losses in a layered agricultural soil. Agric Ecosystems Environ 18: 333–356
 (7) Rijtema PE, Roest CWJ and Kroes JG (1990) Formulation of the nitrogen and phosphate behaviour in agricultural soils, the ANIMO model. Report 30 (in press), Wageningen, The Netherlands: The Winand Staring Centre
 (8) Grant RF (1989) Simulation of carbon accumulation and partitioning in maize. Agron J 81: 563–571
 (9) Grant RF (1989) Test of a simple biochemical model for photosynthesis of maize and soybean leaves. Agric For Meteorol 48: 59–74
(10) Grant RF (1989) Simulation of maize phenology. Agron J 81: 451–457
(11) Grant RF, Peters DB, Larson EM and Huck MG (1989) Simulation of canopy photosynthesis in maize and soybean. Agric For Meteorol 48: 75–91
(12) Grant RF (1990) Dynamic simulation of water deficit effects upon maize yield. Agric Systems (in press)
(13) Keulen H van, Penning de Vries FWT and Drees EM (1982) A summary model for crop growth. In: Penning de Vries FWT and van Laar HH (eds) Simulation of Crop Growth and Crop Production, pp 87–97. PUDOC Wageningen
(14) Vereecken H, Vanclooster M and Swerts M (1989) A model for the estimation of nitrogen leaching with regional applicability. Presented at the Int Congress 'Fertilizer and the Environment', pp 13. August 27–30, 1989, Leuven, Belgium
(15) Hansen S, Jensen HE, Nielsen NE and Svendsen H (1990) DAISY: A soil plant system model. Danish simulation model for transformation and transport of energy and matter in the soil plant atmosphere system, 350 pp (in press)
(16) Groot JJR (1987) Simulation of nitrogen balance in a system of winter wheat and soil. Simulation Reports CABO-TT no. 13, Centre for Agrobiological Research and Department of Theoretical Production Ecology, Wageningen, 69 pp
(17) Kersebaum KC (1989) Die Simulation der Stickstoff-Dynamik von Ackerböden. Ph. D. Thesis University Hannover
(18) Addiscott TM and Whitmore AP (1987) Computer simulation of changes in soil mineral nitrogen and crop nitrogen during autumn, winter and spring. J Agric Sci, Camb 109: 141–157
(19) Whitmore AP and Parry LC (1988) Computer simulation of the behaviour of nitrogen in soil and crop in the Broadbalk continuous wheat experiment. In : Jenkinson DS and Smith KA (eds) Nitrogen Efficiency in Agricultural Soils, pp 418–432. Elsevier, London
(20) Matthäus E, Mirschel W, Kretschmer H, Künkel K and Klank I (1986) The winter wheat crop model TRITSIM of the agroecosystem AGROSIM-W. In: Tag.-Ber. der ADL der DDR (Computer Aided Modelling and Simulation of the Winter Wheat Agroecosystem (AGROSIM-W) for Integrated Pest Management) (242), pp 43–74

(21) Mirschel W, Kretschmer H, Klank I, Matthäus E and Künkel K (1987) Dynamisches Ertragsbildungs- und Entwicklungsmodell TRITSIM für Winterweizen. 2. Mitteilung: Modellvalidierung und -verifizierung. Arch Acker- Pflanzenbau Bodenkd 31: 249–257

(22) Eckersten H and Slapokas T (1990) Biomass production and nitrogen turnover in an irrigated short rotation forest. Agric For Meteor 50: 99–123

first author and a reference to one or more publications in which a detailed description of the model can be found. The order of the models in Table 1 corresponds to the sequence in which they appear in this volume. Both the model and the paper in which its results are shown will be indicated by [A] . . . . [N].

The crop-soil system has as its upper boundary the top of the canopy. As lower boundary can be taken the maximum depth of the root system or a barrier such as the water table which roots cannot penetrate. However, as was pointed out in [3], canopy height is not very important for aboveground resource availability and plant production of a cropped field. Rooting depth, on the other hand, influences the size of the pool of resources available to the crop. Accordingly none of the models presented here explicitly recognizes canopy height as an upper boundary of the system. However, all define a certain depth in the soil profile as a lower boundary. Indeed, many of them deal with the soil only, describing the transfer of mass from the soil to the crop and the atmosphere as sink or source terms in balance equations of the soil water and nitrogen content. All models deal with one-dimensional geometry of the soil only, but [D] contains a pseudo two-dimensional transport module to calculate solute transport to different drainage systems. Except for [A], all models subdivide the soil profile into a number of horizontal layers. Many processes are involved in the nitrogen turnover in the soil-crop system. These influence directly and indirectly the distribution of nitrogen in its component fractions throughout the soil and the crop. As a primary distinction here, we discriminate between soil processes, processes pertaining to aboveground plant growth, and nitrogen and water uptake.

Soil processes

These comprise physical (transport of water, solutes, heat, and gases), biological (mineralization/immobilization, nitrification, and denitrification), and physicochemical (volatilization, adsorption, fixation) processes.

Physical processes

The transport of water, carrying solutes, either vertically, downwards to deeper layers or upwards to the soil surface, or horizontally to the roots, affects the concentration and amount of nitrogen in the root zone. Moreover, the soil water content reflects the transport, extraction and replenishment of water. It has a large influence upon the rates of many of the processes involved in the nitrogen cycle. Therefore, the inclusion of data on modelling of the time course of water content and soil water flux distribution is a prerequisite for models aiming at a description of the dynamics of soil nitrogen in the field. All models, except two, contain a submodel which calculates water flux and water content. The two exceptions ([C] and [D]) do not model soil water dynamics as such. They are meant to be used in conjunction with a hydrological model which produces the information required.

Six of the models calculate the water flux in a fundamental way using Richards' equation ([B], [E], [F], [G], [H] and [N]), the other six use simplified treatments ([A], [I], [J], [K], [L], [M]). The former are called 'mechanistic' and the latter 'functional' [1]. The functional models define a constant upper and lower limit to the water content. The upper limit usually corresponds to a pressure head of -10 to $-30\,kPa$, while the lower limit is given by the water content at the permanent wilting point ($-1.5\,MPa$). The water content cannot have values beyond these limits. Model [N] includes the possibility of calculating bypass flow in macropores.

Transport of solutes is calculated either by simply taking the product of flux and concentration (in [A], [C], [I], [J], [K], [L] and [M]), where concentration is sometimes (in model [I] and [J]) computed as the weighted average of current concentration and incoming concentration. Models [K] and [M] consider mobile and immobile phases of water and solutes. In all the other models, solute transport is calculated by means of the classical convection/dispersion equation. It should be mentioned that model [F] has two options to describe the transport of solutes, one using the convection/dispersion equation and the other making a distinction between mobile and immobile phases.

There are other soil physical processes which create physical conditions which are – in addition to the water content – paramount in influencing the rate and occurrence of important processes in the nitrogen cycle: transport of heat and oxygen.

Heat transport is considered in seven models ([B], [D], [E], [F], [G], [H] and [N]), model [C] requires the results of a heat flow model – similar to that used in model [N] – as input. The other models use measured soil temperatures, if available, or calculate them as a function of the measured air temperatures.

Only two models, [D] and [E], calculate the oxygen concentration in soil air, the other models ignore possible effects of soil oxygen status, although models [C] and [G], in calculating denitrification rate, utilize a function of soil moisture content as an implicit measure of soil oxygen status.

Biological processes
All models include the mineralization of organic forms of nitrogen to ammonium. Most contain a procedure to calculate immobilization as well. There are considerable differences in the description of mineralization/immobilization. Models [B], [L] and [M] describe mineralization as a zero-order process having a constant rate for constant environmental conditions. The other models all use first-order rate kinetics but consider different fractions of organic nitrogen. Model [A] and [F] treat the organic N as if it consists of one pool only, the models [C], [G], [I], [J] and [N] identify two pools – at least in the application to the data provided at the workshop – , whereas models [D], [E], [H], [I] and [K] divide organic nitrogen into three or more pools. Models [E] and [H] employ the most extensive description and distinguish more than five pools. Only four models, [D], [E], [H] and

[K], take explicitly account of one or more pools of biomass.

The microbial oxidation of ammonium, nitrification, is considered by nine models (i.e. [A], [B], [C], [D], [E], [F], [G], [H] and [N]). The others assume mineralization to proceed much slower than nitrification so that within the integration time-step the resulting inorganic form of nitrogen can be assumed to be nitrate. None of the models consider the formation of nitrite as intermediate product of nitrification.

Denitrification – the reduction of nitrate to gaseous nitrogen oxides and eventually to elemental nitrogen – is treated by seven models ([B], [C], [D], [E], [G], [H] and [N]). Models [B] and [G] describe denitrification as a first-order process with respect to nitrate concentration. The other models take into account the oxygen concentration in the soil air, either by calculating the oxygen concentration of soil air from oxygen transport and consumption (models [D] and [E]), or by using a function of the soil moisture content as a measure of anaerobiosis (models [C], [L] and [N]).

In all models except model [B] the rate at which the processes proceed is corrected for temperature and soil moisture content. Table 2

Table 2. Soil biological processes considered by the different models

Model	(a1)	(a2)	(a3)	(b)	(c)
[A]	1	1	–	+	–
[B]	1	0	–	+	+
[C]	2	1	–	+	+
[D]	4	1	+	+	+
[E]	6	1	+	+	+
[F]	1	1	–	+	–
[G]	2	1	–	+	+
[H]	7	1	+	+	+
[I]	2	1	+	–	–
[J]	2	1	–	–	–
[K]	4	1	+	–	–
[L]	1	0	–	–	–
[M]	1	0	–	–	–
[N]	2	1	–	+	+

(a1): number of pools in mineralization/immobilization reaction
(a2): order of the rate equations
(a3): + one or more pools of biomass; – no biomass
(b): + nitrification considered
(c): + denitrification considered

gives an overview of the soil biological processes considered by the different models.

Physicochemical processes

The physicochemical processes relevant for the description of the nitrogen balance in soil (adsorption, volatilization, and fixation within clay minerals) pertain to ammonium. Obviously those models which do not consider the occurrence of ammonium (i.e. [I]–[M]) ignore these processes. Adsorption is treated in six models ([A], [B], [D], [E], [F] and [G]) and volatilization in five models ([B], [D], [E], [G] and [J]). Fixation is not treated by any of the models.

Nitrogen and water uptake

The uptake of mineral nitrogen by the crop is treated very differently. In discussing the formulation of uptake of nitrogen as well as that of water, we shall treat also the transport of nitrogen and water from the bulk soil to root surface. Strictly speaking, however, this process should be discussed under the heading 'Soil Processes'. Models [B] and [E] consider uptake to be determined by the concentration in the soil solution. Model [B] assumes it to be proportional to the concentration, whilst model [E] uses a Michaelis-Menten relation to calculate uptake as a function of concentration. However, most of the models ([A], [C], [F], [G], [H], [I], [J], [K], [L], [M], [N]) consider uptake as driven principally by plant demand. For the models [A] and [F] plant demand must be given as input. Models [C], [K] and [M] calculate demand with a logistic curve. Here the maximum amount of nitrogen to be taken up must be given or estimated a priori. The other models ([G], [H], [I], [J], [L] and [N]) calculate demand from growth rates and the nitrogen content required by the various plant organs. To calculate the actual uptake, models [A] and [F] use a Michaelis-Menten function. Model [C] assumes that only a certain fraction of mineral nitrogen in soil is available for uptake. Models [K] and [M] also assume that a fraction of the soil mineral nitrogen is available, but this fraction decreases with depth. Model [L] assumes that all mineral nitrogen in the root zone exceeding 10 kg ha^{-1} is available for uptake. Except for model [A] and [L], all models take root

Table 3. Features of submodels dealing with N uptake

Model	(a)	(b)	(c)	(d)
[A]	−	+	−	−
[B]	−	−	+	?
[C]	−	+	+	+
[D]	−	?	−	+
[E]	+	−	+	−
[F]	−	+	+	−
[G]	+	+	+	+
[H]	+	+	+	+
[I]	+	+	+	−
[J]	+	+	+	−
[K]	−	+	+	−
[L]	−	+	−	−
[M]	−	+	+	−
[N]	−	+	+	+

(a): + considers transport from bulk soil to root surface
(b): + uptake principally demand-driven
    − uptake determined by concentration
(c): + vertical root density distribution taken into account
(d): + considers both ammonium and nitrate uptake
    − considers nitrate uptake only

distribution into account in describing N uptake. Model [D], however, using transpiration fluxes as input, relies in this respect on the hydrological model it uses. As far as transport through the soil is concerned, models [A], [B], [C], [D], [F], [K], [L], [M] and [N] assume the concentration at the root surface to be the same as in the bulk soil. The other models account for transport from bulk soil towards the root, either by using a steady-state ([E], [G], [H], [J]) or a steady-rate approach ([I]). Table 3 summarizes the discussion of nitrogen uptake.

The uptake of water is driven by atmospheric demand in all models. However, there is provision to correct the potential uptake rate when water is not freely available. Availability of water is considered by defining a limiting value of the water content below which availability decreases with water content ([A], [I], [J], [K], [M], [N]). Models [B], [E], [F], [G] and [H] take transport from bulk soil to root surface into account, using steady-state equations. Rooting density is taken into account for calculating water uptake in models [B], [C], [E], [F], [H] and [J].

Aboveground plant growth

Some of the models ([I], [L]) were developed

from others that originally sought to simulate the growth and development of wheat. Therefore, they have a detailed description of processes associated with growth and development. The majority of the models, however, emanated from models primarily designed to calculate either the nitrogen and/or the water balance of the soil: the models [A], [B], [C], [D], [F], [G], [J], [K] and [M]. These models concentrate on soil processes and deal with plant processes only in a sketchy way. More recently, some of these (i.e. [G] and [J]) have been extended to include a submodel describing plant growth in a more detailed way. Models [E], [H] and [N] appear to have been set up primarily for the purpose of integrating plant and soil processes.

Models [A], [C], [D] and [F] do not consider plant growth in any explicit way. Models [B], [K] and [M] simply use a logistic curve to describe dry matter production and root growth. In the models [E], [G], [H], [I], [J], [L] and [N], plant growth and development is dealt with in detail. These models account for gross photosynthesis, respiration, distribution of dry matter and nitrogen over the different organs, and the phenological stage of the wheat crop. Three of these models (i.e. [G], [I] and [J]) use essentially the same growth and development submodel which is based upon the SUCROS model [6] for describing growth and development.

Results

Data from 18 cases were provided (three locations – Bouwing, Eest and PAGV –, two seasons – 1983 and 1984 –, and three treatments, see [4]). It was left to the participants to choose a case and how to present the results. It turns out that not one case was used by all participants. This makes a straightforward comparison of the performance of the models impossible. Nevertheless there is sufficient overlap in the participants' choice to make it possible to compare the results. For various reasons the majority of the models dealt with the data of the season 1983–1984, so we confine our attention to results pertaining to that season only.

For assessing model performance four variables were chosen, namely the soil water content, soil mineral nitrogen content, dry matter production and nitrogen uptake.

Soil water content

Information about the time course of soil water content is given by only 5 contributions.

For the Bouwing location, papers [A], [B], [F] and [G] show results on the simulated water content. Models [B], [F] and [G] overestimate the water content (see Table 3 in [B], Fig. 4 of [F] and Fig. 1 of [G]). These models are mechanistic in the definition proposed in [1], using Richards' equation, and in the relationship between water content and pressure head, as well as between water content and hydraulic conductivity. This type of models is quite sensitive to the parameters that describe the form of these non-linear relations. Models [F] and [G] fitted to the pressure head and water content data a function, first proposed by Van Genuchten [7]. Different results were obtained: the saturated water content of the upper 40 cm was estimated to be $0.533 \, cm^3 \, cm^{-3}$ in [F] and $0.494 \, cm^3 \, cm^{-3}$ in [G]. The functional model presented in [A] seems to perform reasonably well (Fig. 4 in [A]), although in summer it underestimates the water content. This was probably due to an overestimation of evapotranspiration.

Results for the Eest are shown in [A] (Fig. 7), [B] (Table 3), [D] (Fig. 4) and [N] (Fig. 4). Again model [A] underestimates and [B] overestimates the water content. In [N] (Figs. 4a and b) results are shown both when bypass flow is taken into account and when it is ignored. For the simulated water contents in the unsaturated soil this does not make much difference, but groundwater recharge is much better simulated when bypass flow is considered.

The presentation of the results for the location of PAGV in [B] (Table 3), [G] (Fig. 2) and [I] (Fig. 7) makes it possible to compare the performances of a functional model and a mechanistic model. All three models generally overestimate the water content in the winter period but, overall, the functional model [I] appears to do better.

As the soil water retention data for the PAGV site were not derived for this experimental field but pertain to a 'comparable' soil, part of the deviations between model predictions and measurements can be due to incorrect parameterization of the models.

Soil mineral nitrogen content

The results for the N1 treatment on the Bouwing location in 1984 are shown by [A] (Fig. 5), [B] (Table 4), [F] (Fig. 7), [G] (Figs. 4–7), [J] (Fig. 5), [K] (Fig. 1) and [M] (Fig. 2). All models appear to give satisfactory results, although the second option of model [F] for describing solute transport, where mobile and immobile phases are taken into account, produced much better results than the option in which solute transport is described by convection/dispersion. The results for treatments N2 and N3, shown in [A] (Fig. 5), [F] (Fig. 8), [K] (Fig. 1) and [M] (Fig. 2) are less satisfactory, mainly due to the fact that the measured data did not reflect the effects of the second application of fertilizer (60 and 120 kg ha^{-1}, respectively). This phenomenon of 'apparent loss' of fertilizer nitrogen has also been observed in other treatments.

The papers [A], [B], [C], [D], [G], [H], [K], [L] and [M] furnish information about the simulated results on the soil mineral nitrogen content at the Eest location in 1984. The measured data, again, show an unaccountable loss of fertilizer nitrogen. The first application (on 17 February) of fertilizer nitrogen resulted in a corresponding rise in soil mineral nitrogen, whereas the second and third applications did not. In the simulated results all the applications manifest themselves as peaks in the soil mineral nitrogen, at least for the models [D] (Figs. 7a and b), [H] (Fig. 8b), [K] (Fig. 3), [L] (Fig. 11) and [M] (Fig. 4). Model [C] assumes that no less than 20% of the nitrogen taken up is found in the roots. It therefore calculates a very high uptake rate and consequently the second and third application only lead to relatively small peaks, which is also due to a serious underestimation of the increase in soil mineral nitrogen following the first application. In [C] (Table 1), [D] (Fig. 5), [G] (Table 5) and [H] (Table 2) the components of the nitrogen balance are presented. Though, unfortunately, different periods are involved, still some comparisons can nonetheless be made. According to all models, leaching played but a minor role. Model [C] predicts that the amount leached in the period 1 February-31 December 1984 is at most 19 kg ha^{-1}, model [D] dealing with the period 16 February-31 August predicts no leaching at all, model [H] dealing with approximately the same period predicts a loss of 1 kg ha^{-1}, and model [G] predicts a loss of at most 7 kg ha^{-1} during a 300 day period starting in October 1983. There is a striking difference in the loss by denitrification predicted by model [D] as compared with the other models ([C], [G] and [H]), which predict denitrification losses in the order of 15–20 kg ha^{-1}, whereas model [D], considering only a relatively short period, arrives at a loss of 85 kg ha^{-1}.

Results for the PAGV location are shown by [D], [H], [I], [K] and [M]. As at the other locations, the measured data show the expected increase in soil mineral nitrogen content after the first fertilizer application in February but fail to do so for the second and third applications (see Figs. 13a–c in [D], Figs. 5a–b and 6a–b in [I], Fig. 2 in [K] and Fig. 6 in [M]).

Nitrogen uptake by the crop

The papers [A] (Fig. 6), [G] (Table 8), [K] (Fig. 4), [L] (Fig. 7) and [M] (Fig. 2) show the simulated crop nitrogen uptake at the Bouwing site, season 1984. Generally speaking the results of the simulations seem to be satisfactory, though model [G] overestimates uptake, especially in the case of the N1 treatment.

Results pertaining to the Eest location are shown as follows: [A] (Fig. 9), [C] (Fig. 3, Table 1), [D] (Figs. 5, 6 and 10), [G] (Tables 5 and 8), [H] (Table 2, Figs. 9 and 10), [I] (Fig. 4b), [K] (Fig. 3), [M] (Fig. 4) and [N] (Fig. 6b). Again, the simulations yielded acceptable results. Paper [H] (Fig. 9) and [N] (Fig. 6b) compare the combined measured and simulated data from the locations Eest and PAGV. Both show a good correlation between measured and simulated results, although both models tend to overestimate nitrogen uptake (see Fig. 10 in [H]). From the nitrogen balance presented in [C] (Table 1), [D] (Fig. 5), [G] (Tables 5 and 8) and [H] (Table 2) the calculated amounts of nitrogen taken up can be compared. The measured crop uptake by the aboveground parts in the N1 treatment was 169 kg ha^{-1}, model [C] predicts an uptake of 177 kg ha^{-1}, model [D] of 176 kg ha^{-1}, model [G] of 168 kg ha^{-1}, and finally model [H] of 185 kg ha^{-1}. These differences are not very large, but because of a rather different estimate of nitrogen stored in the roots, the total uptake calculated for model [C] is much higher than that for the other models. Model [C] assumes, as mentioned above, that 20% of the crop nitrogen is in the roots, which amounts to 44 and 53 kg ha^{-1} for treatments N1 and N3, respectively. Model [D] assumes 15 kg ha^{-1} in the roots independent of the treatment, model [G] predicts 23 and 26 kg ha^{-1}.

Results pertaining to the location PAGV are presented in [D] (Figs. 12a–c), [G] (Tables 6 and 8), [H] (Fig. 9), [I] Fig. 5, [K] (Fig. 2), [M] (Fig. 6) and [N] (Fig. 6b). Here again model [G] rather overestimates uptake at the lowest N application, the other models appear to give good estimates. Fig. 5b in [I] illustrates that uptake is calculated quite well, whereas the soil mineral nitrogen content is overestimated during the later part of the season by 60–70 kg ha^{-1}. In [N] it is also shown that correlation between simulated and measured uptake (Fig. 6b) is much better than between simulated and measured soil mineral nitrogen content (Fig. 6c).

Dry matter production

Predicted dry matter production is shown by model [G] (Table 7), [H] (Figs. 9 and 10), [I] (Fig. 3b), [J] (Fig. 5), [K] (Figs. 1–3), [M] (Figs. 1–6) and [N] (Fig. 6a). As with nitrogen uptake, the simulation of dry matter production yielded better results than the simulation of soil mineral nitrogen. Compare, for instance, Fig. 9b with Figs. 7a and 7b in [H], Figs. 6a and 6c in [N], Figs. 1–3 in [K] and Figs. 1–6 in [M].

Discussion and conclusions

In general, simulation of the aboveground variables yielded better results than that of the belowground variables (soil water and mineral nitrogen content). When the objective is prediction of nitrogen uptake by the crop and dry matter production, models such as [H], [I], [J] and [N], which all calculated dry matter production and nitrogen uptake satisfactory in an independent way, i.e. without parameter fitting, seem to be an appropriate choice. But as already mentioned in the introduction, research for adequate fertilization recommendations should give as much attention to prevention of pollution as to production of good yields. If simulation models are to be of any assistance in achieving this goal, a sufficiently accurate prediction of the soil mineral nitrogen content and its dynamics is required.

When the results on soil mineral nitrogen are reviewed the most striking phenomenon is that of the loss of mineral nitrogen immediately after the second and third application of fertilizer (in late spring and early summer) [5]. None of the models could satisfactorily account for this phenomenon. In [J] and [K] some suggestions are made as to the possible causes of these losses. In [J] attention is drawn to the course of biomass nitrogen during the growing season, fluctuations being of the order of 80 kg nitrogen per ha. According to [K] a possible explanation is fixation of ammonium-nitrogen. The fixed ammonium would not be extracted by the extraction method currently in use in the Netherlands, but would nevertheless be largely available for plant uptake, as Wehrmann and Coldewey-Zum-Eschenhoff [8] found. If fixation is the explanation for the apparent loss, there is no need to incorporate such a mechanism in a model meant to predict nitrogen uptake by a crop, in a period when losses by denitrification and leaching are of minor importance since this type of fixation would not affect the availability of nitrogen for the crop. But, again, where interest is mainly in the soil mineral nitrogen content, a quantitatively satisfactory explanation of the apparent loss should be found.

The prediction of the soil water content by the mechanistic models is not better and sometimes worse than that of the simpler functional models. The former require detailed information about soil hydraulic properties and, usually, the results are very sensitive to the parameter values. Functional models require fewer parameters, but are applicable to a limited range of conditions. The upper limit of water content, for instance, implies that saturation never occurs and that prediction of denitrification based on oxygen deficiency is virtually impossible with such models. Thus functional models cannot be used for extreme conditions. Generally, mechanistic models apply to a wider range of conditions. However, they have the severe disadvantage of requiring detailed knowledge of many parameters. When the main interest is the modelling of the nitrogen cycle, functional models seem to be more advisable than mechanistic models.

Also for mineralization is the conclusion valid that the more complicated models did not give better results than the simpler ones. This can best be illustrated by comparing the results of model [K] with those of model [M]. The difference between these two models mainly concerns the way in which mineralization is calculated. Model [K] distinguishes 4 pools using first-order rate kinetics, while model [M] uses one pool only and zero-order kinetics (Table 2). However, the results of the two models with respect to prediction of soil mineral content are very similar.

Seven years ago the Institute for Soil Fertility Research organized a similar workshop, but with fewer participants (6 models). Then, as now, the participants ran their model with a data set provided by the Institute. Then the data set pertained to an uncropped soil during the winter period. In discussing the results, De Willigen & Neeteson [2] came to the conclusion that the main problems were caused by the modelling of the soil microbiological processes. For the present workshop it seems that a similar conclusion can be drawn: the main difficulties in modelling the nitrogen turnover in the soil-crop system lie in the description of the soil processes, and of these the biological processes appear to pose the most serious problems.

References

1. Addiscott TM and Wagenet RJ (1985) Concepts of solute leaching in soils: a review of modeling approaches. J Soil Sci 36: 411–424
2. De Willigen P and Neeteson JJ (1985) Comparison of six simulation models for the nitrogen cycle in the soil. Fert Res 8: 157–171
3. De Willigen P and Van Noordwijk M (1987) Roots, plant production and nutrient use efficiency Ph D Thesis Agricultural University Wageningen the Netherlands, 282 pp
4. Groot JJR and Verberne ELJ (1991) Response of wheat to nitrogen fertilization, a data set to validate simulation models for nitrogen dynamics in crop and soil Fert Res 27: 349–383
5. Neeteson JJ, Greenwood DJ and Habets EJMH (1986) Dependence of soil mineral N on N-fertilizer application. Plant and Soil 91: 417–420
6. Spitters CJT, Van Keulen H and Van Kraalingen DWG (1989) A simple and universal crop growth simulator: SUCROS87. In: Rabbinge R, Ward SA and Van Laar HH (eds) Simulation and Systems Management in Crop Production, pp 147–181. Simulation Monographs, PUDOC, Wageningen
7. Van Genuchten MTh (1980) A closed-form equation for predicting the hydraulic conductivity of unsaturated soils. Soil Sci Soc Am J 44: 892–898
8. Wehrmann J and Coldewey-Zum Eschenhof H (1986) Distribution of nitrate, exchangeable and non-exchangeable ammonium in the soil-root interface. In: Lambers H, Neeteson JJ and Stulen I (eds) Fundamental, Ecological and Agricultural Aspects of Nitrogen Metabolism in Higher Plants, pp 447–450. Martinus Nijhoff, Dordrecht

Fertilizer Research **27**: 151–160, 1991.

Quantitative aspects of nitrogen nutrition in crops

H. van Keulen & W. Stol
Centre for Agrobiological Research (CABO), P.O. Box 14, 6700 AA Wageningen, The Netherlands

Key words: Nitrogen nutrition, simulation models, fertilizer response, light use efficiency

Abstract

The processes affecting the response of crops to nitrogen fertilizers are reviewed, with special emphasis on quantitative relationships that enable the development of deterministic models. Total dry matter production is described in terms of length of the growing period and average growth rate, and the effect of nitrogen status of the vegetation on both is discussed. Attention is paid to nitrogen influence on stomatal response and crop water use. It is shown that the effect of differential nitrogen nutrition can in some cases be described in terms of light interception, but not always. It is concluded that the present insight in the relevant processes is insufficient to use deterministic models directly for application in fertilizer practices at the farm level, but that further development of such models is an important aid in structuring thinking about the system.

Introduction

Nitrogen is an indispensible element for optimum functioning of crops, as was recognized early in literature on agricultural research, where Roberts [42] states 'as to the application of nitrate of soda to wheat when it has a yellow or sickly appearance in the spring ... it will in a few days alter the sickly hue to a luxuriant green'. Some dozens of years later, the point was made more convincingly at about the same time by Gilbert and Lawes in the UK and von Liebig [30] in Germany. Since then, thousands of experiments have been carried out in which the relation between application rate of nitrogen fertilizer and yield has been established. The results of such experiments are, however, so strongly influenced by the environmental conditions during execution, that they can hardly be used for predictive purposes, such as the formulation of fertilizer recommendations under field conditions. There is, however, an urgent need for better recommendations, especially in view of the contribution of excess nutrients to environmental pollution, such as the nitrate load of

drinking water and the concentration of nitrous oxides and ammonia in the atmosphere.

Increasingly, therefore, it is suggested to use simulation models in which the effects of nitrogen status of the crop on the processes underlying growth, yield and development are described quantitatively, in combination with the processes of the nitrogen balance in the soil, that determine the availability of nitrogen to the crop [15].

However, it often appears that not all relevant processes and causal relations are sufficiently understood to permit their incorporation in deterministic models. In this paper a review is presented of the relevant processes underlying crop response to nitrogen nutrition, using, as much as possible, the wheat crop as an example, to pinpoint the gaps in our knowledge, that are constraints in the development of such predictive models.

Nitrogen status and dry matter production

Dry matter production, either the total for a crop, or of particular plant parts, such as the tubers or the grains can, schematically, be described as the product of the mean rate of dry matter accumulation and the length of the period of growth. Assessment of the effects of nitrogen nutrition on yield and production should thus consider both aspects.

Length of the growing period

The phenological development of a crop cultivar, i.e. the order and rate in which the vegetative and reproductive plant organs appear is governed by both genetic properties and environmental conditions, notably temperature and daylength [10]. For cultivars adapted to the daylength of the environment in which they are grown, the temperature of the stem apex is the main driving force, which can be approximated by either air or canopy temperature.

In the literature conflicting evidence with respect to the effect of nitrogen nutrition on phenological development is reported [20, 47], i.e. nitrogen deficiency can either hasten or delay development, or have no effect. Nitrogen deficiency in the crop may lead to partial stomatal closure as a reaction to impaired photosynthesis [12, 13]. That in turn will lead to a change in the energy balance of the crop and hence to increased canopy temperatures. In the field, differences of up to 4°C have been recorded between crops well supplied with nitrogen and crops under nitrogen stress [43]. Under certain environmental conditions, an indirect effect of nitrogen shortage on the rate of phenological development and hence on the length of the growing period can thus be expected, leading to shorter growing periods. Severe stress can stop phenological development of the crop completely, so that the relative importance of both processes in particular situations determines the overall effect, which may be one of the reasons for the variable results reported.

Rate of dry matter production

The rate of dry matter production by a crop is the result of the balance between carbon dioxide assimilation and respiration, modified by the distribution of the dry matter formed. Quantification of the effect of nitrogen supply on the rate

of dry matter production requires therefore description of the effects on these individual processes, which have been studied in far less detail than the overall effect on cumulative dry matter production.

Carbon dioxide assimilation

The relation between the light-saturated rate of CO_2 assimilation and nitrogen content of individual leaves has been studied for many species, as reviewed by van Keulen and Seligman [26]. In practically all cases a strong positive correlation between the two was established, sometimes linear over the full range of concentrations studied, sometimes showing a saturating behaviour at higher nitrogen concentrations. Statistical analysis of the pooled data suggested a linear relationship at least up to a nitrogen content of $0.06 \, kg \, kg^{-1}$ on a dry weight basis. Net assimilation became zero at a nitrogen content of $0.0038 \, kg \, kg^{-1}$, and the slope of the line was $2.01 \cdot 10^{-3} \, kg \, CO_2 \, m^{-2} \, s^{-1}$ per unit $(kg \, kg^{-1})$ increase in nitrogen content.

As carbon dioxide assimilation is expressed per unit leaf surface, it has been suggested that the nitrogen content should also be expressed on an area basis rather than on a dry weight basis [9]. The analysis referred to earlier did, however, not show any improvement in the goodness of fit when nitrogen content was expressed on an area basis.

The statistical analysis of the relation between maximum carbon dioxide exchange rate and nitrogen content shows that about 80 percent of the variation is explained. However, the residual variability may not only be due to inaccuracies in the measurements or inherent variability in the plant material, but may at least partially be functional and related to the level of nitrogen nutrition in the period prior to the measurements. In a study on carbon exchange rate of flag leaves of field-grown wheat at different levels of nitrogen nutrition, Marshall [32] found also a linear relation with nitrogen content, with a slope similar to that derived from the pooled data, but the nitrogen level at zero assimilation was higher for the high N treatment than for the low N treatment. Groot and van Dijk [19], studying barley leaves in the grain filling phase, showed that also the nitrogen use efficiency, expressed as carbon exchange rate per unit nitrogen (the slope of the regression line) differed for leaves exposed to different nitrogen supplies. Both observations point in the direction of differences in the distribution of nitrogen between an active form, contributing to CO_2 assimilation and a non-active form, referred to as structural or storage protein [22]. Under abundant nitrogen supply a larger proportion of total leaf N would then be in the form of storage proteins [31].

Hence, for most situations a linear relation between nitrogen content and carbon exchange rate at light saturation exists, although the actual parameters describing that relation may vary appreciably, depending on pre-treatment. This phenomenon may also explain the reported curvilinearity in the relation between maximum carbon exchange rate and nitrogen content if leaves with a different history with respect to nitrogen nutrition are combined in the analysis [49].

The initial light use efficiency at the light compensation point seems not to be affected significantly by the nitrogen content of the leaves [19, 49]. However, as this parameter is difficult to determine accurately, it may well be that small differences in measured assimilation rate between canopies with different leaf nitrogen concentrations are due to differences in the initial slope.

Assimilate utilization
The assimilates fixed in the assimilation process are partly used in respiratory processes to provide energy for biological functioning and partly for growth of structural plant tissue. Respiration can schematically be distinguished in a maintenance component and a growth component.

Maintenance respiration of the various plant parts is a function of their weight, their chemical composition and ambient temperature. The major processes involved are the resynthesis of degraded proteins, maintenance of concentration differences across membranes and metabolic activity [38]. Different proteins in plant tissue may have very different life times, but an average rate of turnover in active leaf, stem and root tissue of $0.1 \, d^{-1}$ may be assumed. Hence, tissue with higher nitrogen contents requires more energy for maintenance [28, 35, 49]. However, exact quantitative information is rather scarce.

At low nitrogen contents, protein turnover is apparently very low [21], probably because most of the residual nitrogen is in the form of storage proteins, and in that situation the energy requirements for protein turnover are very small compared to those for maintaining concentration gradients and metabolic activity.

The assimilates not used in maintenance respiration are available for increase in dry weight of the various plant organs. The partitioning pattern changes in the course of the plant's life cycle as a result of variations in sink strength of the various organs, which is probably related to the number of growing cells [48]. Under optimum growing conditions this pattern can be described as a function of phenological development only [26]. Under sub-optimum growing conditions, generally the partitioning pattern changes. Whether this is an active process or the result of a differential effect of stress on different organs is difficult to deduce from the available evidence. The often observed shift to lower shoot/root ratios under water stress led Brouwer [4, 5] to suggest that the effect of water stress is stronger on the conversion of primary photosynthates than on assimilation, so that the level of carbohydrate reserves in the plant increases, resulting in greater availability to the root system and stimulated root growth. He referred to this as the 'functional balance'. Although the concept has been questioned [29] an alternative explanation seems to be lacking sofar.

Nitrogen deficiency also often leads to lower shoot/root ratios [6, 8, 9], which may be the result of the same functional balance. In addition, the partitioning between leaf blades and other above-ground organs is affected, resulting in a lower leaf weight ratio [7, 33, 36]. Practically all these results refer, however, to integrated results over a longer growing period, and the instantaneous effect of sub-optimum nitrogen concentrations in the tissue on assimilate partitioning is poorly documented. The degree of nitrogen stress is both difficult to quantify and hard to relate to assimilate partitioning using existing experimental data. Hence, more attention for these aspects is warranted.

The primary assimilates allocated to the various growing organs consist of a mixture of simple carbohydrates and nitrogenous compounds.

These products are converted into structural plant material and the energy required for transport and conversion, as well as the 'waste' products have to be taken into account [38]. In a schematized way growth respiration can be expressed in the carbohydrate requirement for biosynthesis, i.e. the amount of carbohydrate required to form one unit biomass of a specific composition. This requirement can be derived from a theoretical analysis of the pathways of biosynthesis [39], but it can relatively easily be determined from the composition, in terms of carbon and ash, of the material being formed [51]. In Table 1 the carbohydrate requirements for a number of plant components, as derived by Penning de Vries et al. [40] are given. The main uncertainty in quantifying growth respiration arises from the question which proportion of the energy costs of nitrate reduction have to be taken into account, and which proportion can be covered directly from sun energy in the leaves.

Hence, nitrogen affects substrate utilization by increasing the respiratory costs with increasing nitrogen content and changing the allocation, such that the shoot/root ratio and the leaf weight ratio increase.

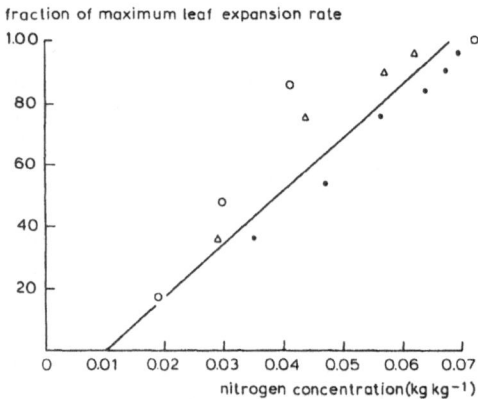

Fig. 1. The relation between the nitrogen content in the leaf and the relative rate of leaf area expansion; dots: Lolium rigidum; open circles: Lolium perenne; triangles: Triticum aestivum (Source: [16, 17]).

Table 1. Carbohydrate requirements (g CH_2O/g dry weight) for biosynthesis of organic plant components (Source: [40])

Component	Carbohydrate requirement
Carbohydrates	1.275
Proteins	1.887* (2.784**)
Fats	3.189
Lignins	2.231
Organic acids	0.954

* synthesis from amides, no nitrate reduction costs
** nitrate reduction costs included

Organ formation

The direct effects of nitrogen status of the vegetation on organ formation are often difficult to disentangle from the indirect effects through the supply of assimilates [26]. Greenwood [16], Greenwood and Titmanis [17] and Wilson [53] in a series of experiments on wheat, *Lolium rigidum* and *Lolium perenne* found that the relative rate of leaf area expansion of the youngest expanding leaf was linearly related to total leaf nitrogen content (Fig. 1). A problem in quantitatively describing these effects is that optimum

leaf nitrogen content declines with leaf age, hence a direct relation with absolute nitrogen content cannot easily be established. In the winter wheat experiments reported by Groot [18], leaf area index around anthesis varied between 1.7 and 5.5. Figure 2 shows that the relation between LAI and total nitrogen uptake by the crop at that moment is linear, suggesting a constant leaf area production per unit nitrogen uptake. This seems an interesting concept, which might be useful in describing the effects of nitrogen availability on leaf area dynamics. However, more detailed analysis is necessary to confirm these findings, also for other crops. Moreover, there are some conspicuous outliers (Bouwing 83/84), which need further explanation.

Tiller formation in gramineous crops, which is comparable to leaf formation, also appears to be directly influenced by the nitrogen content of the crop [1, 58]. To what extent the same holds for branching and formation of side shoots in other crops is difficult to judge from available experimental evidence. Again, these effects are confounded through indirect effects of assimilate availability.

Nitrogen nutrition and water use

Many studies have indicated that water use efficiency, i.e. the amount of dry matter produced per unit input of water, increases with higher

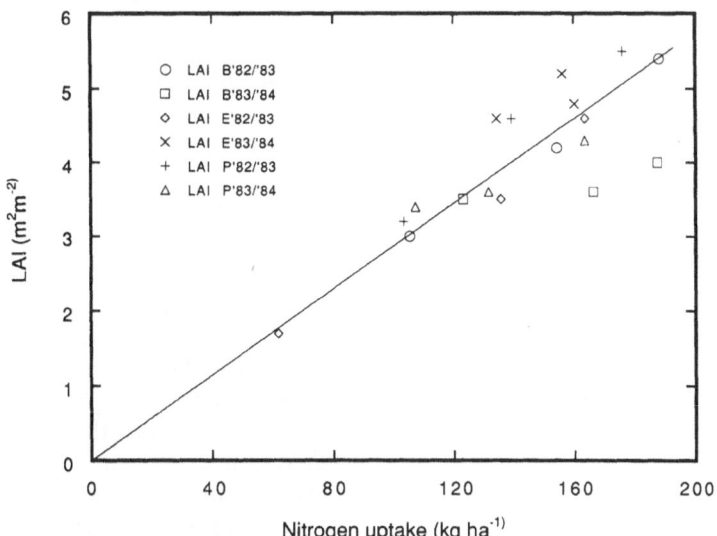

Fig. 2. The relation between total crop nitrogen uptake at anthesis and leaf area index (LAI) at that moment; B is Bouwing; E is Eest; P is PAGV (Source [18]).

nitrogen availability to the crop [2, 25, 52]. Interpretation of these results is hampered when no distinction is made between transpiration by plants and evaporation from the soil surface. Crops growing under nitrogen-deficient conditions generally have a much smaller leaf area than those optimally supplied with nitrogen, hence canopy closure occurs much later, if at all. Consequently, bare soil is exposed for a much longer period of time, which leads to a larger proportion of total water loss from the soil surface, resulting in reduced water use efficiency. This may be one of the reasons that in experiments on water use, where soil surface evaporation was prevented, nitrogen nutrition hardly affected water use efficiency [50, 54].

To examine the effects of nitrogen nutrition on plant water use, a better measure is the transpiration efficiency, the amount of carbon dioxide fixed per unit water transpired. Experiments where assimilation and transpiration were measured simultaneously on leaves of different nitrogen contents have not provided conclusive evidence. Transpiration efficiency was not affected in maize [13, 56] and *Panicum maximum* [3], but increased in *Festuca arundinacea* and *Panicum milioides*, in the latter species almost two-fold over a range in leaf nitrogen contents from 0.01 to 0.05 $kg\,kg^{-1}$ [3].

Shimshi [44, 45] has shown that transpiration from nitrogen-deficient plants was lower when well-supplied with water, but that the situation reversed at soil water contents near wilting point, hence in water-stressed plants. That could be the result of the higher proportion of cell wall constituents in nitrogen-deficient plants, which reduces stomatal flexibility: the opening is restricted under favourable soil moisture conditions and full closure is prevented under water stress. Greater stomatal opening at higher leaf nitrogen contents has also been reported for rice [23, 57], wheat [46], sunflower and maize [13]. The relation between leaf conductance for water vapour, including boundary layer conductance, and leaf nitrogen content as derived from Yoshida and Coronel [57] is given in Fig. 3. The relation between stomatal conductance and leaf nitrogen content would also be linear and to the left of the calculated regression line, as boundary layer conductance under the measuring conditions would be practically constant. This linear relation suggests stomatal control through the CO_2 concentration in the sub-stomatal cavity, in such a way that the CO_2 concentration remains either constant or has a fixed ratio to the external concentration. Any impairment in assimilation will then lead to proportional stomatal closure and lower transpiration. Under what conditions this phenomenon is also operational in field conditions is not clear.

156

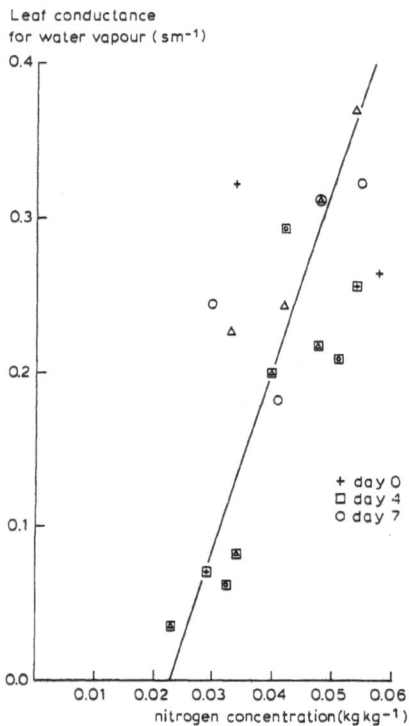

Fig. 3. The relation between the nitrogen concentration in the leaf blade and total leaf conductance for water vapour exchange for individual rice leaves (Source: [57]).

Hence, the quantitative effect of nitrogen status of the crop on water use and water use efficiency is difficult to predict and careful consideration in each situation is therefore necessary.

Crop response to fertilizer application

Countless nitrogen fertilizer experiments have been carried out, often with emphasis on determining response curves with the aim of establishing the optimum application rate, often expressed in economic terms, i.e. the quantity where the costs of the last unit of nitrogen applied is equal to the price of the additional yield obtained. In the last years the emphasis has shifted, as attention for environmental issues increased and restriction of losses of nitrogen to the environment has become a major consideration. Therefore, more attention is being paid to the physical aspects of the fate of applied fertilizer. The use of response curves provides only

limited information on those aspects, and moreover they show a large variability as illustrated in Fig. 4 for three sites from the winter wheat data set of Groot [18]. An alternative method of analysis [27, 55] that permits a distinction to be made between the response of crop uptake to increased nitrogen application and the response of the crop to increased uptake, is therefore more useful. The method is illustrated in Fig. 5, using the same data as in Fig. 4, with addition of the results of chemical analysis of the material harvested. These results show that the relation between nitrogen uptake and yield is invariable for the three sites that showed large differences in response curves. At low nitrogen availability grain yield is proportional to total nitrogen uptake, indicating constant limiting nitrogen con-

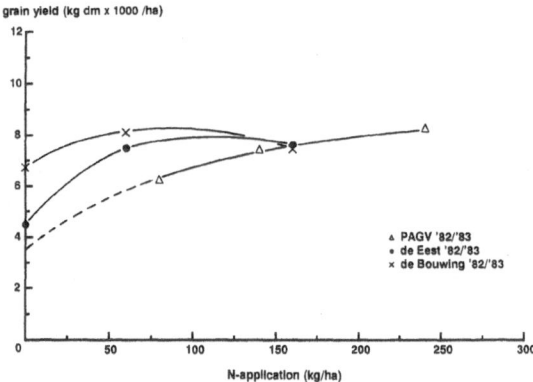

Fig. 4. The relation between nitrogen application and grain yield for winter wheat at three sites in The Netherlands (Source: [18]).

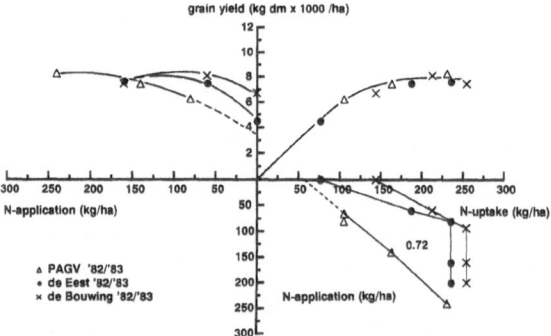

Fig. 5. The relation between total nitrogen uptake and grain yield (right upper quadrant), that between nitrogen application and nitrogen uptake (right lower quadrant) and that between nitrogen application and grain yield (left upper quadrant) for winter wheat at three sites in The Netherlands (Source: [18]).

centrations in both grain and straw. With increasing nitrogen uptake the concentration in the tissue at harvest increases, leading to deviation of the line from the linear and finally a plateau level is reached where some growth factor other than nitrogen availability is yield-determining.

The variability in the response curve thus arises from differences in the relation between nitrogen application and nitrogen uptake (lower right hand side of Fig. 5). Both, the uptake at zero fertilizer application (supply from natural sources) and the recovery fraction (ratio of additional uptake and application) vary substantially. In general this relation appears linear within the relevant range of application rates [24, 27] which suggests that the processes that determine availability of the element to the crop, including those that cause losses, are first-order processes, i.e. that their rates are proportional to the concentration of the element in the soil solution. This also holds for the PAGV data (Fig. 5), but cannot be judged for the two other sites, as at

the highest application rates the crops were probably 'saturated' with nitrogen throughout their life cycle, resulting in active exclusion of nitrogen, as was also observed in nitrogen fertilizer experiments on permanent pasture [41].

The recovery fraction appears 0.72 for the PAGV site, for the other two sites the value exceeds 1 for the lowest application rate. It is not clear what the reason is for this 'aberrant' behaviour.

Leaf area development in winter wheat appears closely related to nitrogen uptake (Fig. 2). Analyses of field crop behaviour in W.Europe suggest that cumulative dry matter production is often proportional to accumulated intercepted radiation [34], which is directly related to the intercepting green surface. In Fig. 6 the relation between cumulative intercepted photosynthetically active radiation (PAR) and cumulative dry matter production is given for the winter wheat crop at PAGV in the '83/'84 growing season. Intercepted photosynthetically active radiation

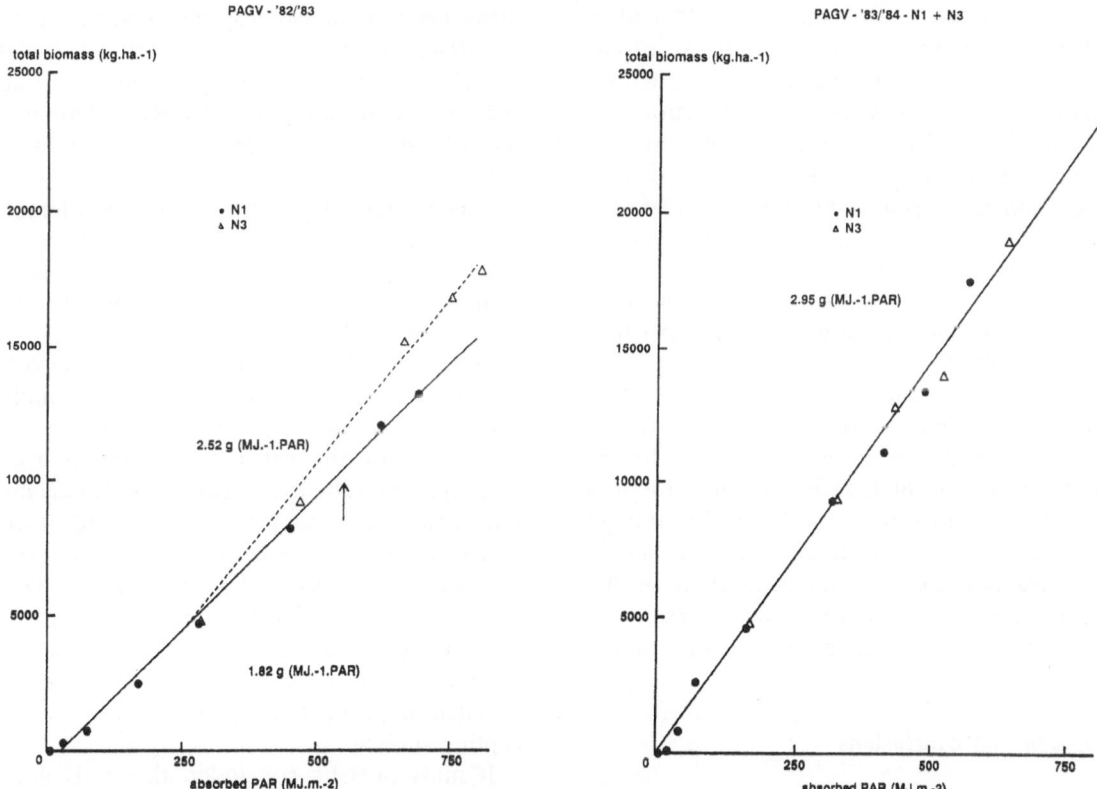

Fig. 6. The relation between cumulative absorbed photosynthetically active radiation and cumulative dry matter production for winter wheat at PAGV, The Netherlands in '82/'83 (left) and '83/'84 (right).

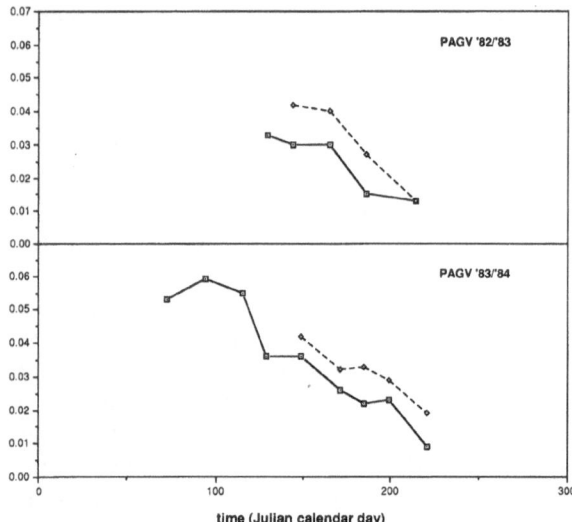

Fig. 7. Time course of nitrogen content in the green leaves of winter wheat at PAGV, The Netherlands, in '82/'83 (upper part) and '83/'84 (lower part); solid lines low N; broken lines diamonds high N (Source: [18]).

was calculated by interpolation in measured leaf area index, assuming an extinction coefficient of 0.6 for global radiation and assuming a proportion of 50% PAR. The results show that the difference in total dry matter production between the low and high nitrogen treatments can be fully explained by the differential radiation interception as a result of differences in leaf area development. The difference in nitrogen concentration in the green leaf mass (Fig. 7) apparently does not influence the radiation use efficiency, i.e. the amount of dry matter produced per unit PAR intercepted, which has a constant value of 2.95 g/MJ. This value is very similar to that reported for wheat and barley crops in the UK [11]. The situation is different for the '82/'83 season, where the high N treatment exhibits a clearly higher radiation use efficiency, although the differences in nitrogen content in the green leaves are not more pronounced than in the '83/'84 season. The latter results agree with those presented for winter wheat crops in the UK [14, 11].

Discussion and conclusions

The basic processes underlying crop response to nitrogen nutrition are not all fully understood quantitatively, as discussed in the first part of this paper. The relation between radiation absorption and dry matter production (Fig. 6) is the integrated result of these processes, and differences in this relation under the influence of differential nitrogen nutrition would have to be explained in terms of these processes. It has been shown that light saturated assimilation rate is linearly related to nitrogen content in the leaves, but the ratio of 'active protein' to 'storage protein' may vary considerably, thus leading to variations in both the nitrogen content at the CO_2 compensation point and the nitrogen use efficiency. Moreover, as initial light use efficiency at the light compensation point seems to be far less affected by nitrogen content, the overall effect of leaf nitrogen content on canopy assimilation rate also depends on the prevailing radiation climate. Taking all of these factors into account, the effect appears to be difficult to predict in general terms.

Even if gross assimilation is higher at higher leaf nitrogen contents, part of that gain will be offset again by higher respiratory losses, as both maintenance and growth respiration are higher for tissues with higher nitrogen contents. Again, generally applicable quantitative relations are difficult to derive, as for instance for maintenance requirements also the ratio of active to storage protein plays a role. Moreover, temperature is important as maintenance respiration is much more sensitive to temperature than is gross assimilation, so that net assimilation is higher at lower temperatures.

Differences in nitrogen content may also lead to differences in distribution of assimilates among various plant parts. This will have consequences for the calculated radiation use efficiency, especially if a larger proportion is invested in the root system as a reaction to nitrogen deficiency, because generally only above-ground dry matter is considered in the analysis. Differences in the distribution between leaves and stem will only have a minor effect on that conversion efficiency, although it may affect total dry matter production by influencing the intercepting surface.

It must thus be concluded that it is not yet possible to quantitatively predict the effects of nitrogen nutrition on radiation use efficiency.

A large amount of work has been done on effects of nitrogen on crop performance and on the effects of nitrogen on individual processes. However, a systematic analysis also reveals that still major gaps in knowledge and insight exist, which makes it difficult to use the existing knowledge at the practical level, for instance in formulating fertilizer recommendations that result for the farmer in the required yield of the required quality and at the same time minimizes effects on the environment. Systems analysis and simulation in which the available knowledge is combined into a consistent framework, as illustrated with a number of examples in this volume, is a research tool that may help in making the gaps in knowledge explicit [26], and design experiments specifically geared to filling these gaps. At the moment it still seems rather premature to claim that the existing models could be of direct practical application for fertilizer practices at the farm level (cf. Otter-Nacke and Kuhlmann, this volume), but they certainly help in structuring the thinking of scientists and thus pointing to the most important processes and parameters.

References

1. Aspinall D (1961) The control of tillering in the barley plant. I. The pattern of tillering and its relation to nutrient supply. Aust J biol Sci 14: 493–503
2. Black CA (1966) Crop yields in relation to water supply and soil fertility. In: Pierre WH, Kirkham D, Pesek J and Shaw R (eds.) Plant environment and efficient water use. Am Soc Agron, Soil Sci Soc Am, Madison, Wisc: 177–206
3. Bolton JK and Brown RH (1980) Photosynthesis of grass species differing in carbon dioxide fixation pathways. V. Response of *Panicum maximum*, *Panicum milioides* and tall fescue (*Festuca arundinacea*) to nitrogen nutrition. Plant Physiol 66: 97–100
4. Brouwer R (1965) Root growth of cereals and grasses. In: Milthorpe FL and Ivins JD (eds.) The growth of cereals and grasses. Butterworths, London: 153–166
5. Brouwer R (1963) Some aspects of the equilibrium between overground and underground plant parts. Jaarb Inst Biol Scheik Onderz Landb 1962: 31–39
6. Brouwer R, Jenneskens PJ and Borggreve GJ (1962) Growth responses of shoots and roots to interruptions in the nitrogen supply. Jaarb Inst Biol Scheik Onderz Landb 1961: 29–36
7. Campbell CA, Davidson HR and McCaig TN (1983) Disposition of nitrogen and soluble sugars in Manitou spring wheat as influenced by N fertilizer, temperature, and duration and stage of moisture stress. Can J Plant Sci 63: 73–90
8. Colman RL and Lazenby A (1970) Factors affecting the response of some tropical and temperate grasses to fertilizer nitrogen. In: Norman MJT (ed.) Proc 11th Intern Grassld Congr. Univ Queensld Press, St Lucia: 392–397
9. Cook MG and Evans LT (1983) Nutrient responses of seedlings of wild and cultivated *Oryza* species. Field Crops Res 6: 205–218
10. Dobben WH van (1962) Influence of temperature and light conditions on dry-matter distribution, development rate and yield in arable crops. Neth J agric Sci 10: 377–389
11. Gallagher JN and Biscoe PV (1978) Radiation absorption, growth and yield of cereals. J agric Sci (Camb) 91: 47–60
12. Goudriaan J and Laar HH van (1978) Relations between leaf resistance, CO_2-concentration and CO_2-assimilation in maize, beans, lalang grass and sunflower. Photosynthetica 12: 241–249
13. Goudriaan J and Keulen H van (1979) The direct and indirect effects of nitrogen shortage on photosynthesis and transpiration in maize and sunflower. Neth J agric Sci 27: 227–234
14. Green CF (1987) Nitrogen nutrition and wheat growth in relation to absorbed solar radiation. Agric Forest Meteorol 41: 207–248
15. Greenwood DJ (1982) Modelling of crop response to nitrogen fertilizer. Phil Trans Roy Soc B 24: 279–288
16. Greenwood EAN (1966) Nitrogen stress in wheat – its measurement and relation to leaf nitrogen. Plant Soil 24: 279–288
17. Greenwood EAN and Titmanis ZV (1966) The effect of age on nitrogen stress and its relation to leaf nitrogen and leaf elongation in a grass. Plant Soil 24: 379–389
18. Groot JJR (1987) Simulation of nitrogen balance in a system of winter wheat and soil. Simulation Reports CABO/TT no. 13, CABO, Wageningen
19. Groot JJR and Dijk W van (1990) The relationship between leaf nitrogen content and carbon exchange rate in leaves of winter barley (*Hordeum vulgare* L.) during senescence. (In press)
20. Halse NJ, Greenwood EAN, Lapins P and Boundy CAP (1969) An analysis of the effect of nitrogen deficiency on the growth and yield of a western Australian wheat crop. Aust J agric Res 20: 987–998
21. Hanson AD and Hitz WD (1983) Whole-plant response to water deficits: Water deficits and protein turnover In: Taylor HM, Jordan WR and Sinclair TR (eds.) Limitations to efficient water use in crop production. Am Soc Agron, Madison, Wisconsin: 331–343
22. Huffaker RC (1982) Biochemistry and physiology of leaf proteins. In: Boulter D and Parthier G (eds.) Encycl Plant Physiol (New Series Vol 14A) Nucleic acids and proteins in plants. I. Proteins. Springer Verlag, Berlin: 370–400
23. Ishihara K, Ebara H, Hirawasa T and Ogura T (1978) The relationship between environmental factors and behaviour of stomata in the rice plants. VII. The relation

160

between nitrogen content in leaf blades and stomatal aperture. Japan J Crop Sci 47: 664–673

24. Keulen H van (1986) Crop yield and nutrient requirements. In: Keulen H van and Wolf J (eds.) Modelling of agricultural production: weather, soils and crops. Simulation Monographs, PUDOC, Wageningen: 155–181

25. Keulen H van (1975) Simulation of water use and herbage growth in arid regions. Simulation Monographs, PUDOC, Wageningen.

26. Keulen H van and Seligman NG (1987) Simulation of water use, nitrogen nutrition and growth of a spring wheat crop. Simulation Monographs, PUDOC, Wageningen

27. Keulen H van and Heemst HDJ van (1983) Crop response to the supply of macronutrients. Agr Res Rep 916, PUDOC, Wageningen

28. Khan MA and Tsunoda S (1970) Evolutionary trends in leaf photosynthesis and related leaf characters among cultivated wheat species and its wild relatives. Japan J Breed 20: 133–140

29. Lamberts H (1983) 'The functional equilibrium', nibbling on the edges of a paradigm. Neth J agric Sci 31: 305–311

30. Liebig J von (1855) Die Grundsätze der Agrikultur-Chemie mit Rücksicht auf die in England angestellten Untersuchungen. Braunschweig, 2nd Ed.

31. Makino A, Mae T and Ohira K (1984) Relation between nitrogen and Ribulose-1,5-biphosphate carboxylase in rice leaves from emergence through senescence. Plant Cell Physiol 25: 429–437

32. Marshall B (1978) Leaf and ear photosynthesis of winter wheat crops. Ph D Thesis, Univ Nottingham

33. McNeal FH, Berg MA and Watson CA (1966) Nitrogen and dry matter in five spring wheat varieties at successive stages of development. Agron J 58: 605–608

34. Monteith JL (1977) Climate and the efficiency of crop production in Britain. Phil Trans Roy Soc London B: 277–294

35. Murata Y (1961) Studies on the photosynthesis of rice plants and its culture significance. Bull Nat Inst Agric Sci D 9: 1–69

36. Os AJ van (1967) The influence of nitrogen supply on the distribution of dry matter in spring rye. Jaarb Inst Biol Scheik Onderz Landb 1966: 51–65

37. Otter-Nacke S and Kuhlmann H (1991) A comparison of the performance of N simulation models in the prediction of N_{min} on farmers' fields in the spring. Fert Res 27: 341–347

38. Penning de Vries FWT (1975) The costs of maintenance processes in plant cells. Ann Bot 39: 77–92

39. Penning de Vries FWT, Brunsting AHM and Laar HH van (1974) Products, requirements and efficiency of biosynthesis: a quantitative approach. J theor Biol 45: 339–377

40. Penning de Vries FWT, Jansen DM, Berge HFM ten and Bakema A (1989) Simulation of ecophysiological processes of growth in several annual crops. Simulation Monograph 29, PUDOC, Wageningen

41. Prins WH, Rauw GJG and Postmus J (1981) Very high application of nitrogen fertilizer on grassland and residual effects in the following season. Fert Res 2: 309–327

42. Roberts E (1847) On the management of wheat. J Roy Agric Soc England 8: 60–77

43. Seligman NG, Loomis RS, Burke J and Abshahi A (1983) Nitrogen nutrition and phenological development in field-grown wheat. J agric Sci 101: 691–697

44. Shimshi D (1970a) The effect of nitrogen supply on transpiration and stomatal behaviour of beans (*Phaseolus vulgaris* L.). New Phytol 69: 405–413

45. Shimshi D (1970b) The effect of nitrogen supply on some indices of plant-water relations of beans (*Phaseolus vulgaris* L.). New Phytol 69: 413–424

46. Shimshi D and Kafkafi U (1978) The effect of supplemental irrigation and nitrogen fertilisation on wheat (*Triticum aestivum* L.). Irr Sci 1: 27–38

47. Spiertz JHJ and Ellen J (1978) Effects of nitrogen on crop development and grain growth of winter wheat in relation to assimilation and utilization of assimilates and nutrients. Neth J agric Sci 26: 210–231

48. Sunderland N (1960) Cell division and expansion in the growth of the leaf. J exp Bot 11: 68–80

49. Takano Y and Tsunoda S (1971) Curvilinear regression of the leaf photosynthetic rate on leaf nitrogen content among strains of *Oryza* species. Japan J Breed 21: 69–76

50. Tanner CB and Sinclair TR (1983) Efficient water use in crop production: Research or re-search. In: Taylor HM, Jordan WR and Sinclair TR (eds.) Limitations to efficient water use in crop production. Am Soc Agron, Madison, Wisc: 1–27

51. Vertregt N and Penning de Vries FWT (1987) A rapid method for determining the efficiency of biosynthesis of plant biomass. J Theor Biol 128: 109–119

52. Viets FG Jr (1962) Fertilizers and the efficient use of water. Adv Agron 14: 223–264

53. Wilson JR (1975) Comparative response to nitrogen deficiency of a tropical and temperate grass in the interrelation between photosynthesis, growth, and the accumulation of non-structural carbohydrate. Neth J agric Sci 23: 104–112

54. Wit CT de (1958) Transpiration and crop yields. Agric Res Rep 64.8, PUDOC, Wageningen

55. Wit CT de (1953) A physical theory on placement of fertilizer. Agr Res Rep 59.4, Staatsdrukkerij, 's Gravenhage

56. Wong SC, Cowan IR and Farquhar GD (1979) Stomatal conductance correlates with photosynthetic capacity. Nature 282: 424–426

57. Yoshida S and Coronel V (1976) Nitrogen nutrition, leaf resistance and leaf photosynthetic rate of the rice plant. Soil Sci Plant Nutr 22: 207–211

58. Yoshida S and Hayakawa Y (1970) Effects of mineral nutrition on tillering of rice. Soil Sci Plant Nutr 16: 186–191

Fertilizer Research **27**: 161–169, 1991.
© 1991 *Kluwer Academic Publishers.*

Modelling of the nitrogen cycle in farm land areas

F. Cabon, G. Girard & E. Ledoux
Centre d'Informatique Géologique, Ecole Nationale Supérieure des Mines de Paris, 35, rue Saint Honoré, 77305 Fontainebleau, France

Key words: Modelling of the nitrogen cycle, leaching of nitrate, hydrologic system, Dutch experimental farms

Abstract

The authors propose a method of modelling the cycles of water and nitrogen in the soil in order to simulate the drainage flow of water and the flow of leached nitrate, the main inputs into a model of transfer in aquifers. The processes of the N cycle included here are: mineralization, nitrification, ammonium adsorption-desorption, immobilization of N from ammonium and nitrate, N uptake by plants and leaching of nitrate. The results of a limited test of the model using data from Dutch experimental farms showed that, on the whole, measured and simulated values were in reasonably good agreement, in spite of the relative simplicity of the model. However, the simulations do not give detailed information concerning the dynamics of nitrogen in the various layers in the soil.

Résumé

Les auteurs proposent une modélisation du cycle de l'eau et de l'azote dans le sol, afin de simuler le flux d'eau drainée et le flux de nitrates lessivés, principales entrées d'un modèle de transfert en aquifère. Les processus du cycle de l'azote considérés sont: la minéralisation, la nitrification, l'adsorption-désorption de l'ammonium, les réorganisations à partir de l'ammonium et des nitrates, la consommation en azote par les plantes et le lessivage des nitrates. Quelques résultats d'applications sur des fermes expérimentales néerlandaises montrent que malgré la relative simplicité du modèle, tant au niveau du transfert de l'eau qu'au niveau de l'approche des processus du cycle de l'azote, les résultats obtenus reflètent dans l'ensemble les observations, mais ne prétendent pas fournir d'informations détaillées sur l'évolution de l'azote dans les différents niveaux pédologiques.

Introduction

The contamination of water resources by nitrogenous compounds is becoming an increasingly serious problem, especially in areas of intensive farming. Nitrogenous compounds may pollute surface waters and groundwater due to runoff and leaching.

A model of the N cycle and the leaching of nitrate was developed to estimate the flow of nitrates that pollute aquifers, and their transport inside a watershed basin. In view of the many heterogeneities encountered at this scale (various types of soil, varying quantities of nitrogen introduced into the soil, different kinds of crops, etc.), only the essential processes have been included in the model. These heterogeneities are accounted for by discretizing the space of the domain [2, 5].

The practical application to data from Dutch experimental farms [4] offers an opportunity to test the part of the model that deals with the

162

water balance and the N balance in the soil and to determine if the results obtained in the root zone agree with the observed data.

Description of the models

Model of the water balance

The modelling of the migration of water in a soil is taken into account by the water production function of the coupled hydrologic model [3, 5], which is available at the Centre d'Informatique Géologique of the Paris School of Mines.

This production function includes:

a) a soil or balance-production reservoir in which the water balance is calculated through a distribution of the precipitation between infiltration, runoff and evaporation, and

b) a certain number of reservoirs that regulate the percolation of the water toward the surface and groundwater domains.

The water balance is calculated at each time step (day) from the amount of precipitation P, the potential evapotranspiration ETP and the water reserve R in the soil reservoir as shown in Fig. 1.

The water production function depends on the following parameters (see Fig. 1):

– DCRT: minimum level (in mm) of the water reserve in the soil, below which no water is available. This parameter mainly governs the role of the first rainfalls after a period of drought.

– RMAX: maximum level (in mm) of the water reserve in the soil, above which all rainfall contributes to the available water (EAU).

The actual evapotranspiration (ETR) increases as a function of these two parameters which thus condition the overall water balance.

– FN: maximum height of water (in mm) infiltrated at each time step. This parameter governs the distribution of the available water between runoff (QR) and infiltration (QI).

– QRMAX and CQR: maximum capacity (in mm) of the runoff reservoir and the corresponding emptying coefficient.

– QIMAX and CQI: maximum capacity (in mm) of the infiltration reservoir and the corresponding emptying coefficient.

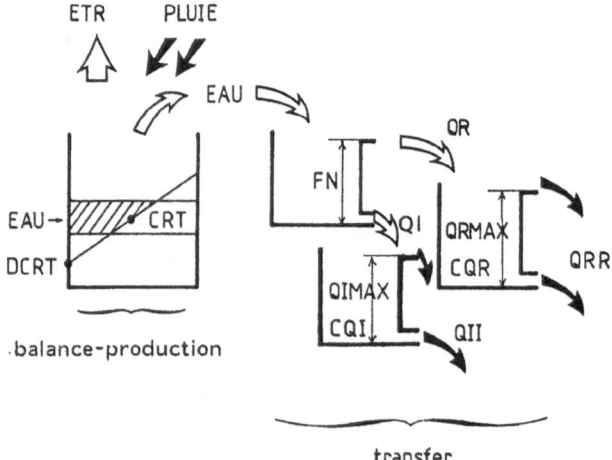

```
RMAX = 2(CRT - DCRT) + DCRT          QR   = max(O, EAU - QI)
RBA  = max(DCRT, R) - DCRT           ETR  = min(R + P - EAU, ETP)
RHA  = min(R + P, RMAX) - DCRT       R    = R + P - EAU - ETR
DR   = max(O, RHA - RBA)
EAU  = max(RP - RMAX, O) + DR(2 RBA + DR)/(4(CRT - DCRT))
QI   = min(EAU, FN)
```

Fig. 1. A schematic outline of the water production function in the model.

In practice, DCRT and RMAX are the major parameters of the model since the amount of water flowing out depends only on the balance-production reservoir, and therefore these parameters describe the soil and its use.

The model requires meteorological data (precipitation, potential evapotranspiration) and gives firstly the quantity of runoff fed into the surface domain and infiltrated into the soil. Secondly, the model makes it possible to describe the actual evapotranspiration and the overall water content in the soil.

The parameter values are determined by fitting in order to obtain the best possible agreement between the measured and the simulated values.

Modelling of the nitrogen cycle and the leaching of nitrate

The various processes involved in the N cycle of the soil-water-plant-atmosphere system can be divided into the following categories:
◇ processes governing the input of the system:
  – biological fixation
  – contribution from the atmosphere
  – artificial additions
◇ processes governing the output of the system:
  – denitrification
  – volatilization of gaseous nitrogen
  – leaching of nitrate
  – plant uptake
◇ processes governing internal transformations:
  – mineralization
  – nitrification
  – immobilization of N from ammonium, immobilization of N from nitrate
  – adsorption-desorption of ammonium

The interaction of the various processes and the existence of several chemical forms of nitrogen make the N cycle extremely complex. Insufficient knowledge and the predominance of certain processes over others have caused us to adopt the following simplifying hypotheses [2]:
1. Soil organic nitrogen is considered as one single pool.
2. Input into the system by biological fixation and from the atmosphere is compensated by output of the system by volatilization and denitrification.

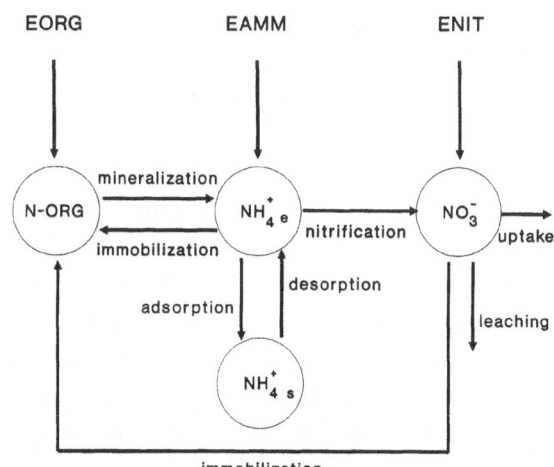

Fig. 2. Diagram of the simplified N cycle for the modelling (EORG, EAMM and ENIT are the inputs of organic nitrogen, ammonium nitrogen and nitrate nitrogen, respectively).

3. Because of the rapidity of the process of oxidation to nitrate versus that of oxidation to nitrite, nitrification is considered as a one-step process governed by the slowest rate (oxidation to nitrite). Therefore the nitrite nitrogen form is disregarded.
4. The processes of plant uptake and leaching only concern the nitrate nitrogen form.

Figure 2 represents a diagram of the simplified N cycle used.

Quantitative methods

The quantitative methods used for each process of the N cycle are the following:

1. *Adsorption-desorption of ammonium*

We use a linear and instantaneous equilibrium between the ammonium dissolved in the liquid phase and that adsorbed on the solid phase [1, 9, 12] according to:

$$(NH_4^+)_s = K_d(NH_4^+)_e \tag{1}$$

where $(NH_4^+)_s$: adsorbed ammonium (kg N ha^{-1}); $(NH_4^+)_e$: dissolved ammonium (kg N ha^{-1}); K_d: distribution coefficient.

2. *Uptake by plants*

The method used to simulate N uptake by plants is of the Michaelis–Menten type. The N uptake rate function can be expressed in the following form [2]:

$$f(t) = B f_{pp}\left(\frac{t}{T}\right) \frac{(NO_3^-)}{(NO_3^-) + K_{ab}} \qquad (2)$$

where $f(t)$: N uptake rate function (kg N ha^{-1} day^{-1}); t: time counted from the sowing date (day); T: life of the crop (time between sowing and harvesting) (day); $f_{pp}\left(\frac{t}{T}\right)$: reduced function (day^{-1}) which calculates the daily fraction of total N uptake by plants; B: potential total N uptake by the crop (kg N ha^{-1}); K_{ab}: Michaelis–Menten coefficient (kg N ha^{-1}).

3. Other biological transformations

For mineralization, nitrification and immobilization of N from ammonium and nitrate we use, in common with many authors [1, 6, 7, 8, 9, 12], first-order kinetics:

$$\frac{dM}{dt} = -kM \qquad (3)$$

where M: nitrogen in the compartment in question (kg N ha^{-1}); and k: first-order rate coefficient(day^{-1})

4. Correction of biochemical parameters according to moisture and temperature

The following equations, based on experimental results from the literature [8, 10, 11] are used:

$$k = k_0 Q_0^{(T-T_m)} \frac{\theta}{cc} \quad \text{if} \quad \theta < cc \qquad (4a)$$

$$k = k_0 Q_0^{(T-T_m)} \frac{cc}{\theta} \quad \text{if} \quad \theta > cc \qquad (4b)$$

where k_0: optimal first-order rate coefficient or biochemical parameter (day^{-1}); Q_0: correction coefficient as a function of the temperature T ($Q_0 = 1.071$ for mineralization [10] and for nitrification [8] and $Q_0 = 1.05$ for immobilization [8]); T: soil temperature (°C); T_m: optimum temperature, usually taken as 35°C; θ: moisture content (cm^3 cm^{-3}); and cc: field capacity (cm^3 cm^{-3}).

The four biochemical parameters: k_{mi} is the mineralization coefficient, k_{ni} is the nitrification coefficient, k_{rn} is the immobilization coefficient

for nitrate, k_{ra} is the immobilization coefficient for ammonium.

Owing to a lack of data we assume that the distribution coefficient (K_d) and the Michaelis–Menten coefficient (K_{ab}) have a constant value for a given soil and for a given crop.

The origin of the method used in the model is an empirical description of the water cycle to which a certain number of phenomenological compartments have been added to account for the N cycle. The nitrate production function is built from balance and transfer reservoirs in the same way as the water production function. The balance of the various nitrogenous forms is achieved by means of equations, and a numerical solution of this system of equations can be obtained by transforming differentials into finite differences, which makes it possible to calculate the flow of nitrate migrating toward the aquifer.

In addition to soil characteristics (type, bulk density, depth, field capacity, temperature) the model also requires knowledge of nitrogenous additives (chemical form, date of introduction) and crop characteristics (type of crops, date of sowing and harvesting, total N uptake by plants).

The simulation provides the flow of leached nitrate and the total N uptake by plants, which data are verified at various scales. It also gives the different flows relative to the biochemical transformations and the nitrogen content in the different soil compartments.

The values of the biochemical parameters (k_0) are determined by fitting in order to obtain the best possible agreement between the measured and the simulated leachate flow.

Materials and methods

On three experimental farms wheat was grown on different soils: a silty clay, a clay loam and a sandy loam on the farms of Bouwing, Eest and PAGV respectively. For each farm and for three treatments of increasing fertilization (N1, N2, N3) data were collected every two or three weeks during each growing season (1982–1983 and 1983–1984); they are described in detail elsewhere [4].

For practical use of the model the fitting methods had to be modified.

For the water balance we had no information concerning the drainage flow of water, but only about the water content in the soil at different depths. From these data we estimated the water reserves in the soil so as to be able to compare them with the reserve R of the water production function (Fig. 1). The parameter values were determined by fitting the measured and simulated reserves to each other.

For the N balance we have taken as a basis the nitrate contents versus the depth and the root length density profiles according to [4]. From these we estimated a reserve of nitrate nitrogen from which the plant can feed itself, and which was then compared with the contents of the balance reservoir of nitrate nitrogen in the model. The introduction of N fertilizers is simulated by the model by assuming that progressive dissolution takes place as a function of daily rainfall and maximum solubility.

For N uptake by plants a function was established assuming a nonlimiting nutrient supply in the soil and absence of moisture stress in the plant [13]. Figure 3 shows the function of total N uptake versus time in nondimensional coordinates. This function, developed for the growing of corn, could not be used in the present case. We therefore worked out a new function on the basis of the data from Bouwing N3, considering that in the year 1983–1984 the conditions of moisture and nitrogen were optimum for development of the plant.

The calculations were made with a time step of one day.

Results

The results of the modelling include the variation in the water reserve in the soil, the variation in nitrate content in the root zone and the total N uptake by plants.

In particular, the results from the experimental farms of Bouwing and Eest are presented. We chose the results of 1983–1984 for which we had the most complete data set, especially the root length density and nitrate profiles in the soil.

Farm 'the Bouwing'

Figure 4 shows the comparison of measured and simulated water contents in the soil as a function of time for the layer 0–100 cm (treatment N1).

There is a good agreement between the observations and the simulation except for the summer season where the model underestimates the water reserve.

The values of the parameters DCRT [minimum level of the water reserve in the soil, below which no water is available] and RMAX [maximum level of the water reserve in the soil, above which all rainfall contributes to the available water] of the water production function obtained during the fitting are the following:

$$DCRT = 310 \, mm \text{ and } RMAX = 380 \, mm$$

The values of the parameters of the water balance were identical for the three treatments

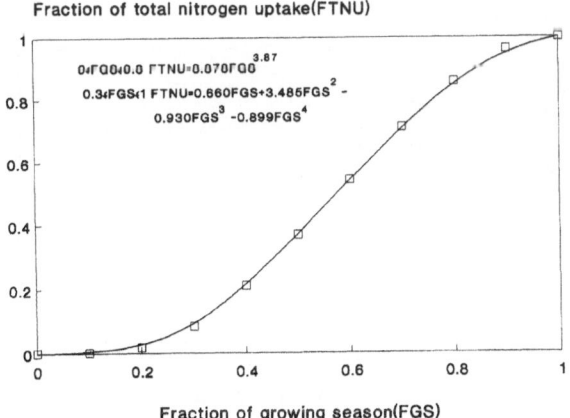

Fraction of total nitrogen uptake(FTNU)

Fig. 3. Dimensionless N uptake function ($f_{pp}(t/T)$) [13].

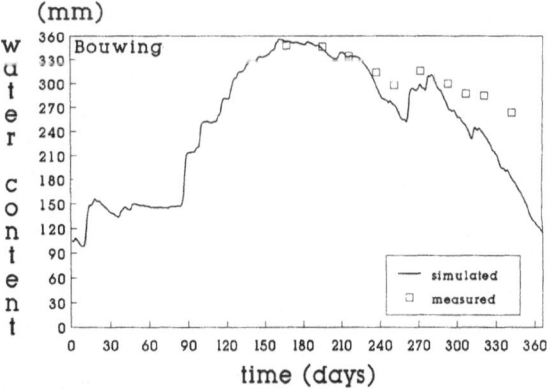

Fig. 4. Comparison of measured and simulated water contents in the soil (0–100 cm, treatment N1) in 1983–1984 (1/09/83 to 31/08/84).

(N1, N2, N3). Although the fertilization rate increased, the differences in water reserves in the soil along treatments were small and only apparent at the end of the growing season, during which period the model underestimated the water reserve in the soil.

The comparison of measured and simulated nitrate contents in the root zone is shown in Fig. 5.

The model correctly simulated the course of the nitrate content except for:

– the two specific points marked by an asterisk. These points are characterized by a higher level of nitrate in the entire layer as well as by a high N uptake by plants compared to the previous measurement. This high level was all the more marked when fertilization was in-

creased. This phenomenon cannot be explained by the model;

– the summer season of treatment N2 where the model underestimated the nitrate contents. We obtained similar results for treatment N3.

The biochemical parameters involved in simulation of the soil N balance were obtained by fitting and had the following values: k_{mi} (mineralization coefficient) = 0.002 day^{-1}, k_{ni} (nitrification coefficient) = 0.8 day^{-1}, k_{rn} (immobilization coefficient for nitrate) = 0.02 day^{-1}, k_{ra} (immobilization coefficient for ammonium) = 0.02 day^{-1}.

The simulation of total N uptake by plants (Fig. 6) can only be satisfactory since the N uptake function was calculated from the measurements.

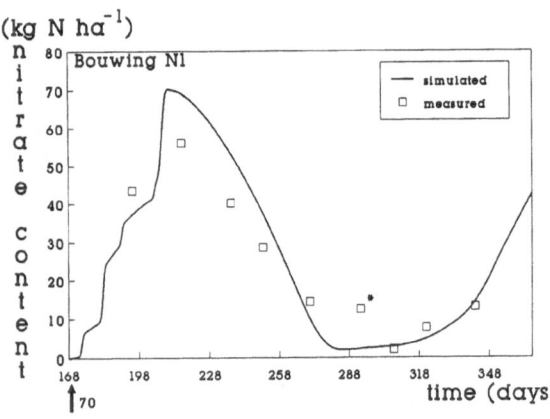

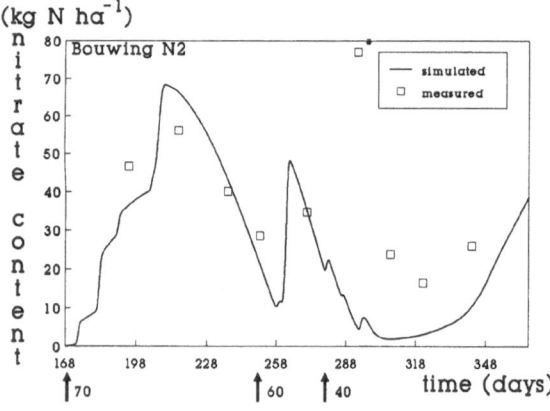

Fig. 5. Comparison of measured and simulated nitrate contents in the root zone in 1984 for treatments N1 and N2 (14/02 to 31/08). The numbers near arrows represent the application of fertilizers (kg N ha^{-1}).

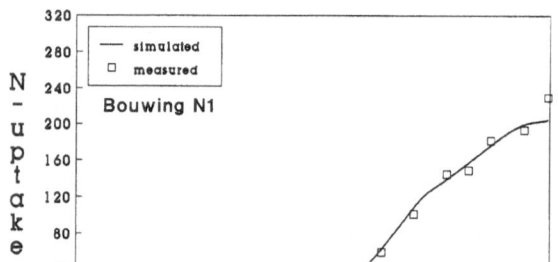

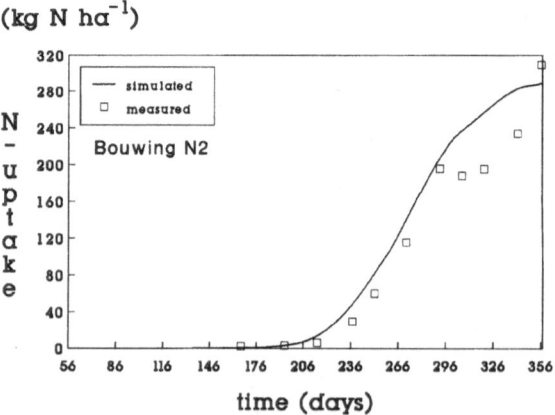

Fig. 6. Comparison of measured and simulated N uptake by wheat in 1983–1984 (27/10/83 to 21/08/84).

In order to follow N uptake by plants versus time, small plots ($0.5 \, m^2$ in 8 replicates) of wheat were harvested periodically and nitrogen content in the different parts were measured. We found that in the growing season the measured nitrogen contents did not always increase (treatment N2). Necessarily, the model simulates an increasing uptake and thus overestimates the plant consumption. This excessive uptake of nitrate in the balance reservoir causes an underestimation of the nitrate content in the root zone (Fig. 5).

Farm 'the Eest'

Figure 7 shows the measured and simulated variations in the water content in the soil as a function of time for the layer 0–100 cm (treatment N1).

The simulation results were less satisfactory than those of the Bouwing farm and the simulated reserve was already underestimated in spring.

The values of the parameters DCRT and RMAX of the water production function are the following:

$$DCRT = 405 \, mm \text{ and } RMAX = 480 \, mm$$

These values were also maintained during the three fertilization treatments for the same reasons as before.

Figure 8 shows the comparison of measured and simulated nitrate contents in the root zone.

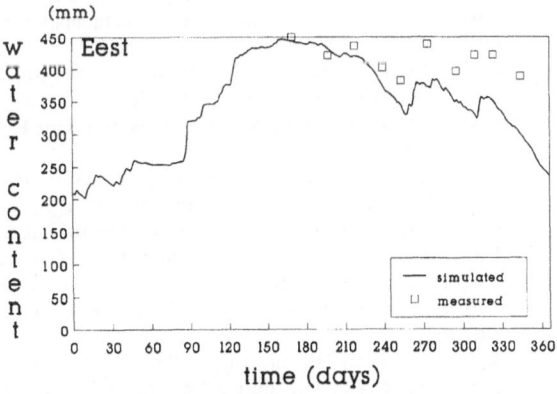

Fig. 7. Comparison of measured and simulated water contents in the soil (0–100 cm, treatment N1) in 1983–1984 (1/09/83 to 31/08/84).

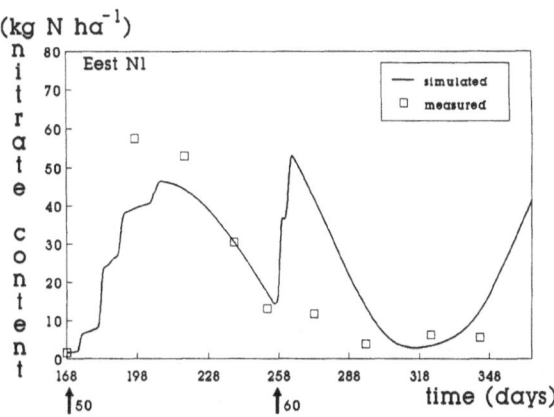

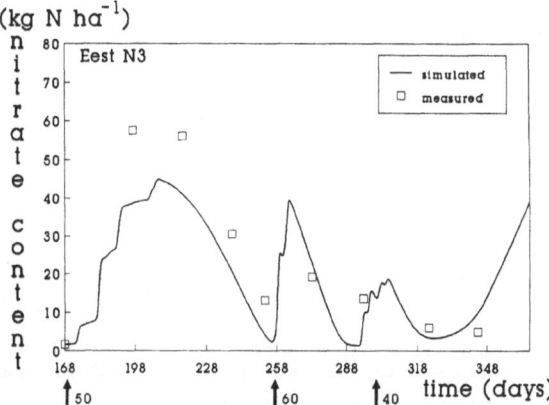

Fig. 8. Comparison of measured and simulated nitrate contents in the root zone in 1984 for treatments N1 and N3 (17/02 to 31/08). The numbers near arrows represent the application of fertilizers ($kg \, N \, ha^{-1}$).

In general, the simulated course of nitrate content in the root zone reflects the measurements, but in two cases (N1 and N3) the model underestimates the reserve of nitrate nitrogen in the soil at the beginning of the simulation.

The values of the biochemical parameters, obtained by fitting, are the following: k_{mi} (mineralization coefficient) = $0.002 \, day^{-1}$, k_{ni} (nitrification coefficient) = $0.5 \, day^{-1}$, k_{rn} (immobilization coefficient for nitrate) = $0.02 \, day^{-1}$, k_{ra} (immobilization coefficient for ammonium) = $0.02 \, day^{-1}$.

For N uptake by plants the function developed on the basis of the Bouwing data was used for the simulation. The results of total N uptake by plants perfectly agreed with the measurements (Fig. 9).

168

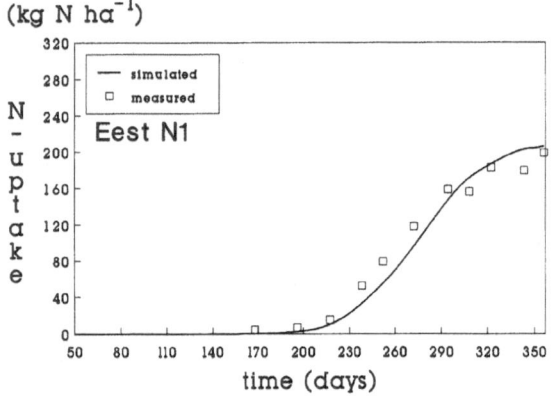

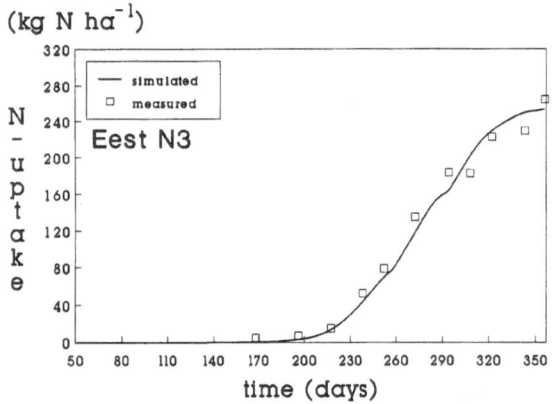

Fig. 9. Comparison of measured and simulated N uptake by wheat in 1983–1984 (21/10/83 to 22/08/84).

For all the practical applications the Michaelis-Menten coefficient (K_{ab}) was assumed to be equal to 3 kg N ha^{-1} for the wheat crop [2]. The values of the biochemical parameters suggested for these simulations are close to those found in the literature [2, 6, 8].

Conclusion

One must remember that the model presented and tested here was developed for the purpose of obtaining a tool for the problems of aquifer contamination and that the soil only represents a small part of the hydrologic system.

Because of a lack of data the calculation of nitrate leaching, the main purpose of the model, could not be validated.

As a consequence certain improvements concerning the N uptake by plants and the organic nitrogen compartment in the soil are necessary. It would be useful to find a function of N uptake which could either be generalized or be attributed specifically to each group of crops (winter, spring, fodder, etc.). A division of the organic nitrogen compartment into different forms that are more or less easily mineralized also seems necessary.

The use of the model on soil data such as those from the Dutch experimental farms [4] represents a further stage in the complete validation of the model.

Acknowledgements

This research project has received financial support from the French Secrétariat d'Etat auprès du Premier Ministre chargé de l'Environnement, Service de la Recherche des Etudes et du Traitement de l'Information sur l'Environnement and from the European Economic Community.

References

1. Cho CM (1971) Convective transport of ammonium with nitrification in soil. Can J Soil Sci 52: 339–350
2. Geng QZ (1988) Modélisation conjointe du cycle de l'eau et du transfert des nitrates dans un système hydrologique. Thèse de doctorat, Ecole Nationale Supérieure des Mines de Paris, Paris
3. Girard G (1975) Modèle global ORSTOM 1974. Première application du modèle à discrétisation spatiale sur le bassin versant de la crique Grégoire en Guyanne. Atelier hydrologique sur les modèles mathématiques, ORSTOM
4. Groot JJR and Verberne ELJ (1991) Response of wheat to nitrogen fertilization, a data set to validate simulation models for nitrogen dynamics in crop and soil. Fert Res 27: 349–383
5. Ledoux E (1980) Modélisation intégrée des écoulements de surface et des écoulements souterrains sur un bassin hydrologique. Thèse de Docteur-Ingénieur, Ecole Nationale Supérieure des Mines de Paris, Paris
6. Mehran M and Tanji KK (1974) Computer modeling of nitrogen transformations in soils. J Environ Qual 3: 391–396
7. Misra C, Nielsen DR and Biggar JW (1974) Nitrogen transformations in soils during leaching. I Theoretical considerations. II Steady state nitrification and nitrate reduction. III Nitrate reduction in soils columns. Soil Sci Soc Am Proc 38: 289–304

8. Prat M (1982) Simulation numérique du transport de produits réactifs dans les sols. Cas de l'azote dans les relations bassin versant rivière. Thèse de Docteur-Ingénieur, Institut National Polytechnique, Toulouse

9. Selim HM and Iskandar IK (1981) Modeling nitrogen transport and transformations in soils. Soil Sci 131: 233–241

10. Stanford G, Frere MH and Schwaninger DH (1973) Temperature coefficient of soil nitrogen mineralization. Soil Sci 115: 321–323

11. Stanford G and Epstein E (1974) Nitrogen mineralization. Water relations in soils. Soil Sci Soc Am Proc 38: 103–107

12. Tillotson WR and Wagenet RJ (1982) Simulation of fertilizer nitrogen under cropped situations. Soil Sci 133: 133–143

13. Watts DG and Hanks RJ (1978) A soil-water-nitrogen model for irrigated corn on sandy soils. Soil Sci Soc Am J 42: 492–499

Fertilizer Research **27**: 171–180, 1991.
© 1991 *Kluwer Academic Publishers.*

Nitrate leaching and soil moisture prediction with the LEACHM model

C. Ramos & E.A. Carbonell
Instituto Valenciano de Investigaciones Agrarias, Apdo. Oficial, 46113 Moncada, Spain

Key words: Nitrate leaching, soil moisture, soil mineral nitrogen, models

Abstract

The LEACHM model developed by Wagenet and Hutson [1989] was used to predict the mineral nitrogen and water content in the soil under a winter wheat crop from February to April in two years and three locations. The model grossly overestimated soil water content, probably due to the bad fitting of the assumed water retentivity function to the experimental data at high water contents, and to the presence of a relatively shallow water table (1.0–1.5 m). Measured soil hydraulic conductivity varied with water content in a different manner than predicted by the model. By assuming a sandy or gravelly soil layer between the bottom of the measured soil profile and the water table, prediction of soil water content improved considerably. Simulation showed that, under the experimental conditions studied, soil mineral nitrogen varied mainly due to the fertilizer additions, mineralization and denitrification. Nitrogen uptake by plants and leaching were small. Low values of nitrate leaching were predicted by the model because of low drainage. Large differences between predicted and observed values in the mineral nitrogen in the soil occurred in some cases, both in the total amount and its profile distribution.

Introduction

Transport of water and solutes in the soil under field conditions is a research area of continuous interest as some recent publications show [21, 24, 34]. There are different modeling approaches to describe solute leaching in field soils, but testing of these models has, in general, been restricted to those who developed them. As pointed out by some workers [1, 25] there is a need for validating these models under different soil and environmental conditions. Following this idea, we present here an application of the model LEACHM [29], to estimate soil moisture and soil mineral nitrogen under a winter wheat crop in two years (1983, 1984) and three experimental locations (Bouwing, PAGV, EEST), using an existing data set [9].

Description of LEACHM

A detailed description of the model was given by their developers [29]. Here, a general view of the model will be given and the main points relevant to its specific use in our simulations will be emphasized.

LEACHM evolved from modeling efforts in the last twenty years and has successfully been used by several workers to describe pesticide movement in field soils [28, 30]. It is a research type model that can be used for management purposes if some of the required inputs, difficult to obtain, are estimated and not actually measured. It has four modules, all of which with similar schemes, to simulate water and chemical movement. One of these modules, LEACHN, describes nitrogen transport and transformation and it is the one used in the present simulations; a second one, LEACHP, deals with pesticides; a third one, LEACHC, describes flow of inorganic ions (Ca, Mg, Na, K, SO_4, Cl, CO_3, HCO_3) and a fourth one, LEACHW, describes water transport only. LEACHM has also a heat flow subroutine that allows to adjust rate constants according to temperature, although this adjustment is not included in the second version used here. Figure 1 shows a flowchart for LEACHN.

In each of the four modules, there is a main

172

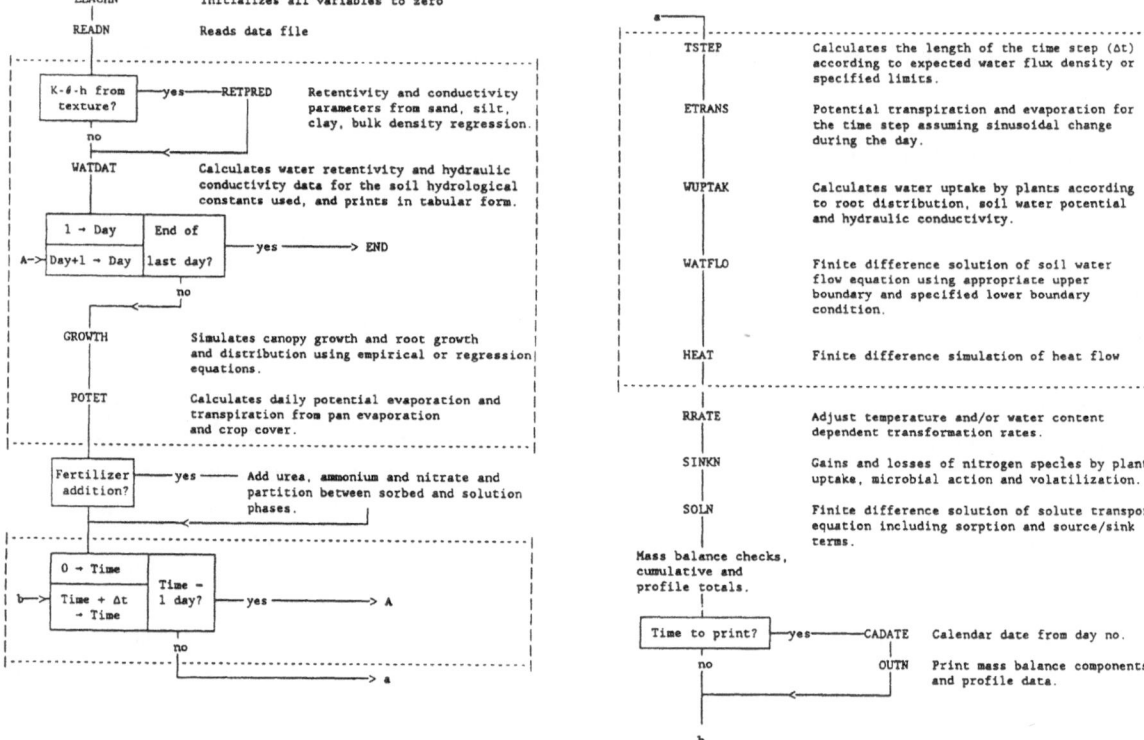

Fig. 1. Flow chart for LEACHN, the nitrogen module of LEACHM (after Wagenet and Hutson [29]). Subroutines are given in capitals.

program that initializes variables, calls subroutines and performs mass balancing. Subroutines deal with the different processes such as water and solute flow, evapotranspiration, sinks, sources, plant growth, heat flow, and also with data input and output, and time step calculation. Main required inputs for LEACHN are:

1. Soil data for the different layers of equal thickness:
   – initial water content or water potential.
   – hydrological constants for the moisture retentivity and hydraulic conductivity curves. The model can also estimate these curves from particle size distribution.
   – initial inorganic nitrogen content.
2. Soil surface boundary conditions:
   – irrigation and rainfall amounts and rate of application.
   – mineral nitrogen fertilizer application rates and dates.
   – mean temperature and diurnal amplitude

for each period regarded as having a constant temperature regime (only if a temperature simulation is required).
   – weekly totals of pan evaporation.
3. Crop data:
   – time of planting.
   – root and crop maturity and harvest dates.
   – root and ground cover growth parameters.
   – pan factor for converting pan evaporation to potential crop evapotranspiration.
   – lower soil and plant water potentials for water extraction by plants.
4. Other constants needed include diffusion coefficients, maximum value for time step, dispersivity coefficient, rate constants for the nitrogen transformations considered (urea hydrolysis, ammonia volatilization, nitrification and denitrification) and the adsorption coefficients for urea, ammonium and nitrate.

LEACHM is not intended to:
   – use unequal depth increments.

– predict runoff water quantity or quality.
– simulate how plants respond to soil or environmental changes.
– predict crop yields.
– predict solute distributions in situations with two- or three-dimensional flux patterns.

Soil water

a) Hydraulic retentivity and conductivity functions

LEACHM solves the Richard's equation using a numerical procedure. Basic to the formulation used are the hydraulic retentivity and conductivity functions. These functions are those proposed by Hutson and Cass [11], in which the retentivity equation used by Campbell [5] was modified in the wet range to remove the discontinuity at the water tension equivalent to the air-entry value. Thus, the retentivity curve consists of two parts: a parabolic function at the wet end, and a different power function for the drier part.

For pressure potentials between 0 and h_c:

$$h = [a(1 - \theta/\theta_s)^{1/2}(\theta_c/\theta_s)^{-b}]/(1 - \theta_c/\theta_s)^{1/2} \quad (1)$$

where h is pressure potential, a and b are constants, θ is volumetric water content and θ_s is θ at saturation. In addition, (h_c, θ_c) is the intersection point of the power and parabolic curves. This point is given by

$$h_c = a[2b/(1 + 2b)]^{-b} \quad (2)$$

$$\theta_c = 2b\theta_s/(1 + 2b) \quad (3)$$

For pressure potentials more negative than h_c the relation used is:

$$h = a(\theta/\theta_s)^{-b} \quad (4)$$

The resulting composite retentivity curve is continuous and has a differential water capacity of zero at saturation.

The hydraulic conductivity function used in the model was proposed by Campbell [5]:

$$K(\theta) = K_s(\theta/\theta_s)^{(2b+2+p)} \quad (5)$$

where $K(\theta)$ is hydraulic conductivity at θ, K_s is K

at $\theta = \theta_s$, b is as in (1), and p is a pore interaction parameter, set to 1 in the model.

Matric potentials higher (i.e. closer to zero) than that corresponding to a θ equal to 0.9999 θ_s are not allowed in order to prevent mathematical overflow errors when calculating and using the differential water capacity.

b) Evapotranspiration

At the start of each day, the daily potential evapotranspiration is calculated from the weekly pan evaporation and the pan factor. Then, potential transpiration and evaporation are estimated from the crop cover. An option exists to increase transpiration by a certain amount whenever evaporation is limited by soil conductivity. During the day, it is assumed that evapotranspiration starts at 0.3 day and ends at 0.8 day and that during this period it varies sinusoidally.

The absorption of water by roots is estimated following the method proposed by Nimah and Hanks [18] in which water uptake in each soil layer is a function of potential transpiration, fraction of total root length in the layer, matric and osmotic soil water potential, layer depth, soil hydraulic conductivity, and a root resistance term. No water flow from roots to soil is allowed. There is also a water content below which roots cannot extract water. Therefore, actual transpiration can be lower than potential transpiration.

Solute movement

Solute transport is estimated by a numerical solution to the diffusion-convection equation once water fluxes are estimated with the subroutine WATFLO (Fig. 1). The general transport equation is:

$$\frac{\partial(\theta + \rho k_d + \varepsilon K_H^*)c}{\partial t}$$

$$= \frac{\partial}{\partial z}\left[\theta D(\theta, q)\frac{\partial c}{\partial z} - qc\right] \pm \phi \quad (6)$$

where c is solution concentration of the particular solute in mg l^{-1}, ρ is soil bulk density in kg l^{-1}, k_d is the solute partition coefficient between the liquid and solid phases in l kg^{-1}, ε is the gas filled soil porosity, K_H^* is a modified Henry's law constant, q is the macroscopic water flux in mm day^{-1}, $D(\theta, q)$ is the apparent diffusion coefficient in mm^2 day^{-1} that includes mechanical dispersion and chemical diffusion, z is the soil depth in mm, and ϕ indicates sources and/or sink terms in mg l^{-1} day^{-1}. The chemical diffusion component of $D(\theta, q)$ depends on two empirical constants m and n [20].

Since the finite difference solution of equation (6) may cause considerable numerical dispersion, a correction for this is made. The model developers found that, for time steps less than 0.1 day and for layer thicknesses (Δz) less than 100 mm, numerical dispersion was equivalent to an increase in dispersivity of $0.16\,\Delta z$. This numerical dispersion correction is applied to the mechanical dispersion coefficient only.

Nitrogen transformations and plant uptake

The nitrogen transformation reactions considered in LEACHN are:

$$
\begin{array}{c}
\text{ammonia (gas)} \\
\uparrow k_4 \\
\underset{c_1}{\text{urea}} \xrightarrow{k_1} \underset{c_2}{\text{ammonium}} \xrightarrow{k_2} \underset{c_3}{\text{nitrate}} \xrightarrow{k_3} \text{gas}
\end{array}
$$

where k_1, k_2, k_3, and k_4 represent the rate constants of the different processes and c_1, c_2 and c_3 are the concentrations of urea, ammonium and nitrate, respectively. Concentrations and rates can apply to the liquid phase only or to the liquid + solid phases.

Nitrogen uptake by plants is calculated as [19]:

$$I = 2\pi\alpha rc$$

where I is the rate of nutrient uptake per unit length of root in mg m^{-1} day^{-1}, r is the mean root radius in mm, c is nutrient concentration in solution in mg l^{-1}, and α is a root uptake coefficient in m day^{-1}.

Simulation objectives

There were two main reasons why LEACHN could not be applied directly to the data set provided: 1) plant growth description is for corn, and 2) plant growth is not limited by any environmental factor (temperature, water or nitrogen supply) and depends only on time. Therefore, the simulation objectives were limited to the evaluation of LEACHN in a situation where the effects of plant growth on soil water and mineral nitrogen could be considered negligible in comparison to other nitrogen or water sinks or sources such as drainage, fertilizer additions, or chemical transformations and transport. Thus, the option available in the model, in which root distribution and crop cover are assumed constant, was used. The period in which these conditions were likely to be met was from the first soil sampling (early February) to the third soil sampling (late March or early April). Inspection of the data set on nitrogen uptake by the crop in this period confirmed that, in general, it was quite small (8 kg ha^{-1} on average) compared to the initial mineral nitrogen in the soil or the fertilizer additions.

Input data

a) Soil-water parameters

Details on soil moisture content, water retentivity curves and hydraulic conductivity measurements are given elsewhere [9].

To calculate the parameters a and b required by the model, $\log |h|$ was plotted against $\log \theta$. The points corresponding to low water content followed approximately a straight line. Then, a and b were estimated by linear regression. The complete retentivity curve was obtained by applying eqs. (1) to (4).

Table 1 shows the soil hydraulic parameters used in the simulations. K_s and θ_s were given in the data set, except for the 40 – 100 cm layer of the Bouwing plot where K_s had to be obtained by extrapolation, using the log-linear relationship between K and θ. The hypothetical 'sandy' layer was introduced for convenience and will be discussed later.

b) Potential evapotranspiration estimation

LEACHM calculates crop evapotranspiration by multiplying the required pan evaporation data by

Table 1. Soil hydraulic parameters

Plot	Layer (cm)	K_s (mm day^{-1})	θ_s (cm^3 cm^{-3})	Campbell's parameters a (kPa)	b
Bouwing	0–40	636	0.517	−6.61	4.93
	40–100	856	0.517	−8.32	5.06
	(100–110)(1)	856	0.517	−1.00	2.00
EEST & PAGV	0–20	1780	0.411	−17.0	2.70
	20–40	264	0.499	−9.8	3.15
	40–100	2300	0.523	−10.7	3.50

(1) hypothetical 'sandy' layer (see text)

a pan factor. Since the data set did not provide pan evaporation data, reference (i.e. potential) evapotranspiration, ET_0, was calculated following the FAO modified Penman method [6], using a computer program in which the empirical coefficients required in the FAO procedure were those derived by Frevert et al. [7], and the relationship between air temperature and saturated water vapor pressure was taken from Weiss [31]. In the program input, ET_0 was then substituted for pan evaporation assuming a pan factor of 1.

For comparative purposes two additional methods of ET_0 estimation were used: 1) the FAO radiation method [6], and 2) the Makkink method, originally calibrated for Dutch conditions [6]. Daily climate data from nine weeks taken at random from the two sites and years, during the February-March period, were used in the comparison.

In LEACHM, crop cover fraction is numerically equal to the ratio of plant transpiration to evapotranspiration when the soil surface is not limiting evaporation. In the simulation, a constant value for the crop cover fraction was taken for the whole study period. This fraction was estimated using results by Ritchie and Burnett [22]. They found that in a subhumid climate and under nonlimiting soil water conditions, the transpiration to evapotranspiration ratio was a common function of the leaf area index for cotton and grain sorghum. In the present case, the leaf area index for the different plots and years in the selected periods varied mainly from about 0.15 to 0.40. These values correspond, using the relationship found by these workers, to a transpiration/evapotranspiration ratio of 0.06

and 0.23, respectively. Therefore, a constant value of 0.15 for the crop cover fraction was selected for the whole study period.

c) Rate constants and other parameters

Rate constants for the nitrogen transformations and other parameter values were chosen after a review of literature data. A summary of these values is presented in Table 2.

Most of the rate constant values found in the literature had been obtained at higher temperatures than those occurring in the upper soil layers during the simulation period (estimated to be about 5°C on average). A temperature adjustment was made using a Q(10°C) of 2.3, an average value found by different workers [2, 23, 32].

Although the model does not consider mineralization, it was modified to include a constant mineralization rate of 0.15 and 0.075 kg N ha^{-1} day^{-1} for layers 0–30 cm and 30–60 cm, respectively. These values are somewhat lower than those reported by others [2, 17, 33] for temperatures 5 to 15°C higher.

Results and discussion

a) Soil water

Before comparing field soil moisture predictions with measured data, the adequacy of the soil hydraulic functions will be discussed. Figure 2 shows the fitting of the Hutson and Cass model [11] used in LEACHM to the experimental data for the 0–40 cm layer in the Bouwing plot. It is

Table 2. Some parameter values used in the simulations

Parameter	Value	Observed range	References
Diffus. coef. in water (mm^2/day)	120	86–173	3, 19
Apparent diff. coef. constants			
m	0.005	0.005–0.010	20
n	10	10	20
Dispersivity (mm)	40	1–300	4, 8, 13
NH$_4^+$ partition coef. (dm^3 kg^{-1})	5	1–9	14, 15, 17
Nitrific. rate const. (day^{-1}) (1)		0.05–0.30 (2)	12, 15, 16, 26, 27
0–30 cm	0.10		
30–60 cm	0.05		
60–100 cm	0.02		
Denitrif. rate const. (day^{-1}) (1)		0.0003–0.02 (2)	10, 12, 15, 16, 27
0–30 cm	0.005		
30–60 cm	0.002		
60–100 cm	0.0001		

(1) Assuming first order kinetics
(2) Values adjusted to $T = 5°C$ assuming $Q(10°C) = 2.3$

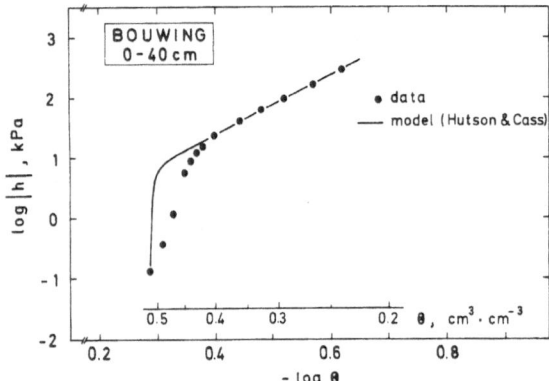

Fig. 2. Fitting of experimental data to the Hutson and Cass soil retentivity function [11] for the 0–40 cm layer at the Bouwing site.

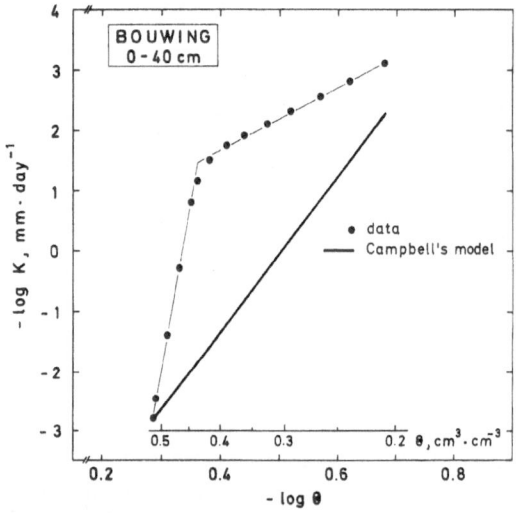

Fig. 3. Fitting of experimental data to the Campbell soil hydraulic conductivity function [5] for the 0–40 cm layer at the Bouwing site.

apparent that in the wetter part of the curve (i.e. $\theta > \theta_c$), corresponding to the parabolic relationship, the model predicted higher θ values than measured. The good fitting for the drier part of the curve is because here the relationship between θ and h follows well a power function and a and b were actually obtained from these points by linear regression. Results for the 40–100 cm depth and for all depths in the EEST and PAGV plots showed a similar type of curves to that in Fig. 2.

Figure 3 shows the relationship between log K and log θ. In Campbell's model, used in LEACHM, a linear relationship is proposed for the whole range of θ. However, data clearly show two linear relationships. Changes were made in LEACHM so that by entering appropriate p values for each of the linear parts of the log K vs. log θ relationship, a good fitting was obtained in the whole range of θ.

Soil water content was first simulated in the Bouwing plot and although no data were available on the water table depth in 1983, it was assumed at 1.5 m depth, guessing from 1984

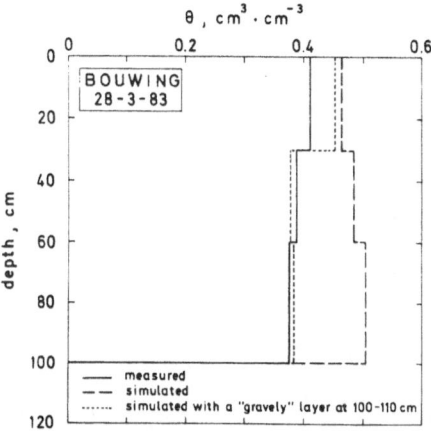

Fig. 4. Predicted and measured volumetric soil water profile at the Bouwing site on 28-3-83, with or without a hypothetical sandy or gravelly layer at 100–110 cm (see text).

data. Simulation results indicated a high upward water flux and an increase in soil moisture. This contrasted very much with measured values (Fig. 4). This difference is probably due to the bad fitting of the soil water retentivity function at low water tension, and to the imposed bottom boundary condition. Under equilibrium conditions, having a water table at 0.5 m deeper than the bottom of the soil profile is equivalent to maintaining a constant water suction of 5 kPa at that depth; at this suction the simulated water content was much higher than that measured (Fig. 2).

Since at the time of the first soil sampling each year the soil water profile was probably close to equilibrium with the water table, at least in the deeper layers, an attempt was made to improve the simulated values by trying to decrease the predicted upward water flux from the water table. Lowering water table levels was successful only when set at unrealistic values (i.e. >5 m). Another way of preventing the predicted upward water flow from the water table was to assume a soil layer at 100–110 cm with a sandy or gravely texture since small water tensions would decrease its water content and its hydraulic conductivity would become very small, reducing the wetting influence of the water table on the soil above. This procedure was quite successful. Differences between measured and simulated soil water profiles were similar for all samplings and years in the Bouwing plot, and the introduction of the hypothetical 'sandy' layer considerably improved the simulation (Fig. 4). Information provided later by the farmer confirmed the presence of layers of gravel at approximately 1.00–1.20 m depth (JJR Groot, personal communication).

Table 3 summarizes soil water content predictions and measurements. It is very apparent that LEACHM overestimated soil water content in all cases. Although not presented in Table 3, differences between predicted and measured soil

Table 3. Soil water balance (mm) for different plots and dates

Plot	Date (1)	Initial content (2)	Rainfall (3)	Predicted (3)			Measured soil water	Soil water predic.-measur.
				ET	Drainage	Soil water		
Bouwing	28–2–83	377	8	18	−109	476	385	91
Bouwing	28–3–83	377	64	51	−96	486	391	95
Bouwing	13–3–84	348	20	24	−129	474	346	128
Bouwing	3–4–84	348	48	60	−140	477	334	143
EEST	2–3–83	443	6	18	−44	475	368	107
EEST	30–3–83	443	57	58	−34	476	466	10
EEST	15–3–84	447	26	28	−29	473	421	52
EEST	5–4–84	447	56	66	−40	476	436	40
PAGV	1–3–83	356	6	16	−109	456	436	20
PAGV	29–3–83	356	57	44	−110	477	373	104
PAGV	14–3–83	371	26	25	−89	462	367	95
PAGV	3–4–84	371	50	53	−102	471	361	110

(1) Date of the second and third sampling each year.
(2) Measured at the first (initial) sampling each year.
(3) Cumulative since the first sampling each year.

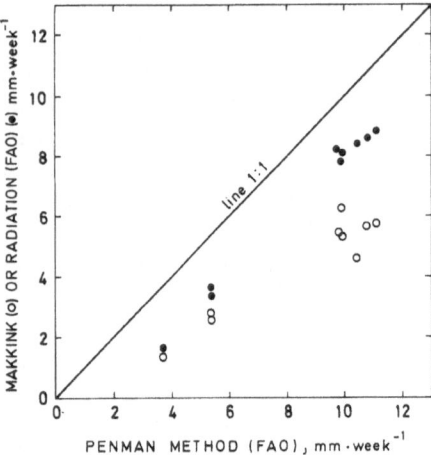

Fig. 5. Comparison of ET0 estimates by three different methods for the February-March period in the two experimental years and locations.

water content for the whole profile Bouwing plot, assuming the 'sandy' layer, ranged from -12 to $19\,\mathrm{mm}$, indicating a better simulation. Upward flow (i.e. negative drainage) was simulated in all cases (Table 3).

Predicted soil water in the whole profile tended to be quite constant for all plots and years, irrespective of the soil water content at the first sampling each year and of rainfall or *ET*. This suggests that the water table had a dominant influence on the soil water content.

Although calculated *ET* was small in comparison to drainage, and therefore errors in its estimation would have a small influence on predicted soil water, it is interesting to compare the three methods of estimating *ET*0 (Fig. 5). The two radiation methods gave lower *ET* values than the Penman-FAO method. This difference may be due to the quite low total radiation values measured during the two months used in the comparison. These results point out some of the uncertainties that still remain when trying to estimate potential crop evapotranspiration.

b) *Soil mineral nitrogen*

A comparison of the predicted and measured soil mineral nitrogen is presented in Table 4. The main components considered in the nitrogen balance are also shown. There was a tendency to underestimate soil mineral nitrogen with

Table 4. Soil mineral nitrogen (kg ha^{-1}) for different plots and dates.

Plot	Date (1)	Initial (2)	Additions (3)		Losses (3)			Predict.	Measured	Pred.-Measur.
			fert.	mineraliz.	leach.	denitr.	pl. uptake			
Bouwing	28-2-83	113	0	5	0	3	2	112	130	-18
Bouwing	28-3-83	113	0	11	0	7	5	110	86	24
Bouwing	13-3-84	29	70	6	0	7	3	96	98	-2
Bouwing	3-4-84	29	70	11	0	14	6	91	109	-18
EEST	2-3-83	94	0	5	1	3	2	93	116	-23
EEST	30-3-83	94	0	11	8	6	4	86	61	25
EEST	15-3-84	36	50	6	0	6	3	83	133	-50
EEST	5-4-84	36	50	11	1	11	6	79	110	-31
PAGV	1-3-83	46	80	5	0	3	2	126	47	79
PAGV	29-3-83	46	80	11	0	15	6	115	54	61
PAGV	14-3-84	16	80	6	0	8	3	91	124	-33
PAGV	3-4-84	16	80	11	0	15	6	85	124	-39

(1) Date of the second and third sampling each year.
(2) Measured at the first sampling each year.
(3) Cumulative since the first sampling each year.

LEACHM, although in PAGV 1983 predicted values were more than twice those measured. Disregarding these results, the relative errors ranged from −38 to 41%, with an average value of −11%.

Nitrate leaching prediction was extremely low, undoubtedly due to the lack of water drainage. Although predicted drainage was negative, some losses of nitrate nitrogen were estimated, probably because of an occasional downward flux of nitrate below the soil bottom boundary and then an upward flow, the latter with an imposed zero nitrate concentration.

Predicted mineralization values were almost the same for all plots and years because of the assumed constant rate and the similar time intervals between soil sampling in all cases. Denitrification losses were comparable to mineralization gains, whereas simulated plant uptake was about half those values.

It is concluded that, under the experimental conditions considered, the soil water retentivity function used in LEACHM was not appropriate and resulted in a overestimation of soil water content. This, in turn, affected nitrate leaching estimation and may partially explain some of the differences observed between predicted and measured soil mineral nitrogen values.

References

1. Addiscott TM and Wagenet RJ (1985) Concepts of solute leaching in soils: a review of modeling approaches. J Soil Sci 36: 411–424
2. Addiscott TM and Whitmore AP (1987) Computer simulation of changes in soil mineral nitrogen and crop nitrogen during autumn, winter and spring. J Agric Sci 109: 141–157
3. Barber SA (1984) Soil Nutrient Bioavailability. New York: Wiley
4. Beese F and Wierenga PJ (1983) The variability of the apparent diffusion coefficient in undisturbed soil columns. Z Pflanzenernaehr Bodenk 146: 302–315
5. Campbell G (1974) A simple method for determining unsaturated conductivity from moisture retention data. Soil Sci 117: 311–314
6. Doorenbos J and Pruitt WO (1984) Crop water requirements. FAO Irrig & Drain Paper No. 24, FAO, Rome
7. Frevert DK, Hill RW and Braaten BC (1983) Estimation of FAO evapotranspiration coefficients. J Irrig Drain Eng 109: 265–270
8. Gelhar LW, Matutoglou A, Welty C and Rehfeldt KR (1985) A review of field-scale physical solute transport processes in saturated and unsaturated porous media, EPRI Topical Report EA-4190, Electric Power Research
9. Groot JJR and Verberne ELJ (1991) Response of wheat to nitrogen fertilisation, a data set to validate simulation models for nitrogen dynamics in crop and soil. Fert Res 27: 349–383
10. Hagin J and Welte E (1984) Nitrogen Dynamics Model Verification and Practical Application. Verlag Erich Goltze, Göttingen, F.R.G.
11. Hutson JL and Cass A (1987) A retentivity function for use in soil-water simulation models. J Soil Sci 38: 105–113
12. Johnsson H, Bergström L, Jansson PE and Paustian K (1987) Simulated nitrogen dynamics and losses in a layered agricultural soil. Agric Ecosystems Environ 18: 333–356
13. Jury WS (1988) Solute transport and dispersion. In: Steffen WL and Denmead OT (eds.) Flow and Transport in the Natural Environment: Advances and Applications. pp 1–16. Berlin: Springer-Verlag
14. Khan MA, Green RE and Cheng P (1981) A numerical simulation model to describe nitrogen movement in the soil with intermittent irrigation. HITAHR Res Ser 010–11.81 University of Hawaii at Manoa
15. Misra C, Nielsen DR and Biggar JW (1974) Nitrogen transformations in soil during leaching: II. Steady state nitrification and nitrate reduction. Soil Sci Soc Am Proc 38: 294–299
16. Myrold DD and Tiedje JM (1986) Simultaneous estimation of several nitrogen cycle rates using 15N: Theory and application. Soil Biol Biochem 18: 559–568
17. Neeteson JJ, Greenwood DJ and Draycott A (1987) A dynamic model to predict yield and optimum nitrogen fertilizer application rate for potatoes. Proceedings 262. The Fertiliser Society, London
18. Nimah MN and Hanks RJ (1973) Model for estimation of soil water, plant, and atmospheric interrelations: I. Description and sensitivity. Soil Sci Am Proc 37: 522–527
19. Nye PH and Tinker PB (1977) Solute Movement in the Soil-Root System. Berkeley: University of California Press
20. Olsen SR and Kemper WD (1968) Movement of nutrients to plant roots. Adv Agron 20: 91–151
21. Richter J (1987) The Soil as a Reactor. Modelling Processes in the Soil. Catena Verlag, Cremlingen, FRG
22. Ritchie JT and Burnett E (1971) Dry land evaporative flux in a subhumid climate: II. Plant influences. Agron J 63: 56–62
23. Stanford G, Frere MH and Van der Pol RA (1975) Effect of fluctuating temperatures on soil nitrogen mineralization. Soil Sci 119: 222–226
24. Steffen WL and Denmead OT (eds.) (1988) Flow and Transport in the Natural Environment: Advances and Applications. Springer-Verlag, Berlin
25. Tanji KK (1982) Modeling of the soil nitrogen cycle. In: Stevenson FJ (ed.) Nitrogen in Agricultural Soils, pp 721–772. Madison, Wisconsin: Am Soc Agron

26. Tanji KK and Mehran M (1979) Conceptual and dynamic models for nitrogen in irrigated croplands. In: Pratt PF (principal investigator) Nitrate in effluents from irrigated lands. Final Report to the National Science Foundation, Univ of California, Riverside

27. Wagenet RJ, Biggar JW and Nielsen DR (1977) Tracing the transformations of urea fertilizer during leaching. Soil Sci Soc Am J 41: 896–902

28. Wagenet RJ and Hutson JL (1986) Predicting the fate of nonvolatile pesticides in the unsaturated zone. J Environ Qual 15: 315–322

29. Wagenet RJ and Hutson JL (1989) LEACHM: Leaching Estimation and Chemistry Model: A process based model of water and solute movement transformations, plant uptake and chemical reactions in the unsaturated zone. Continuum Vol 2 Version 2 Water Resources Inst, Cornell Univ Ithaca, NY

30. Wagenet RJ, Hutson JL and Biggar JW (1989) Simulating the fate of a volatile pesticide in unsaturated soil: A case study with DBCP. J Environ Qual 18: 78–84

31. Weiss A (1983) A quantitative approach to the Pruitt and Doorenbos version of the Penman equation. Irrig Sci 4: 267–275

32. Westerman DT and Crothers SE (1980) Measuring soil nitrogen mineralization under field conditions. Agron J 72: 1009–1012

33. Whitmore AP and Addiscott TM (1986) Computer simulation of winter leaching losses of nitrate from soils cropped with winter wheat. Soil Use Management 2: 26–30

34. Wierenga PJ and Bachelet D (eds.) (1988) Validation of Flow and Transport Models for the Unsaturated Zone. Proc Int Conf, Ruidoso, NM New Mexico State University

Fertilizer Research **27**: 181–188, 1991.
© 1991 *Kluwer Academic Publishers.*

Simulation of soil nitrogen dynamics using the SOILN model

L. Bergström, H. Johnsson & G. Torstensson
Swedish University of Agricultural Sciences, Division of Water Management, P.O. Box 7072, S-750 07 Uppsala, Sweden

Key words: Mineralization, nitrogen model, plant N-uptake, winter wheat

Abstract

A model dealing with transport and transformations of nitrogen in soil is briefly described. The model has a one-dimensional layered structure and considers processes such as plant uptake, mineralization/immobilization, leaching and denitrification. A soil water and heat model provides daily values for abiotic conditions, which are used as driving variables in the nitrogen simulation. In this study, the model was run with data from a polder-soil area in the Netherlands, with winter wheat as the crop. The simulation results showed that if a measured time course of crop nitrogen uptake throughout the growing season is available, mineral-N dynamics in soil can be satisfactorily described with this model. The main problems identified in the simulations were related to the partitioning between above- and below-ground plant-N, and supplying the crop with sufficient N, as given by the measurements.

Introduction

A number of models dealing with transformations and transport of nitrogen in soil have been developed [3]. The degree of sophistication and applicability for the different processes included in these models varies considerably, from simple empirically based formulations to complex mechanistic approaches. The complexity in itself does not necessarily restrict widespread use, however. The applicability is often limited by the need for calibration, since models requiring calibration, both 'simple' and 'sophisticated', are not suited for sites which lack monitoring programs [3]. The SOILN model [11] described and used in this study is an example of a model which was developed with the intention of facilitating numerous applications. It has been applied to several sites and data sets of different areal and temporal scales [1, 2, 6, 7, 8, 11].

This paper briefly describes the SOILN model and the simulation results obtained when using a provided data set [5].

Materials and methods

Structure of the model

The model consists of two parts, which are used in sequence. In the first step, soil water and temperature conditions are simulated by a soil water and heat model [9] of which outputs are utilized as driving variables for the nitrogen model [11]. Common to both models is the vertical structure that facilitates division of the soil into different layers depending on the accuracy required and available information on basic soil physical and biological characteristics.

Water and heat transport. The water and heat submodel is based on two coupled differential equations describing heat and water transport (derived from Fourier's and Darcy's laws, respectively) in a one-dimensional soil profile [10]. Snow dynamics, frost, evapotranspiration, precipitation, groundwater flow, water uptake by plants, and drainage flow are included. The

182

model predicts daily values of soil climate variables (soil temperature, and soil water content and tension) at any level in the soil profile, using standard meteorological data as driving variables.

The model also deals with water transport in the saturated zone. The flow rate from each saturated layer, above the depth of the drainage tiles, is calculated by taking into account the saturated conductivity and a gradient estimated as the difference between the mean depth of the groundwater table and the depth of the tiles divided by the distance between the drainage tiles. Vertical redistribution of water below the groundwater table is then calculated from conservation of mass and the water contents exactly at saturation. In addition to the water flow drained via the drainage tiles, the saturated water flow directed towards a stream or ditch is calculated with an empirical equation.

Surface runoff can occur due to limited infiltration capacity or limited permeability of the soil.

A detailed description of the water and heat model is found elsewhere [10].

Nitrogen transformations and plant N-uptake. Biological N-transformations in the model (Fig. 1) apply to each layer. Mineral-N pools include ammonium and nitrate. Organic-N is distributed over litter, faeces, and humus. Organic carbon pools are included for both litter and faeces to control nitrogen mineralization and immobilization rates. Undecomposed material (*e.g.* crop residues, dead roots, microbial biomass) constitute the litter component, while the humus consists of stabilized decomposition products.

Mineralization of humus-N ($N_{h\text{-}am}$; kg ha^{-1} d^{-1}) is calculated as a first-order rate process:

$$N_{h\text{-}am} = k_h e_t e_m N_h \qquad (1)$$

where k_h is the specific mineralization constant (d^{-1}), e_t and e_m are response functions for temperature and moisture, and N_h is the mass of humus-N. Similarly, decomposition of the or-

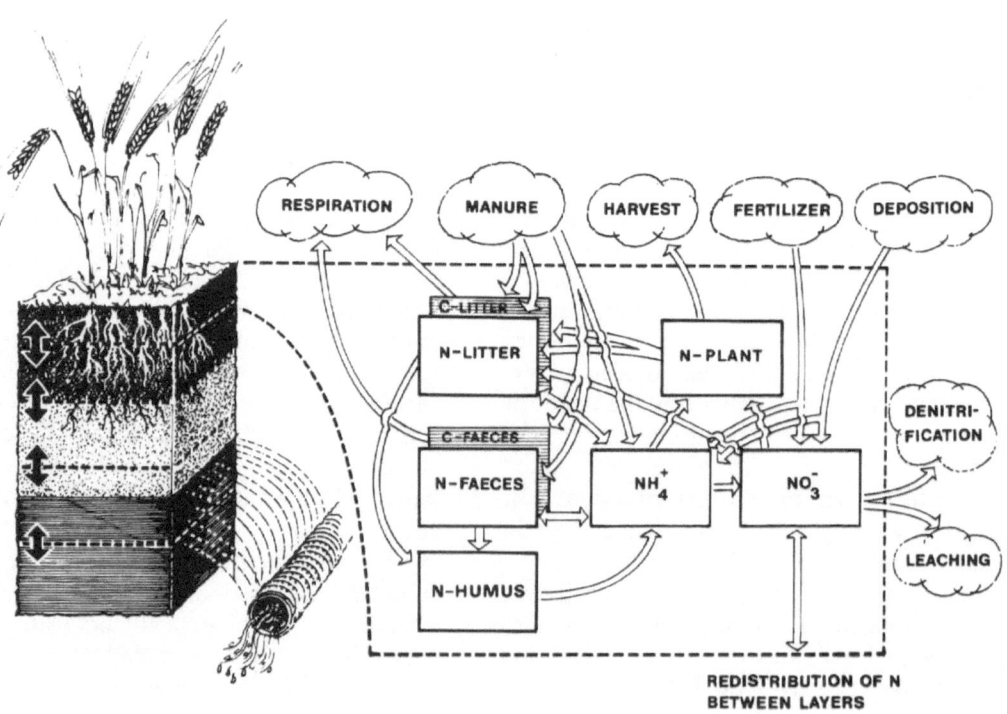

Fig. 1. Structure of the soil nitrogen model, showing state variables (boxes) and flows (arrows) included in the model. The model structure is replicated for each soil layer (From [11]).

ganic carbon pools of litter and faeces are calculated as first-order rate processes controlled by specific mineralization constants and the same abiotic response functions as above. Decomposition products are partitioned into three fractions according to a microbial synthesis efficiency (f_e) and a humification fraction (f_h). One fraction ($1 - f_e$) is lost to the atmosphere as CO_2, the fraction $f_e(1 - f_h)$ is assimilated and recycled within the pool, and the fraction $f_e f_h$ is stabilized as humus (Fig. 2). Corresponding nitrogen flows are calculated assuming a constant C/N ratio of decomposing biomass and humification products (r_0). The net mineralization or immobilization in the litter pool (kg ha^{-1} d^{-1}) is given by the balance between the release of nitrogen during decomposition and the nitrogen immobilized during microbial synthesis and humification, *i.e.*:

$$N_{l\text{-}am} = (N_l/C_l - f_e/r_0)C_{ld} \qquad (2)$$

where N_l and C_l are the actual masses of nitrogen and carbon in the litter pool, and C_{ld} is the decomposition rate of litter carbon (kg ha^{-1} d^{-1}). Mineralization from faeces is handled in the same way. When net immobilization occurs (*i.e.* $N_l/C_l < f_e/r_0$) the immobilization rate is reduced by assuming a maximum fraction of the mineral-N is available. Both ammonium and nitrate can be immobilized, but with preference for available ammonium.

Nitrification of ammonium to nitrate is calculated as a first-order process, modified by the inclusion of a threshold level defined by an equilibrium nitrate/ammonium ratio, which is assumed to be characteristic for a particular soil. The transfer rate of ammonium to nitrate further depends on a specific rate contant (d^{-1}) and the same response functions for temperature and moisture as above.

The abiotic response functions regulating decomposition, mineralization, and nitrification are functions of soil temperature and soil water content. A Q_{10}-expression is used for temperature. An empirical relationship is used for volumetric water content, based on the assumption that the range within which water content is optimum is defined by two thresholds in dry and very wet soil.

Denitrification is calculated as a zero-order process based on a potential rate (kg ha^{-1} d^{-1}) and on response functions of temperature, water content, and nitrate concentration. The temperature response function is the same as that used for other microbial processes, while the response to water content is zero up to a value close to saturation, from where the response increases with increased water content to an optimum level at saturation. The nitrate concentration response is calculated with a Michaelis-Menten expression with a half-saturation constant (mg l^{-1}).

Plant uptake of nitrogen is controlled by a potential demand (logistic growth curve) which is distributed over an assumed root distribution. To limit nitrogen uptake if soil mineral-N concentration is low, a maximum availability fraction (d^{-1}) is assumed which is proportional to the total amount of mineral-N in each layer. This is similar to the situation described for immobilization. Compensatory uptake from any layer (except the uppermost layer) may also occur. When for a given layer the actual uptake is less than the demand, the difference between demand and actual uptake for this layer is added to the total demand of the crop in the layer below. Ammonium and nitrate are taken up by plants in proportion to amounts available in the soil solution.

Only nitrate is considered to be mobile, and nitrate flows are calculated as the product of water flow and nitrate concentration in the soil layer from which the water flow originates. Dif-

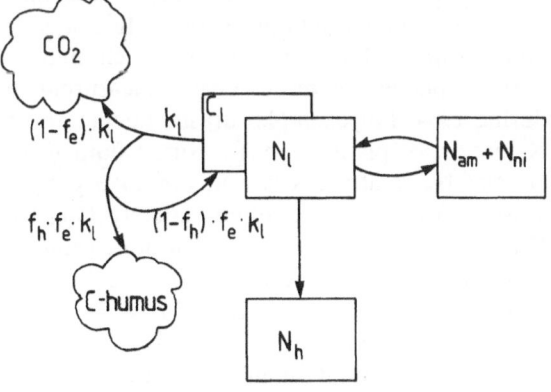

Fig. 2. Flow diagram showing carbon and nitrogen balances associated with litter (or faeces). Explanation of parameter symbols are given in the text (From [11]).

fusion/dispersion is not explicitly accounted for, but due to the division of the soil profile into layers of finite thickness, numerical dispersion will occur.

The nitrogen model is described in more detail elsewhere [11].

Model application

The model was applied to experimental data for winter wheat grown in the Netherlands [5]. Simulations are only reported for the Eest site 1982/83 and 1983/84, because of the better agreement between simulated and measured water conditions for the Eest data compared to the other sites.

The adaptation of the model to the winter wheat crop was based on parameter values previously used for an application with barley [11]. However, some parameter values were changed to better reflect the conditions in the fall sown wheat.

The simulation of soil water and heat processes was the same as that described by Eckersten and Jansson [4], and is therefore not described here.

Since this study was not primarily aimed at simulating crop growth, and since comprehensive measurements of above-ground N-uptake by plants were available, we decided to use these values as target values for the simulation. This was only done, however, for the N_2 treatment in 1984. All the other simulations of plant N-uptake were parameterized in the same way, with the exception of the parameter determining potential uptake which was different between treatments and years. It was assumed that 80% of plant-N was fixed in above-ground parts, while the rest was located to roots. The C/N ratio of above- and below-ground crop residues was set at 50 and 25, respectively, as used in the previous barley study [11]. In order to provide the crop with sufficient N, as indicated by the measurements, the maximum availability fraction of mineral-N in the soil (see above) was doubled (from 8 to 16%) compared to the previous barley simulation.

The soil moisture-response function for microbial activity increased from 0 at the wilting point to an optimal rate at a water content of 13%

above the wilting point, and decreased linearly from 8% below saturation to 0.6 at saturation. The temperature-response function was identical to that used in the barley simulation, *i.e.* using a Q_{10}-relationship with a Q_{10}-value of 3.

The rate constant for humus mineralization was reduced by 28.6% (from 7.0×10^{-5} to 5.0×10^{-5} d^{-1}) compared with the original barley simulation. In fact, the chosen value in this simulation was the average of several earlier model applications [1, 2, 6, 7, 11].

The nitrate/ammonium ratio in the soil was lowered from 8 as used earlier [11] to 6, motivated by site specific measurements.

In the first model application with barley, little emphasis was put on simulation of denitrification dynamics. Recently, this has been done in more detail by Johnsson *et al.* (unpublished), who found a soil moisture activity range (see above) of 18% to be more reasonable, as used in this simulation, rather than 10%, as before.

Dry deposition ($0.05 \text{ kg N ha}^{-1} \text{ d}^{-1}$) and nitrogen concentration in precipitation (3 mg l^{-1}) were adjusted to give reasonable values of total N-deposition, typical for the Netherlands.

Results and discussion

Plant N-uptake

There was a tendency to overestimate simulated N-uptake by plants in comparison to measurements, especially during the first part of the growing period (Fig. 3). A better agreement between simulated and measured values was obtained during 1984 than 1983, which may be partly explained by more reliable measurements during 1984. For example, drastic fluctuations in N-content of plants were measured both in the N_2 and N_3 treatments late in the 1983 growing season, which seemed quite unlikely (Fig. 3). During 1984, simulated and measured values of plant-N were the same in the N_2 and N_3 treatments (Fig. 3), since N-fertilization was identical in the two treatments.

A general limitation in the simulations concerns assumptions related to the partitioning between above- and below-ground plant-N. This was assumed the same for all treatments. Al-

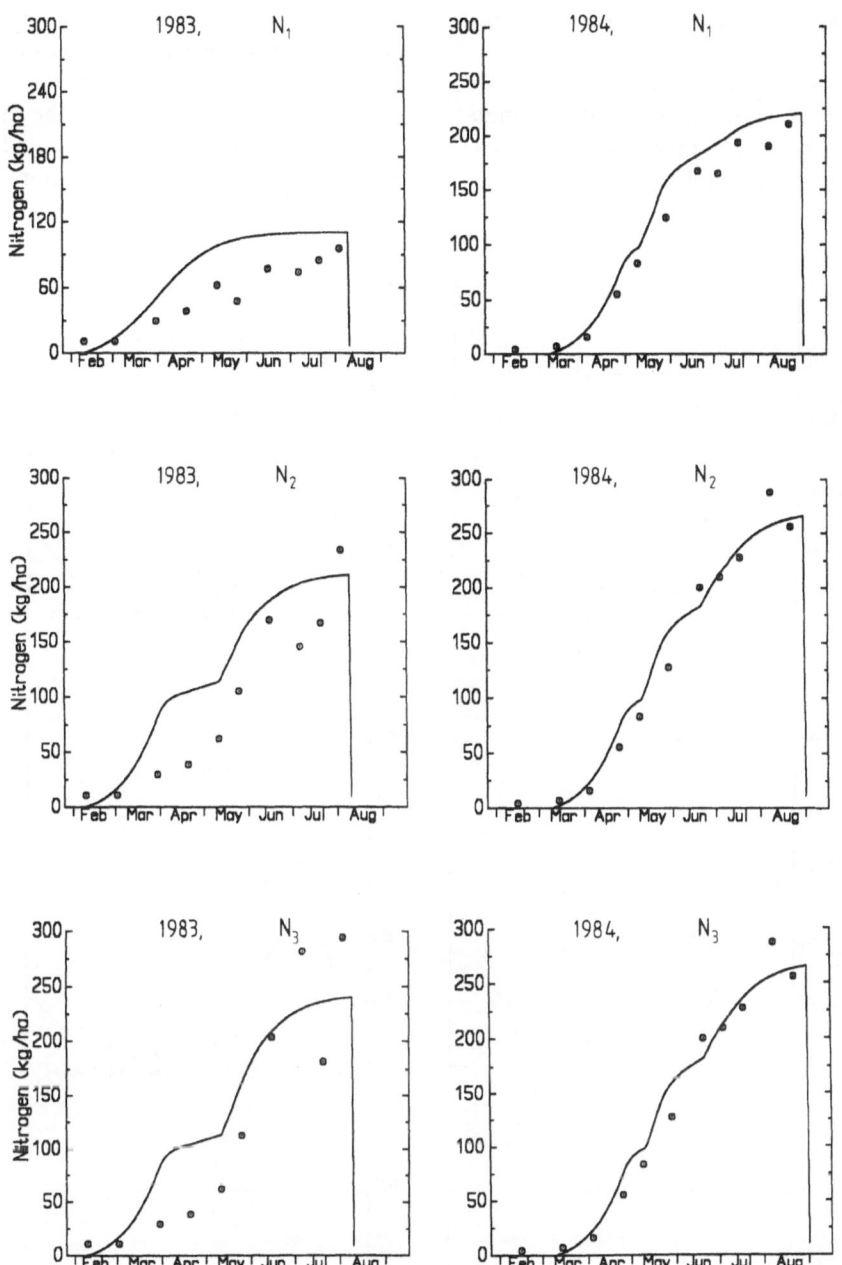

Fig. 3. Simulated (lines) and partly measured (dots) N-uptake by plants accumulated over each season (nitrogen in roots was never measured, but was assumed to be 20% of plant-N).

though no below-ground measurements were available, there is reason to believe that the treatment with low N-input (N_1) had a higher proportion of N located to roots compared with the treatments with higher N-inputs (N_2 and N_3). This may alter the deviations between simulated and measured values of plant-N. However, the export of N will not change by altering the partitioning of N within the plant, since the total N-uptake is increased. In a long-term perspective, higher proportion of roots will increase litter input and thus litter mineralization.

N mineralization and mineral-N dynamics in soil

All simulations of mineral-N content of the soil showed a reasonable agreement with the measurements (Fig. 4). Since plant N-uptake was simulated using measured uptake as target values (see above), this indicates that the chosen rate constants for humus- and litter-mineralization were reasonable values. Over the simulated period (1 Febr.–31 Dec.), the humus-N pool decreased by ca. 60 kg N ha^{-1}, while there was no net change of N in the litter pool (Fig. 5).

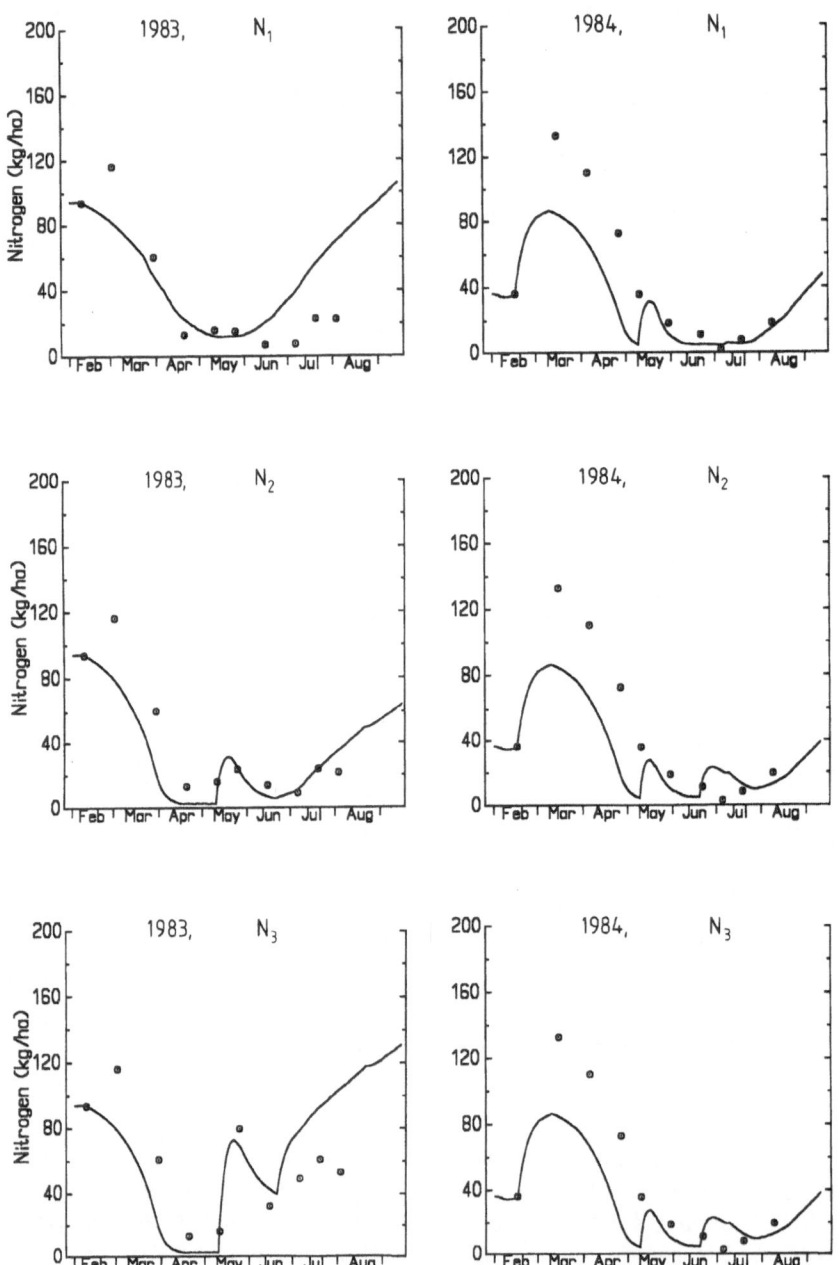

Fig. 4. Simulated (lines) and measured (dots) mineral-N contents in the soil down to a depth of 1 m.

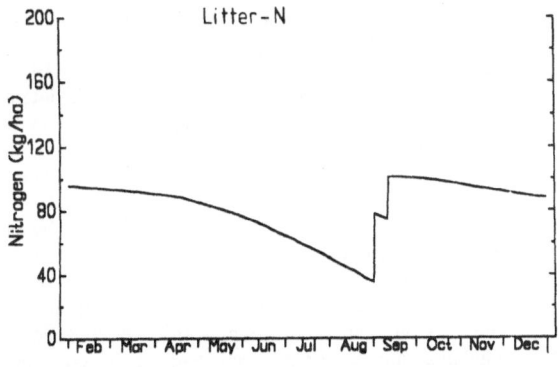

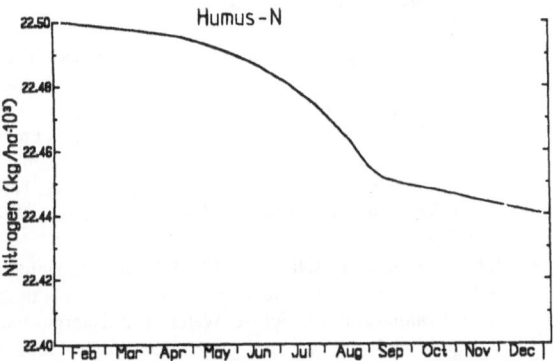

Fig. 5. Simulated variations in the litter- and humus-N pools in the soil (N_2 treatment during 1984).

with sufficient N, as indicated by the measurements, this decrease in humus-N is a likely explanation. There is reason to believe that the 'young' soil used in this study, with a relatively high organic matter content throughout the profile, also depletes its humus-N pool over the short-term, as in this study. Decreases of the humus-N pool during relatively short periods have also been shown in another simulation with the SOILN model, when applied to a perennial grass ley system [1].

As was shown for the plant N-uptake (see above), there was also a better agreement between simulated and measured mineral-N dynamics during 1984 than in 1983 (Fig. 4). This was especially obvious from June onwards, when the increase in simulated mineral-N values was clearly overestimated in 1983. Plant N-uptake slowed down and stopped much earlier during 1983 than during 1984 (Fig. 3), which could partly explain the overestimation. Moreover, the fact that plant N-uptake generally seemed to be simulated better during 1984, would also lead to an improved description of mineral-N dynamics, since N-uptake by plants is a large N-sink; *i.e.* small errors in plant uptake can lead to large deviations between simulated and measured mineral-N dynamics.

These results refer to the N_2 treatment during 1984, for which plant N-uptake was calibrated. Whether this simulated decrease of humus-N is in accordance with actual field conditions could not be verified. However, to supply the crop

Concluding remarks

If measurements of nitrogen contents in plants over the season are available for calibration of

Table 1. Annual simulated nitrogen flows in the different treatments. Partial flows are given in parentheses. All values in kg N ha^{-1}

Flow	1983			1984		
	N_1	N_2	N_3	N_1	N_2	N_3
Fertilization	0	60	120	110	150	150
Deposition	34	34	34	36	36	36
Mineralization	161	156	155	154	137	137
Net litter min	(43)	(38)	(37)	(48)	(31)	(31)
Humus min	(118)	(118)	(118)	(106)	(106)	(106)
Humification	40	49	51	36	47	47
Plant uptake	110	211	240	221	266	266
Harvest	(77)	(148)	(168)	(155)	(186)	(186)
Crop residues	(33)	(63)	(72)	(66)	(86)	(86)
Leaching	11	4	4	19	17	17
Denitrification	9	10	15	20	19	19

188

the simple plant N-uptake approach in this model, mineral-N dynamics in soil can be satisfactorily described. However, quite often, only data on harvested nitrogen are available, which then requires a more sophisticated submodel for N-uptake by plants, including response functions for climatic variables (*cf.* [4]).

Various nitrogen flows, which have earlier been described satisfactorily with the model, were not possible to validate in these simulations. However, in order to get a general idea of the magnitude of these simulated flows, their annual sums are given in Table 1. Leaching and denitrification both reached ca. 20 kg N ha^{-1}. It is notable that the highest simulated leaching losses occurred in the treatment with lowest or no input of fertilizer-N. Total simulated mineralization was ca. 150 kg N ha^{-1}, again, with the highest amounts in the N_1 treatment. This must be due to lower immobilization over the short-term in the N_1 treatment than the other two treatments, depending on less input of crop residues with a high C/N ratio (50). In summary, these simulations indicate an improved nitrogen use efficiency with increased N-fertilization.

References

1. Bergström L and Johnsson H (1988) Simulated nitrogen dynamics and nitrate leaching in a perennial grass ley. Plant Soil 105: 273–281
2. Borg G CH, Jansson P–E and Lindén B (1990) Simulated and measured nitrogen conditions in a manured and fertilised soil. Plant Soil 121: 251–267
3. De Willigen P, Bergström L and Gerritse R G (1990) Leaching models of the unsaturated zone; their potential use for management and planning. In: Decoursey D G (ed.) Proceedings of the International Symposium on Water Quality Modeling of Agricultural Non-Point Sources, pp 105–128. USDA publication
4. Eckersten H and Jansson P–E (1991) Modelling water flow, nitrogen uptake and production for wheat. Fert Res 27: 313–329
5. Groot J J R and Verberne E L J (1991) Response of wheat to nitrogen fertilization, a data set to validate simulation models for nitrogen dynamics in crop and soil. Fert Res 27: 349–383
6. Gustafson A (1988) Simulated nitrate leaching from arable land in southern Sweden. Acta Agric Scand 38: 12–23
7. Jansson P–E and Andersson R (1988) Simulation of runoff and nitrate leaching from an agricultural district in Sweden. J Hydrol 99: 33–47
8. Jansson P–E, Borg G CH, Lundin L–C and Lindén B (1987) Simulation of nitrogen storage and leaching – applications to different Swedish agricultural soils. National Swedish Environmental Protection Board, Report 3356, 63 p
9. Jansson P–E and Halldin S (1979) Model for annual water and energy flow in layered soil. In: Halldin S (ed.) Comparison of Forest Water and Energy Exchange Models, pp 145–163. Copenhagen: International Society for Ecological Modelling
10. Jansson P–E and Halldin S (1980) Soil water and heat model. Technical description. Swedish Coniferous Forest Project. Tech Rep 26, 81 p. Uppsala: Swedish University of Agricultural Sciences
11. Johnsson H, Bergström L, Jansson P–E and Paustian K (1987) Simulated nitrogen dynamics and losses in a layered agricultural soil. Agric Ecosystems Environ 18: 333–356

Fertilizer Research **27**: 189–198, 1991.
© 1991 *Kluwer Academic Publishers*.

Some results of nitrogen simulations with the model ANIMO

P.E. Rijtema & J.G. Kroes
The Winand Staring Centre for Integrated Land, Soil and Water Research, P.O. Box 125, 6700 AC Wageningen, The Netherlands

Key words: Winter wheat, nitrogen, uptake, simulation, model, soil

Abstract

Simulation of the nitrogen behavior in the soil and the nitrogen uptake by winter wheat was performed using the model ANIMO. As input for the model ANIMO simulations of the hydrological conditions in the soil crop ecosystem were executed with the model SWATRE. Compared with measured data the simulation of nitrogen uptake by the crop was satisfactory. The simulation of mineral nitrogen in the soil agreed reasonably well with measured data for one of the experiments used for the analysis. The agreement was less for experiments with additional fertilizer applications in May and June.

Introduction

Intensification of agriculture has led to an increased fertilizer use, increasingly contaminating groundwater and surface waters. This contamination can have negative effects on other activities like municipal water supply and preservation of natural areas. The formulation and execution of a rational and effective policy to reduce the nitrogen load on groundwater reservoirs and surface waters requires a thorough understanding of the behaviour of nitrogen and its compounds in the soil.

There is a need to quantify the sources of nitrogen compounds from rural regions under various conditions of climate, soil type, water management, cropping pattern, and agricultural technologies. Nitrogen in the soil primarily originates from inorganic fertilizers, animal manures, organic matter, precipitation, biological fixation and irrigation water. It is mainly removed by crops, drainage water, denitrification and volatilization of ammonia.

The development of nitrogen models should not only focus on the selection of methods for detailed analysis of the physical, chemical and biological processes constituting the behaviour of nitrogen in the plant-water-soil system. The main aim should be to derive cause-effect relationships between agricultural practices as cropping pattern, fertilizer type, fertilizer rate and application technology, as well as water management and climatic conditions, and the generation of nitrogen compounds leaving the agricultural system. This paper deals with a description of the processes to be considered in nitrogen modelling and the performance of the model ANIMO (Agricultural NItrogen MOdel). Also a discussion of the uncertainties will be given in both measured field data and in the simulation results.

Description of the model ANIMO

A suitable model for regional use, simulating the nitrogen management in both agricultural areas and nature reserves, should be based on a clear and quantitative description of the main processes which are:
- mineralization and immobilization of nitrogen related to processes in the carbon cycle;
- nitrogen uptake by plants under conditions of both excess supply and limited nitrogen availability;
- denitrification related to (partial) anaerobiosis and decomposing organic materials, which im-

190

plicates the modelling of the oxygen and temperature distribution in the soil;
– soil moisture dynamics and nitrogen transport. A complete description of the newest version of the model ANIMO is given by Rijtema et al. [6], the present paper only gives a brief description of the model principles. The central part of the model is the transport and conservation equation. By means of this equation the new concentrations of soluble compounds in all layers distinguished can be calculated after simultaneous transport and transformation processes. Assuming complete mixing in each identified model layer, the transport and conservation equation can be written as:

$$\frac{d\{V(n, t)C(n, t)\}}{dt} + \frac{dQ_a(n, t)}{dt}$$

$$= \sum f_i C_i - \sum f_o C(n, t) - f_e S C(n, t)$$

$$+ k_0 L + k_1 V(n, t) C(n, t) \qquad (1)$$

in which: n = layer number ($-$); t = time (d); $V(n, t)$ = moisture volume of layer n at time t (m); $C(n, t)$ = concentration in layer n at time t (kg m^{-3}); $Q_a(n, t)$ = nitrogen adsorbed at the soil complex (kg m^{-2}); $\sum f_i C_i$ = total incoming flux of nitrogen (kg m^{-2} d^{-1}); $\sum f_o C(n, t)$ = total outgoing flux of nitrogen (kg m^{-2} d^{-1}); f_e = transpiration flux (m d^{-1}); S = selectivity coefficient for crop uptake ($-$); k_0 = production rate coefficient of zero order (kg m^{-3} d^{-1}); k_1 = production rate coefficient of first order (d^{-1}); L = layer thickness (m).

Different analytical solutions for this equation have been introduced, depending on the change of the moisture content with time.

The model ANIMO can be linked in practice to any hydrological model for the calculation of fluxes and changes in moisture content per layer. In all cases the hydrological calculations should be executed completely before the model ANIMO can be applied. The output file of the hydrological model was used as input for ANIMO. We used the model SWATRE for the hydrological simulation using the given soil physical data as input. A description of this model is given by Belmans et al. [1] and Feddes et al. [3].

Though processes and transport of solutes in the unsaturated zone can be described one-dimensionally, this is not the case in the saturated zone, where also horizontal transport to different drainage systems should be considered. The model ANIMO has a pseudo two-dimensional transport module to calculate solute transport to different drainage systems.

The different organic materials added to or present in the soil contain nitrogen as well as carbon, so the transformations in the carbon cycle correspond with the transformations in the nitrogen cycle. To understand the processes in the nitrogen cycle it is necessary to quantify the processes in the carbon cycle too, because of the many interdependences between organic material and nitrogen. For the quantification of the carbon cycle a Jenkinson and Rayner [5] type of model formulation has been used.

The carbon cycle as used in ANIMO is given in Fig. 1. The input of carbon by organic plant parts and manure can be given by a number of fractions depending upon the decomposition rate and C/N ratio. The choice of the number of fractions is in principle free.

The different organic materials mentioned contain besides carbon also nitrogen. So the transformations in the carbon cycle correspond with transformations in the nitrogen cycle. The schematized nitrogen cycle is given in Fig. 2. It can be seen that part of the cycle corresponds closely with the carbon cycle. In this part the decomposition rate of organic carbon and the formation rate of soil organic material determines the mineralization and immobilization of nitrogen. The other part of the nitrogen cycle

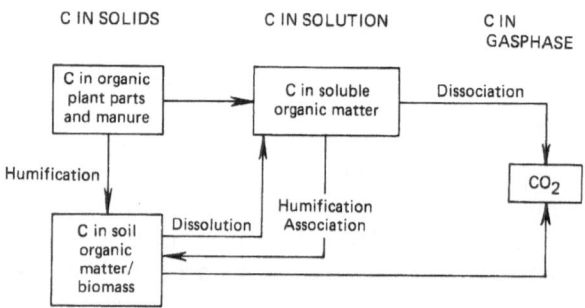

Fig. 1. The carbon cycle in the model ANIMO.

N IN SOLIDS N IN SOLUTION N IN GASPHASE

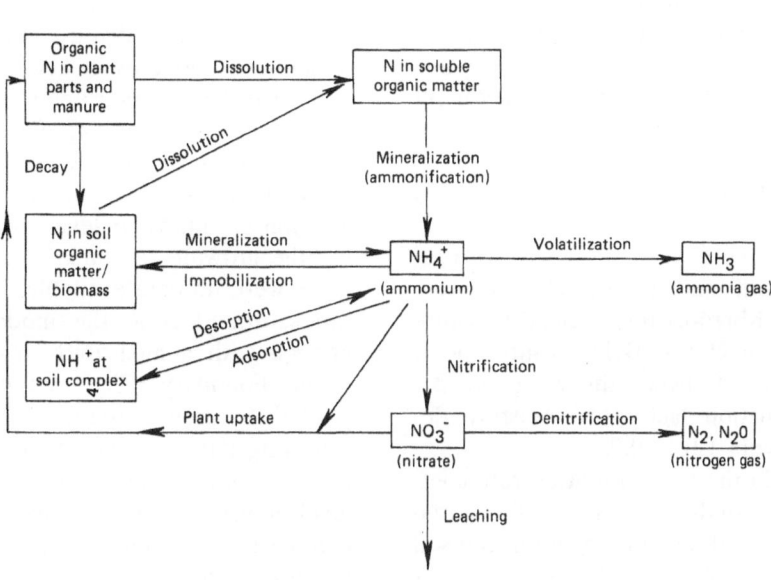

Fig. 2. The nitrogen cycle in the model ANIMO.

describes processes such as nitrification and denitrification. Nitrification is described as a first-order process, denitrification is described as a zero-order process with respect to nitrate, but as a first-order process with respect to organic carbon decomposition.

The ammonium ion with its positive charge can be adsorbed at the soil complex; this is formulated as linear adsorption.

NH_4 ions and NO_3 ions can be taken up from the soil solution by the plant roots using the transpiration flux and/or through diffusion. Both processes were taken into account by a proper choice of the selectivity coefficient S in Eq. (1):

$$S(t+1) = N_m(t)/N_r(t) \qquad (2)$$

in which $S(t+1)$ = selectivity coefficient for timestep $t+1$, $(-)$; $N_m(t)$ = maximum cumulative nitrogen uptake by the crop $(kg\,m^{-2})$; $N_r(t)$ = real cumulative nitrogen uptake $(kg\,m^{-2})$.

In the model the main environmental influences on the transformation processes are temperature, pH, moisture and oxygen.

The transformation rate for chemical and biological processes generally increases with temperature. The soil temperature at a certain depth z (m) from soil surface and at a certain day of the year was calculated using a simple sinus wave submodel with a damping effect for depths below the soil surface:

$$T(z, t) = T_a + A_0 \exp(-z/D_m) \\ \times \cos(\omega t + \phi - z/Dm) \qquad (3)$$

in which $T(z, t)$ = temperature at depth z and time t, (°C); T_a = average yearly temperature, (°C); A_0 = amplitude of temperature wave, (°C); D_m = damping depth, (m); ω = frequency of temperature wave, $(rad\,d^{-1})$; ϕ = phase shift, (rad).

Because the optimum temperature for biological processes is often around 30°C or higher, it can be assumed for the temperatures occurring in soil that the reaction rates increase with temperature. For temperatures from 0 to 26°C this curve has been described by the Arrhenius equation:

$$r(T) = Z \exp\{-a/(T + 273)\} \qquad (4)$$

in which: $r(T)$ = rate at temperature T $(-)$; Z = constant $(-)$; a = constant (°C).

For the effect of pH on reaction rates only one function for the various processes has been introduced in the model. The pH-reduction factor (r) is given as:

$$r = (1 + \exp\{-2.5(pH - 5.0)\})^{-1} \qquad (5)$$

Moisture content and oxygen status are strongly related in the soil and therefore they were treated together. Microorganisms need moisture to perform their functions. Below wilting point these are disturbed. Between the values pF 3.2 and 4.2 the reduction factor for moisture decreases linearly from 1.0 to 0.2.

The reduction in the transformation rate coefficients under wet conditions is generally caused by dilution effects and lack of oxygen in the soil system. In the model ANIMO this reduction has therefore been described on the basis of soil aeration. The aeration of the soil depends on the air-filled pore space, the diffusion coefficient and the oxygen consumption in the soil. The conditions for partial aeration in the soil have been described in the model by vertical diffusive transport through the air-filled pores and a radial horizontal transport through the water phase, while a stochastic approach has been used for the non-homogeneous distribution of the air-filled pores in the soil system. The conditions in the upper unsaturated zone are strongly determined by the results of the calculations of the hydrological model used, which determines the amount of percolating water to the groundwater reservoir, but also the soil moisture distribution and the oxygen transport in the unsaturated zone. A poor simulation of the soil moisture distribution by the hydrological model used automatically results in a poor simulation of the carbon and nitrogen cycle.

The model calculates the total oxygen requirement on the basis of the decomposition of the organic materials and nitrification. The difference between the maximum required quantity of oxygen and the available oxygen by diffusion has been used as a basis for the total denitrification in each layer. If insufficient nitrate is available a nitrate-related reduction factor for the decomposition of organic material will be calculated.

Available data

Data on soil moisture dynamics, nitrogen dynamics, crop growth and nitrogen uptake of extensive field experiments were collected by Groot and Verberne [4]. Fertilizer applications varying from 0–240 kg N per ha were applied between the months February and June. At each location groundwater levels were measured during the growing season of the year 1984. Since there were no data available of vertical fluxes in the saturated zone the model SWATRE was applied with a measured groundwater level as lower boundary. The output of the model SWATRE, used as input for ANIMO, gave the following data for each timestep: precipitation, evaporation, transpiration, runoff, groundwater level, moisture contents, fluxes to/from different compartments, boundary fluxes across the lower boundary and lateral fluxes to/from drainage systems per compartment.

Simulation results and discussion

The description of the simulation results will be limited to the upper meter of the soil profile since the available measured data on nitrogen turnover refer to this part of the soil. Mineral nitrogen is presented for the layers 0–40 cm and 0–100 cm to give an impresssion of the model results in the upper part of the soil.

Simulations were executed for the Eest and PAGV. Calculations were started with the Eest-data as this was the most complete dataset. The simulation of the nitrogen treatment N1 will be discussed in detail.

Eest N1

The water balance for the Eest experiments as a result of simulations with the model SWATRE is given in Fig. 3 for the top soil layer to 1 meter depth. In this balance the seepage and leakage indicate the total flow through the lower boundary at one meter depth. The leakage gives the total outflow to deeper soil layers during drainage conditions whereas the seepage gives the inflow by capillary rise through the lower boundary during periods with a surplus of evaporation

193

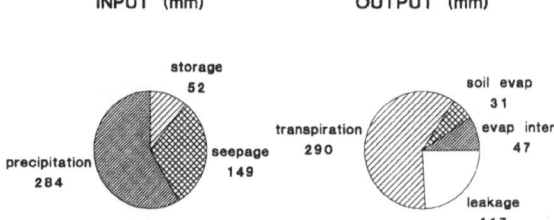

Fig. 3. Water balance of the top soil layer 0–100 cm; evap = evaporation, interc = intercepted water (Eest N1).

and a groundwater table at more than one meter depth. Figure 4 shows the simulated and measured moisture volumes for two layers. Simulated data followed the measured ones reasonably well.

The seasonal nitrogen balance for the Eest N1 field is given in Fig. 5. The results showed that the contribution of the fertilization input and availability of mineral N by mineralization were of the same order of magnitude, each contributing for about 40% of the total mineral N input

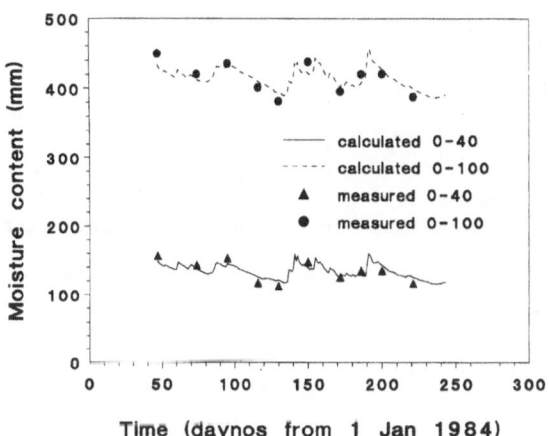

Fig. 4. Simulated and measured moisture volumes for the layers 0–40 cm and for the layer 0–100 cm (Eest N1).

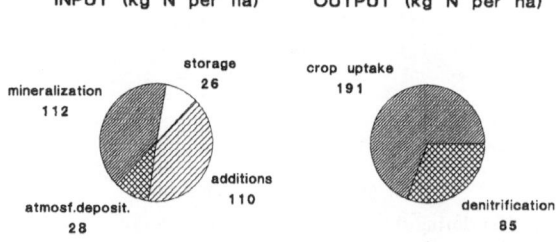

Fig. 5. Nitrogen balance of the top soil layer 0–100 cm (Eest N1).

during the period of growth. The N uptake of the crop was about 69% of the total available nitrogen as simulated by ANIMO. Denitrification losses were 31% of the total available mineral nitrogen.

A more detailed analysis of the nitrogen uptake by the crop is given in Fig. 6. The model ANIMO simulates total nitrogen uptake, including nitrogen present in roots and stubble, whereas the measured data were without this quantity. On the basis of available experimental data the quantity of nitrogen present in roots and stubble can be estimated at about 15 kg per ha. It appeared from Fig. 6 that measured and simulated nitrogen uptake during growth agreed reasonably well.

The total mineral nitrogen present in the soil of the Eest N1 field during growth is given for two layers in Fig. 7. Figure 8 shows the measured and simulated data separately for NH_4–N and NO_3–N for the layer 0–100 cm. It was remarkable that after the fertilizer addition of 60 kg N applied at dayno. 132 an increase in mineral nitrogen in the measured data did not occur. The simulated data, however, showed an increase in total mineral nitrogen immediately after application, followed by a sharp reduction in the days after application. Next to an increased uptake by the crop after this application also an increased denitrification reduced the quantity of mineral nitrogen in the soil, caused by partial anaerobiosis due to heavy summer rains. Figure 9A gives the distribution of precipi-

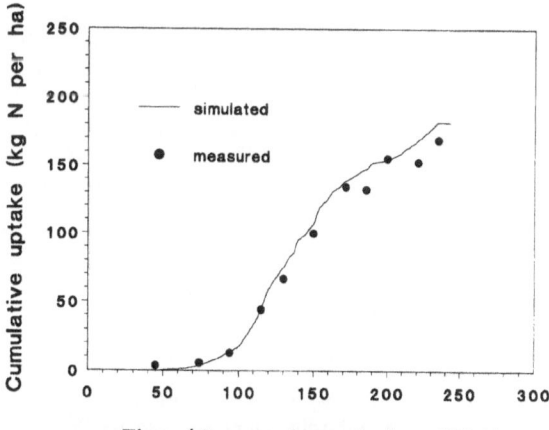

Fig. 6. Simulated and measured cumulative nitrogen uptake by the crop (Eest N1).

194

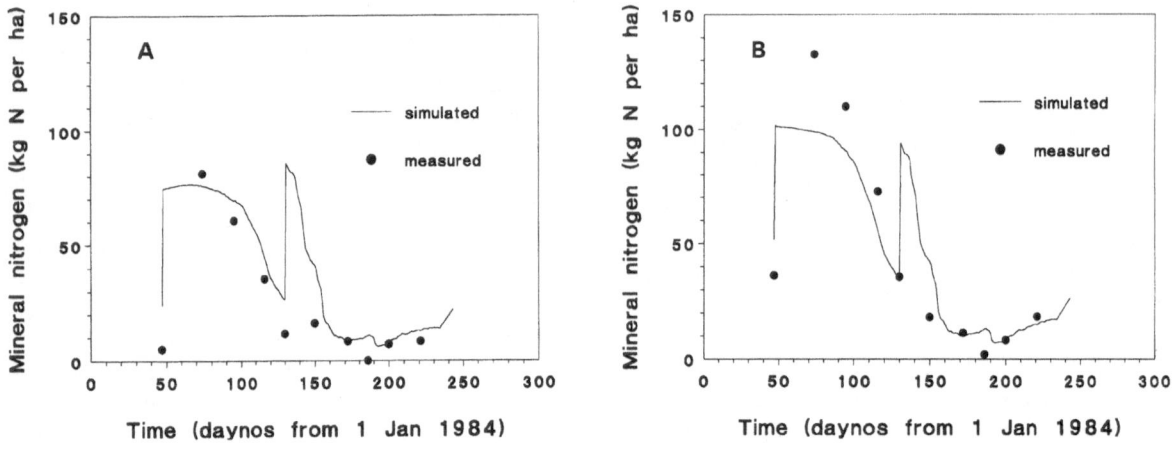

Fig. 7. Simulated and measured soil mineral nitrogen (ammonium + nitrate) for the layer 0–40 cm (A) and for the layer 0–100 cm (B) (Eest N1; applications of 50 and 60 kg N per ha at daynos 48 and 132 respectively).

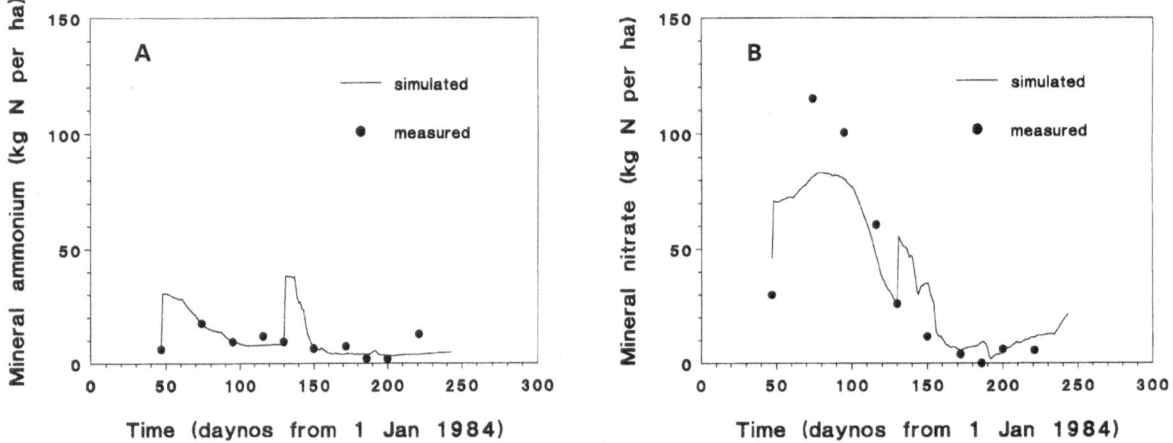

Fig. 8. Simulated and measured soil mineral ammonium (A) and nitrate (B) for the layer 0–100 cm (Eest N1).

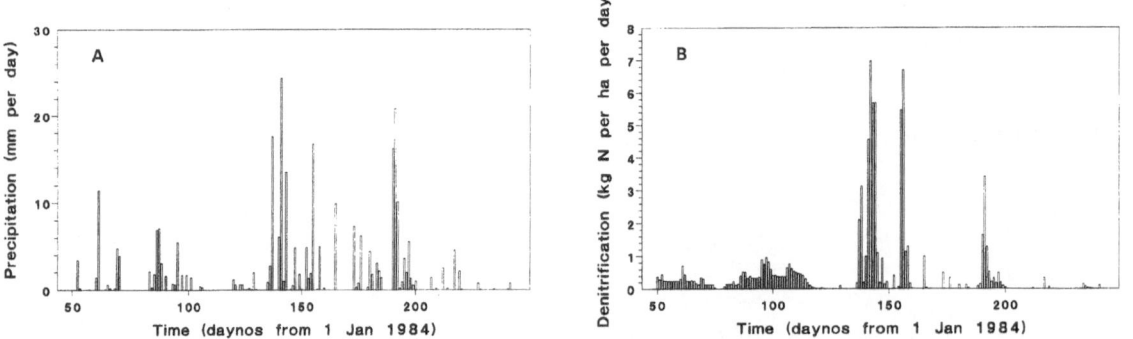

Fig. 9. The distribution of precipitation (A) and denitrification (B) during the growing season (Eest N1).

tation during growth. The simulated denitrification rate is presented in Fig. 9B, showing denitrification peaks coinciding with heavy rains. The calculated total reduction in mineral nitrogen in the soil followed the measured data reasonably well.

Eest N2, N3

Figure 10 shows the results for measured and simulated nitrogen uptake by the crop of the Eest N2 and N3 fields. Both fields received the same amount of nitrogen fertilization and can be considered as an experiment in twofold. The measured data of both fields as plotted in Fig. 10

give an indication of the spatial variability of the measured data. The simulated data followed the measured ones reasonably well. Figure 11 gives the mineral nitrogen in the soil profile for two layers. The duplicated measurements give also an indication of the spatial variability in measured mineral nitrogen data. The simulated data appeared to describe the temporal variation in mineral nitrogen in the soil following the fertilizer additions of 60 kg N at dayno. 132 and of 40 kg N at dayno. 173 satisfying as compared with the measured data.

PAGV N1, N2, N3

Figure 12 gives the comparison between both measured and simulated nitrogen accumulation in the crop. The simulated crop uptake data for the three fertilizer treatments were in good agreement with the measured ones.

The course in mineral nitrogen in the soil for the 0–40 cm top layer is given in Fig. 13. A good agreement between simulated and measured data was present in the N1 experiment. The results were less for the N2 and N3 experiments. The fertilizer additions of 60 and 120 kg N per ha for respectivily the N2 and N3 fields at dayno. 135 and the addition to both fields of 40 kg N per ha at dayno. 160 affected the simulated mineral nitrogen in the 40 cm top layer considerably, whereas they did hardly affect the measured data.

The total mineral nitrogen simulated for the

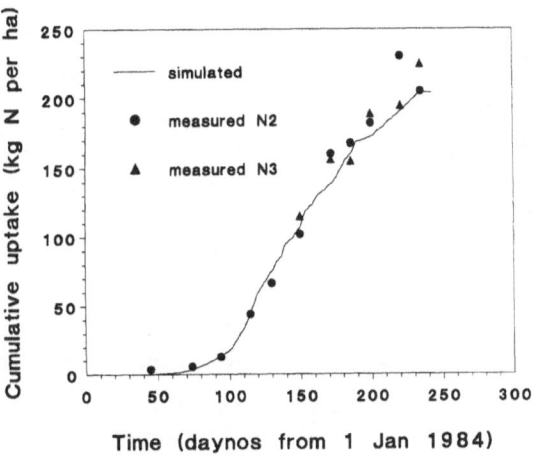

Fig. 10. Simulated and measured cumulative nitrogen uptake by the crop (Eest N2 and N3).

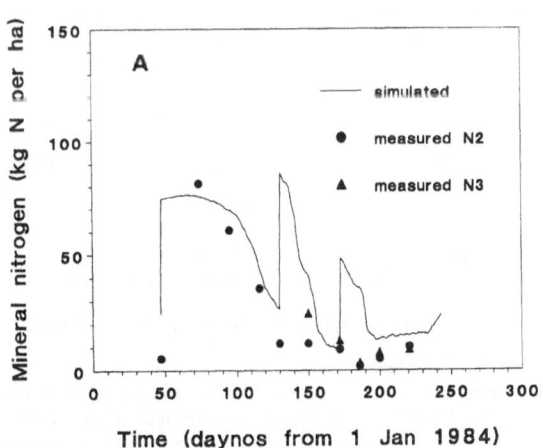

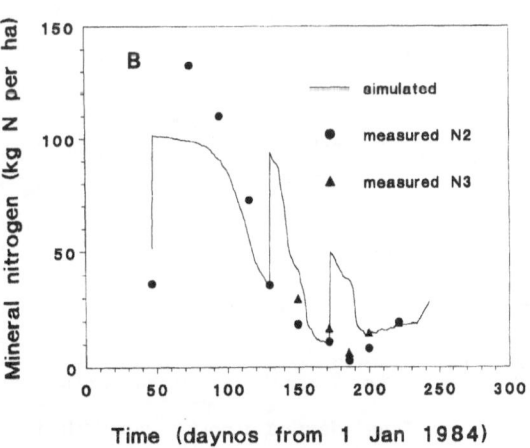

Fig. 11. Simulated and measured soil mineral nitrogen (ammonium + nitrate) for the layer 0–40 cm (A) and for the layer 0–100 cm (B) (Eest N2 and N3; applications of 50, 60 and 40 kg N per ha at daynos 48, 132 and 173 respectively).

196

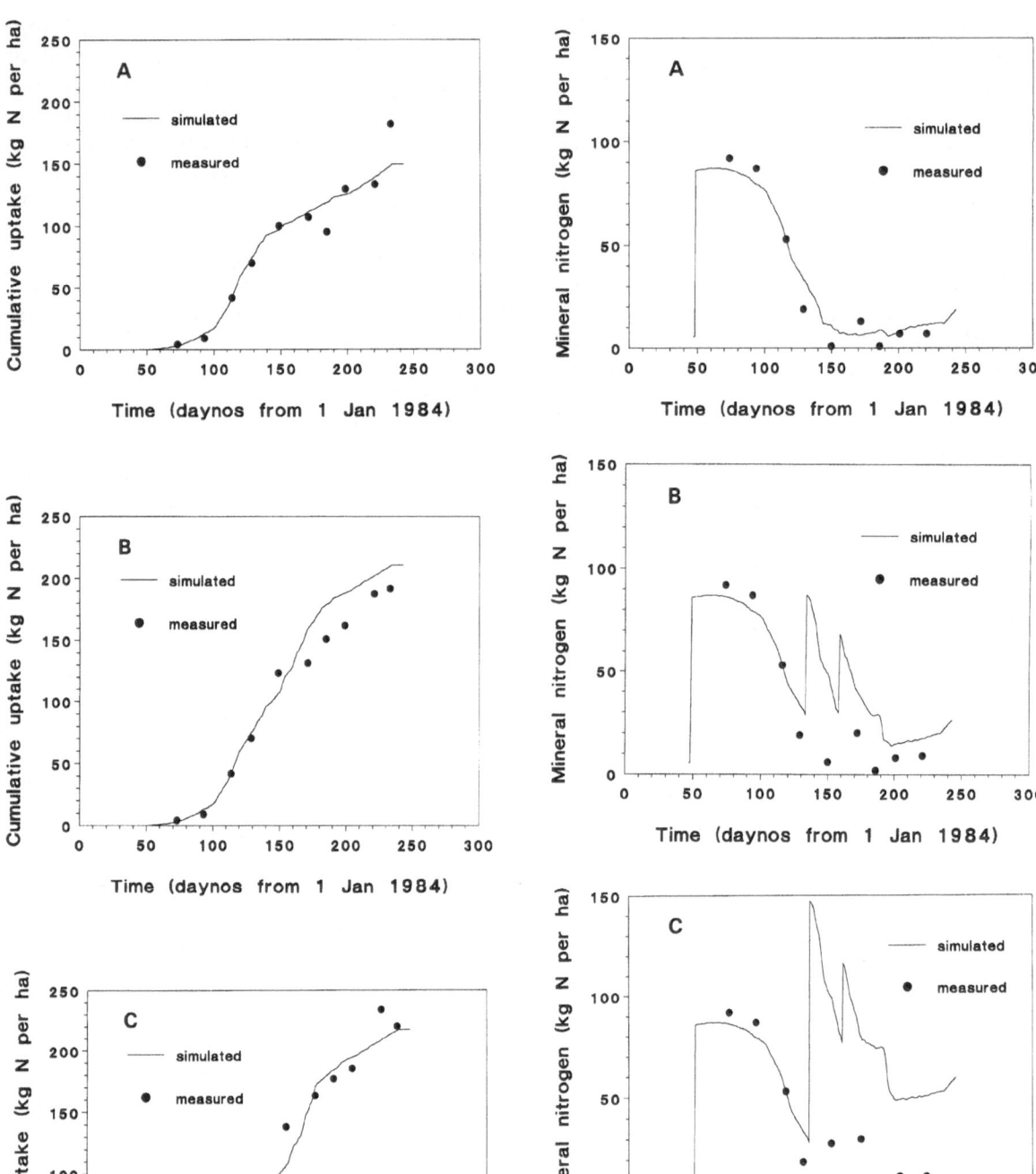

Fig. 12. Simulated and measured cumulative nitrogen uptake by the crop, PAGV N1 (A), N2 (B) and N3 (C).

Fig. 13. Simulated and measured soil mineral nitrogen (ammonium + nitrate) at location PAGV, for the layer 0–40 cm, N1 (A), N2 (B), N3 (C) (N1: application of 80 kg N per ha at dayno. 48, N2: applications of 80 and 60 Kg N per ha at daynos 48 and 135, respectively, N3: applications of 80, 120 and 40 kg N per ha at daynos 48, 135 and 160 respectively).

soil profile from 0–100 cm, as given in Fig. 14, deviated from the measured data. The total increase in measured mineral nitrogen between dayno. 46 and 73 was 108 kg N per ha, while the fertilizer addition at dayno. 48 was only 80 kg per ha. The difference between increase in mineral nitrogen and addition seemed to be mainly due to the variation in mineral nitrogen in the layer 40–100 cm. At daynos 18, 46 and 73 the mineral nitrogen was respectively 52, 15 and 32 kg N per ha for this layer. The precipitation excess during this period was however very small, and denitrification is unlikely to have occurred.

The effects of the nitrogen additions of the N2 and N3 fields at daynos 135 and 160 on nitrogen uptake by the crop were simulated in a correct way as appears from Fig. 12. Also an extra denitrification rate during the summer rains was simulated. However, during the rainfall in the period from dayno. 141 to 155, as given in Fig. 9, a sharp rise of the groundwater table from 160 cm to 116 cm below soil surface was observed. It is not clear whether due to crack formation short-cuts in the soil system were present resulting in a rapid transport of mineral nitrogen below the maximum sampling depth of 100 cm, or not.

General discussion and conclusions

Only one set of soil physical data was available for the Eest and PAGV experiments while the hydrological boundary conditions were derived from measured data of the groundwater table depth. The limited availability of these data prevented to analyse thoroughly the effect of spatial variation on soil physical properties and hydrological conditions. Van der Bolt et al. [2] showed in an analysis of the uncertainties in the performance of the model ANIMO on a field scale that the effects of different bulk densities and organic matter contents on the variation in soil properties did hardly effect the dynamic water balance simulations of the hydrological model. The input of the simulated soil moisture distributions in the model ANIMO, resulted in different aeration conditions. Results of the simulation of mineral nitrogen in the soil by the model

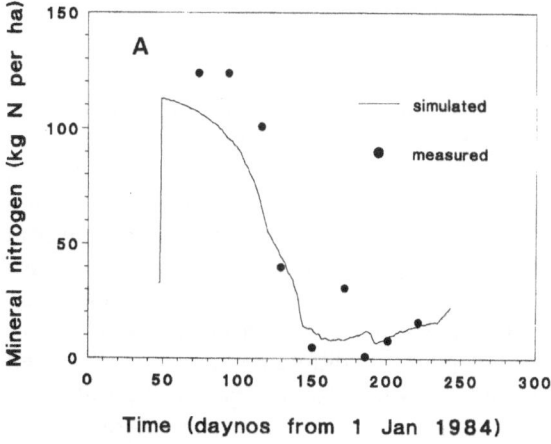

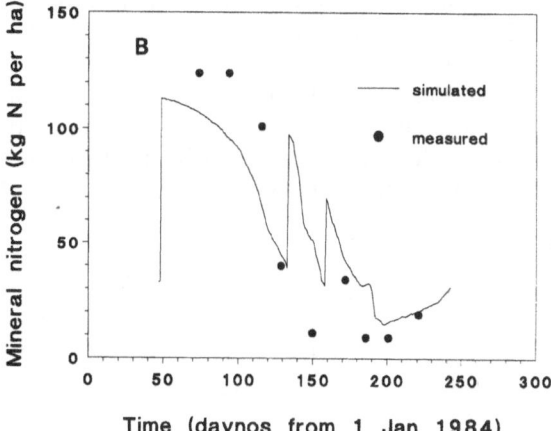

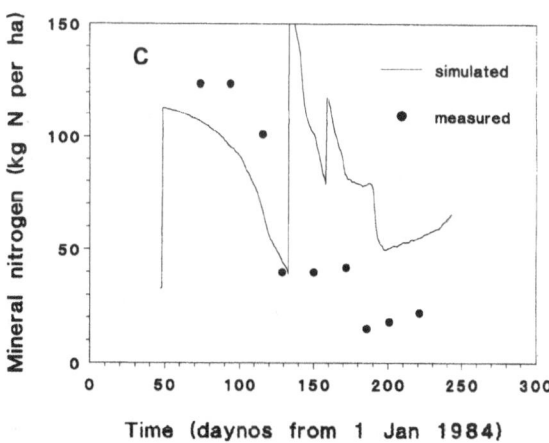

Fig. 14. Simulated and measured soil mineral nitrogen (ammonium + nitrate) at location PAGV, for the layer 0–100 cm, N1 (A), N2 (B) and N3 (C) (N1: application of 80 kg N per ha at dayno. 48, N2: applications of 80 and 60 kg N per ha at daynos 48 and 135 respectively, N3 applications of 80, 120 and 40 kg N per ha at daynos 48, 135 and 160 respectively).

ANIMO did strongly depend on the variation in moisture conditions. The variation in measured mineral nitrogen in the soil profile at different locations in the same field was of the same order of magnitude as the variation in the simulated values of mineral nitrogen due to the variation in soil physical properties. In both the Eest and PAGV experiments the variation in soil physical conditions has not been taken into account. Also the effects of bypass flow due to cracking in the soil profile has not been considered, which might partly explain the differences between measured and simulated mineral nitrogen in the soil.

The formulation of partial anaerobiosis in the model ANIMO gave only partially an explanation for the sharp reduction in mineral nitrogen in the soil profile after additions during the growing season.

The model ANIMO did give a good simulation of the accumulated nitrogen uptake by the crop for all the fertilizer treatments in the Eest and PAGV experimental fields.

References

1. Belmans C, Wesseling JG and Feddes RA (1983) Simulation model of the water balance of a cropped soil: SWATRE. J Hydrol 63, 3/4: 271–286
2. Bolt FJE van der, Pankow J, Roest CWJ, Toorn A van den (1990) De invloed van de stikstofhuishouding in de bodem op de grondwaterkwaliteit in waterwingebied 't Klooster. Rapport 31, 29 pp Wageningen, The Netherlands: The Winand Staring Centre
3. Feddes RA, Kowalik PJ and Zaradny H (1978) Simulation of field water use and crop yield. Simulation Monographs, 189 pp Wageningen, The Netherlands: PUDOC
4. Groot JJR and Verberne ELJ (1991) Response of wheat to nitrogen fertilization, a data set to validate simulation models for nitrogen dynamics in crop and soil. Fert Res 27: 349–383
5. Jenkinson DS and Rayner JH (1977) The turnover of soil organic matter in some of the Rothamsted classical experiments. Soil Science 123: 298–305
6. Rijtema PE, Roest CWJ and Kroes JG (1990) Formulation of the nitrogen and phosphate behaviour in agricultural soils, the ANIMO model. Report 30, (in press) Wageningen, The Netherlands: The Winand Staring Centre

Fertilizer Research **27**: 199–213, 1991.
© 1991 *Kluwer Academic Publishers.*

The distribution of water and nitrogen in the soil-crop system: a simulation study with validation from a winter wheat field trial

R.F. Grant
Department of Soil Science, University of Alberta, Edmonton, Alberta, Canada T6G 2E3

Key words: Water uptake, nitrogen uptake, mass flow, diffusion, rooting distribution, crop growth, nitrogen partitioning

Abstract

The realistic simulation of uptake processes for water and nitrogen, and of partitioning processes for nitrogen, are necessary to the accurate reproduction of water and nitrogen deficit effects on crop growth and yield. Mathematical descriptions of these processes, based on the findings of detailed studies, were used as part of a larger simulation model to calculate the extraction and uptake of water and nitrogen, and the distribution of nitrogen, by a winter wheat crop from a multi-layered soil profile over a growing season. Descriptions of water uptake processes allowed the model to reproduce the hourly dynamics of water uptake and redistribution through the root system, and to estimate the yearly dynamics of water depletion from the soil profile consistent with field data. Descriptions of nitrogen uptake processes allowed the model to reproduce the hourly dynamics of mass flow, diffusion and active uptake, but estimates of the yearly dynamics of mineral nitrogen depletion did not closely follow field data. This inconsistency arose largely from the partial disappearance of fertilizer nitrogen in the field plots shortly after application. This disappearance was not reproduced in the model. The model was able to reproduce the seasonal accumulation and redistribution of dry matter and nitrogen within the crop for fertilizer applications from 0 to $16 \, g \, m^{-2}$. These applications gave dry matter and nitrogen yields of phytomass from 1340 to 1600 and from 13.4 to $23.4 \, g \, m^{-2}$ respectively, and of grain from 673 to 810 and from 11.5 to $20.4 \, g \, m^{-2}$ respectively. However, the model tended to overestimate both dry matter and nitrogen yields under high rates of fertilizer application. This simulation study demonstrates that results from more detailed studies of water and nitrogen uptake may be used to understand the annual dynamics of water and nitrogen distribution in the soil-crop system.

Introduction

The accurate simulation of water and nutrient uptake by root systems is necessary in models used to study the use of irrigation and fertilizers in crop production. Mathematical equations for the uptake of water by root systems have been in use for many years [9]. These equations describe the movement of water along hydraulic potential gradients from the soil through the roots to the plant canopy as determined by hydraulic resistances in the soil and roots. However, these equations are not included in many models used

in the study of water deficit effects on crop growth. Some such models [25] use dimensionless factors to reduce uptake from dry soil while others [4, 10] normalize water uptake according to soil depth. In some crop models [22] water uptake is simulated as radial flow to single roots uniformly distributed through a soil layer, but it is not linked to dynamic estimates of transpiration rates based on canopy water potentials. These models cannot reproduce the entire hydraulic gradient by which extraction rates are determined. More detailed models of water uptake [43] represent the entire hydraulic gradient, allowing dynamic estimates of crop water status under changing atmospheric conditions. Such models need to be incorporated into larger models of agroecosystem behavior if these latter models are to be based on more fundamental processes.

Mathematical equations for the uptake of nutrient ions by root systems have also been available for many years [2, 6, 7, 33]. These equations describe the movement of ions by mass flow and diffusion to root surfaces, and their absorption by active uptake processes at the root surfaces. The translocation, assimilation and storage of uptaken N has also been reviewed in detail [24]. However, in some models of crop growth [18, 22] uptake is simulated from empirical estimates for the supply and demand of inorganic N based on concentrations in the soil and crop, rather than from explicit representations of the component processes. If studies of nitrogen cycling at the ecosystem level of resolution are to be based on studies of nitrogen behavior at higher levels of spatial and biological resolution, then mathematical treatments of these processes need to be included in simulation models.

In this simulation study, the behavior of process-based equations for the movement and uptake of water and nitrogen are examined as part of an agroecosystem model. The objective of this study was (1) to determine whether the observed short-term behavior of root systems in the uptake of water and nitrogen could be reproduced by these equations, and (2) to determine whether the use of these equations in an agroecosystem model would allow the model to reproduce seasonal trends in the uptake of water and nitrogen, and in the yield of nitrogen and dry matter (DM) observed in a fertilizer experiment. An important part of this second objective was the reproduction of the dynamics of nitrogen and DM distribution within the growing crop.

Methods

Simulation technique

Dynamics of soil C and N. The dynamics of water and nitrogen uptake are simulated as part of a larger submodel for the movements and transformations of carbon, nitrogen, and phosphorus in organic and inorganic phases in the soil and crop (Fig. 1). These dynamics are based on explicit finite different approximations to partial differential equations describing movement and transformations of C, N, P, O, water and heat through a horizontally layered soil profile on a fixed time step of 1 h. Subsequent discussion of the model will be limited to C and N.

In the soil organic phase, C and N undergo first order decomposition and mineralization from each of four subcomponents of crop residue [5], and three components of manure [11] into soluble C and NH_4 pools. These pools are the substrates for the formation of two pools of microbial biomass of differing activity, through either aerobic [29] or anaerobic [27] processes based on Michaelis-Menten kinetics. Microbial activity is fully coupled to the movement of O in gaseous and soluble forms through the soil profile. The microbial pools undergo first order decomposition [23], releasing some C into the soluble C pool for recycling to the microbial biomass [30] and some into humads, with associated mineralization and fixation of NH_4^+. Similarly, humads undergo first order hydrolysis [34], with partial solubilization and partial reformation of C and N as humus, again with associated mineralization and fixation of NH_4^+. Humus also undergoes first order hydrolysis and mineralization.

In the soil inorganic phase, soluble NH_4^+ may be adsorbed into exchangeable NH_4^+ [38], volatilized into gaseous NH_3 [8] and nitrified [28] into soluble NO_3^-. When oxygen is deficient at the microbial microsites, soluble NO_3^- is used as an alternative electron acceptor for microbial respiration, causing denitrification for which soluble C is the substrate.

Both soluble NH_4^+ and soluble NO_3^- are fully coupled to submodels for water movement through the soil profile. The vertical movement of each N species i between adjacent layers ($F_{V,N,i}$ in g m^{-2} h^{-1}) is calculated as the sum of convective and diffusive fluxes:

$$F_{V,N,i} = F_{V,W}C_{N,i} + D_e \, \partial C_{Ns,i}/\partial z \qquad (1)$$

where $F_{V,W}$ = vertical water flux (m^3 m^{-2} h^{-1}), $C_{Ns,i}$ = concentration of N in soil water (g m^{-3}), D_e = effective dispersion-diffusion coefficient of

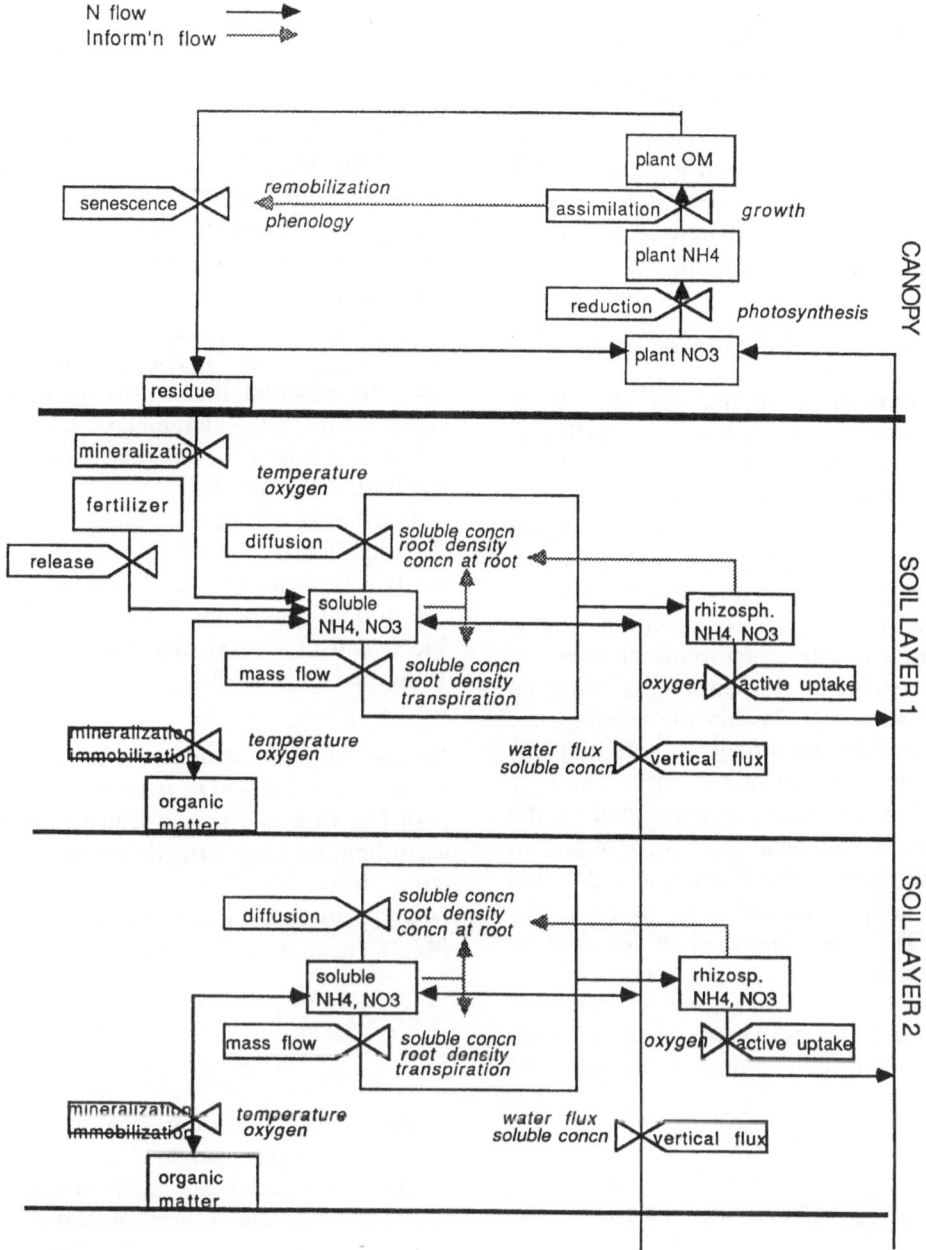

Fig. 1. Flow diagram for nitrogen cycle in the agroecosystem simulation model, showing processes for nitrogen movement and uptake in the soil profile.

N ion (m^2 h^{-1}), and z = soil depth (m). The flux $F_{V,W}$ is calculated from vertical gradients of soil water potential (ψ_s) and hydraulic conductivity between adjacent soil layers as described elsewhere [17]. The diffusive flux $D_e\, \partial C_{Ns,i}/\partial z$ is calculated from vertical gradients of concen-

tration and effective diffusivity between adjacent layers.

Uptake of mineral N. The distribution of root DM growth through a horizontally layered soil profile is calculated recursively from current uptake rates of water and N per unit root surface area in each layer. From this distribution, root length and surface area are calculated from an assumed root radius of $0.15 * 10^{-3}$ m, a DM content of 0.075 g g^{-1}, and an internal porosity of 0.10 m^3 m^{-3} [32]. The radial movement of N to crop roots within each horizontal soil layer ($F_{R,N,i}$ in g m^{-2} h^{-1}) is calculated as the sum of convective and diffusive radial fluxes:

$$F_{R,N,i} = F_{R,W} C_{Ns,i} + D_e \, \partial C_{Ns,i}/\partial r \qquad (2)$$

where $F_{R,W}$ = transpiration flux (m^3 m^{-2} h^{-1}), calculated from radial gradients of hydraulic potential and conductivity between the soil and cylindrical roots in each soil layer. This calculation involves an iterative convergence solution for the canopy water potential (ψ_c) at which $\Sigma\, F_{R,W}$ for all soil layers equals the canopy transpiration rate ($F_{C,W}$). The transpiration rate is calculated from hourly meteorological data and from canopy stomatal resistance (r_c), itself an indirect function of ψ_c. Within the convergence solution, a value for the root water potential (ψ_r) of each soil layer is calculated such that $F_{R,W}$ from the soil to the roots, as controlled by the soil-root potential gradient $\psi_s - \psi_r$, is equal to that from the roots to the canopy, as controlled by the root-canopy potential gradient $\psi_r - \psi_c$, for each layer. The equations used in the convergence solution are given elsewhere [15].

The diffusive flux $D_e \, \partial C_{Ns,i}/\partial r$ in Eq. 2 is calculated from radial gradients of concentration and effective diffusivity between the soil and cylindrical roots:

$$D_e \, \partial C_{Ns,i}/\partial r$$
$$= \{2\pi D_e (C_{Ns,i} - C_{Nr,i})/\ln(r_2/r_1)\} \, L_r \qquad (3)$$

where C_{Nr} = N concentration in the rhizosphere (g m^{-3}), r_2 = mean half-distance between root axes (m) as calculated from root length density (L_d in m m^{-3}), with a maximum value equal to the mean distance for ion diffusion during each time step as calculated from D_e, and where r_1 = root radius and L_r = root length per layer (m m^{-2}).

Active uptake of N by crop roots ($U_{N,i}$ in g m^{-2} h^{-1}) is calculated according to Barber [3]:

$$U_{N,i} = \{U_{N,i\,max}(C_{Nr,i} - C_{Nr,i\,min})\}/$$
$$\{C_{KNr,i} + (C_{Nr,i} - C_{Nr,i\,min})\} \qquad (4)$$

where $U_{N,i\,max}$ = maximum uptake rate at saturating $C_{Nr,i}$ (g m^{-2} h^{-1}) calculated as the product of an experimental rate constant (g m^{-2} root area h^{-1}) and the root surface area ($2\pi r_1 L_r$ in m^2 root area m^{-2} ground area), $C_{Nr,i\,min}$ = minimum $C_{Nr,i}$ for N uptake (g m^{-3}) and $C_{KNr,i} = (C_{Nr,i} - C_{Nr,i\,min})$ at which $U_{N,i} = 1/2\, U_{N,i\,max}$ (g m^{-3}). An Arrhenius equation is used to describe the sensitivity of $U_{N,i\,max}$ to temperature [26]. Parameters for Eq. 4 were taken from the experimental data of Barber [3] for both NH$_4^+$ and NO$_3^-$.

In order to calculate N uptake, $F_{R,N,i}$ from Eq. 2 is equilibrated with $U_{N,i}$ from Eq. 4 by iteratively solving for $C_{Nr,i}$ with a convergence criterion of 0.001. The movement and uptake of soluble P and O are calculated in the same way as those of NH$_4^+$ and NO$_3^-$.

Storage and assimilation of crop N. N taken up as either NH$_4^+$ or NO$_3^-$ is stored in an inorganic pool (N_p in g m^{-2}) from which N is withdrawn according to crop growth requirements calculated from the nitrogen : dry matter ratios ($R_{N,DMo}$ in g g^{-1}) and the DM growth rates (dDM_o/dt in g m^{-2} h^{-1}) of each organ (o):

$$dN_p/dt = \sum_i U_{N,i} - \sum_o \{(dDM_o/dt)\, R_{N,DMo}\} \qquad (5)$$

If the nitrogen concentration in an inorganic pool ($C_{N,p}$ in g(N) g(DM)$^{-1}$), calculated as $N_p/$ {leaf DM + sheath DM + stalk structural DM + root DM}, exceeds a threshold value (set at 0.01 for crop plants), $U_{N,i\,max}$ in Eq. 4 is reduced [37].

The term dDM_o/dt is calculated from the soluble carbohydrate pool of the crop (C_p in g m^{-2}, stored in the leaves as the primary product of photosynthesis), calculated from leaf and canopy level algorithms [13, 16], and from the maintenance respiration rate (R_m in g m^{-2} h^{-1}), the dry matter : carbohydrate (CH$_2$O) ratio of organ synthesis (R_{DM,CH_2Oo} in g g^{-1}) and phenology-dependent partitioning coefficients (P_o) [14]:

$$dDM_o/dt = (dC_p/dt - R_m)\, P_o\, R_{DM,CH_2Oo} \quad (6)$$

If N_p is less than $\Sigma\{(dDM_o/dt)\,R_{N,DMo}\}$ from Eq. 5 (i.e. the N reserves present during a given hour are inadequate to meet N demands during the hour), then dDM_o/dt is reduced to that enabled by N_p, such that all N_p is withdrawn during the hour:

$$dDM_o/dt = P_oN_p / \left\{ t\sum_o (P_oR_{N,DMo}) \right\} \quad (7)$$

where t = time step (1 h).

Remobilization of crop N. During the pre- and post-anthesis period, reserves of soluble CH_2O and N are accumulated in the stalk for subsequent translocation to the grain, as determined by the balance between crop uptake and grain demand [21]. Under conditions of reduced uptake, or of rapid translocation of CH_2O or N, these reserves may decline, under which conditions CH_2O and N in the model may be transferred from C_p and N_p to the reserves, such that C_{Cp}, calculated as $C_p/\{\text{leaf DM}\}$, and C_{Np} also decline. When C_{Cp} declines below a threshold value, additional CH_2O may be remobilized to C_p from leaf DM of the simulated crop, with accompanying remobilization of leaf N at current leaf $R_{N,DMo}$. This remobilization is intended to allow the model to reproduce the accelerated leaf DM senescence observed under conditions of rapid translocation of CH_2O to the grain [39], as may occur during water deficits. Similarly if C_{Np} declines below a threshold value of $1.0*10^{-3}$ g N g^{-1} DM [41], additional N may be remobilized to N_p from leaf N. This remobilization is intended to allow the model to reproduce the accelerated loss of leaf N observed under low N fertilization [40]. Up to 67% of leaf N may be remobilized as simulated amino acids with accompanying remobilization of leaf DM at a $R_{N,DMo}$ of 16%. Remaining leaf N may be remobilized with leaf DM only at current leaf $R_{N,DMo}$ which is much lower, causing accelerated leaf DM senescence. Additional remobilization of leaf DM may occur if the CH_2O production of a leaf is insufficient to meet its maintenance respiration requirement over a 24 h period, as may occur in the lower canopy layers. Only 25% of leaf DM is recovered as C_p during simulated

remobilization of leaf DM, while the rest is lost to crop residue on the soil surface. Remobilization of DM and N occurs first in the lower leaves of the simulated canopy, and proceeds upwards. In the model, remobilization of N proceeds in advance of that of DM, as observed experimentally [21].

Leaf photosynthetic capacity is calculated as the product of RuBP carboxylase activity and superficial N density, assuming a constant relationship between RuBP carboxylase and total leaf N. Reductions in leaf $R_{N,DMo}$ arising from remobilization of leaf N cause reductions in superficial N density, such that CO_2 fixation [42] and hence C_p is reduced. In this way, a dynamic balance between C_p and N_p arises within the simulated crop. Within each hourly time step, Eq. 6 is first calculated for root growth, and then for shoot growth, such that the latter is more sensitive to inadequate N_p. Under these conditions, C_{Cp} and C_p partitioned to the roots increase, such that root:shoot ratios shift in favour of the roots as observed experimentally [21].

Experimental technique

Winter wheat (cv Arminda) was planted at Bouwing (51°57′ N, 5°45′ E) on 21 October, 1982 as part of an N fertilizer response study carried out at three sites over two years. Plant density after emergence was 200 m^{-2}. Soil characteristics at the site are presented in Table 1. Data for maximum and minimum temperature, global radiation, vapor pressure, windspeed and rainfall were recorded daily. Data for the distribution of carbon, nitrogen and water through the soil-crop system were collected in the field every two to three weeks from February through July 1983, for three fertilizer treatments (Table 2) ranging from 0.0 (N1) to 16.0 (N3) g N m^{-2}. Experimental methods and results are presented by Groot and Verberne [20].

Recorded data for soil, weather and management were used in the model to reproduce the experimental conditions under which the field data were recorded. Because weather data were collected daily, hourly values used in the model were estimated [14]. Model runs were begun on 1 October 1982, and ended on 1 August 1983. A simulated manure application of 750 g DM m^{-2}

Table 1. Physical and chemical characteristics of the clay at Bouwing

Layer	1	2	3	4	5	6	7
Depth to bottom (m)	0.10	0.20	0.30	0.40	0.60	0.80	1.00
Sand (%)	2.0	2.0	2.0	3.5	4.0	4.5	4.5
Silt (%)	60.0	60.0	60.0	60.0	60.0	60.0	60.0
Clay (%)	35.0	35.0	35.0	35.0	35.0	35.0	35.0
Organic Matter (%)	2.8	2.8	2.8	1.4	1.4	0.7	0.0
Saturated conductivity (mm h^{-1})	0.28	0.28	0.28	0.28	0.10	0.10	0.10
VWC at 0.03 MPa (m^3 m^{-3})	0.38	0.38	0.38	0.38	0.40	0.40	0.40
VWC at 1.5 MPa (m^3 m^{-3})	0.17	0.17	0.17	0.17	0.18	0.18	0.18
Bulk Density (Mg m^{-3})	1.35	1.35	1.35	1.40	1.40	1.35	1.35

Table 2. Fertilizer treatments (g m^{-2}) at Bouwing. N1 = lowest level, N2 = intermediate level, and N3 = highest level

	Date (ddmmyy)		
Treatment	130583	220683	total
N1	0	0	0
N2	6.0	0	6.0
N3	12.0	4.0	16.0

was added to the uppermost soil layer on 1 October 1982 to enable apparent mineralization rates for the soil profile that were consistent with those estimated for these soils [31]. Output from the model was studied at two levels of temporal resolution: (1) hourly, to study the behavior of the algorithms for water and nutrient uptake by the simulated root system, and (2) seasonal, to compare the longer term distributions of carbon, nitrogen and water arising from these algorithms in the model with corresponding distributions recorded in the field experiment.

Results

Hourly

Water uptake. Hourly simulated profiles of ψ_s, $\psi_s - \psi_r$, L_d, and $F_{R,W}$ are presented at 0300 h, 0900 h, 1500 h and 2100 h on 4 July 1983 for the N1 treatment (Fig. 2). At 0300 h (Fig. 2a), ψ_c was −0.40 MPa and $F_{C,W}$ was very low, while $\psi_s - \psi_r$ was close to zero through the soil profile. Such equilibrium conditions in the model arise from high r_C in the absence of solar irradiance. At 0900 h (Fig. 2b), ψ_c was −0.47 MPa, and $F_{C,W}$ was $0.08 * 10^{-3}$ m^3 m^{-2} h^{-1} of which the greater

part was taken up from the 0.20–0.50 m soil zone. Because ψ_s, and hence hydraulic conductivity, increased with depth, the vertical distribution of $F_{R,W}$ was deeper than that of L_d. Only a very small $\psi_s - \psi_r$ was necessary to generate the $F_{R,W}$ in each soil layer required to meet $F_{C,W}$. In fact, because ψ_s in the uppermost layer was lower than ψ_c, $\psi_s - \psi_r$, and hence $F_{R,W}$, was negative, such that some redistribution of water in the soil profile occurred through the root system. Such redistribution has been observed experimentally [1]. At 1500 h (Fig. 2c), ψ_c was −0.62 MPa, and $F_{C,W}$ was $0.28 * 10^{-3}$ m^3 m^{-2} h^{-1}. In comparison with those at 0900 h, $\psi_s - \psi_r$ and $F_{R,W}$ were increased, most notably in the soil layers above 0.30 m. The small size of $\psi_s - \psi_r$ in comparison with $\psi_r - \psi_c$ indicates that under these conditions, the root hydraulic resistance was the greatest component of the soil – root – canopy resistance pathway in the model. At 2100 h (Fig. 2d), ψ_c was −0.45 MPa, and $F_{C,W}$ was $0.004 * 10^{-3}$ m^3 m^{-2} h^{-1}. Negative $\psi_s - \psi_r$ gradients developed in the upper 0.20 m of the soil profile, causing some redistribution of water to this zone from lower in the profile where $\psi_s - \psi_r$ remained slightly positive. The simulated profile of L_d was similar in magnitude and distribution with that recorded at the same site and crop growth stage during the following year [20].

N uptake. The simulated dynamics of NH_4^+ and NO_3^- uptake through the soil profile are presented in Table 3 for treatments N1 and N3 at 0300 and 1500 h on 4 July 1983. Below 0.10 m, NH_4^+ had been extracted in almost all layers to the minimum concentration (0.012 g m^{-3} [3]) above which uptake can occur (Table 3a). At 0300 h, $F_{R,N,i}$ in treatment N1 was entirely diffu-

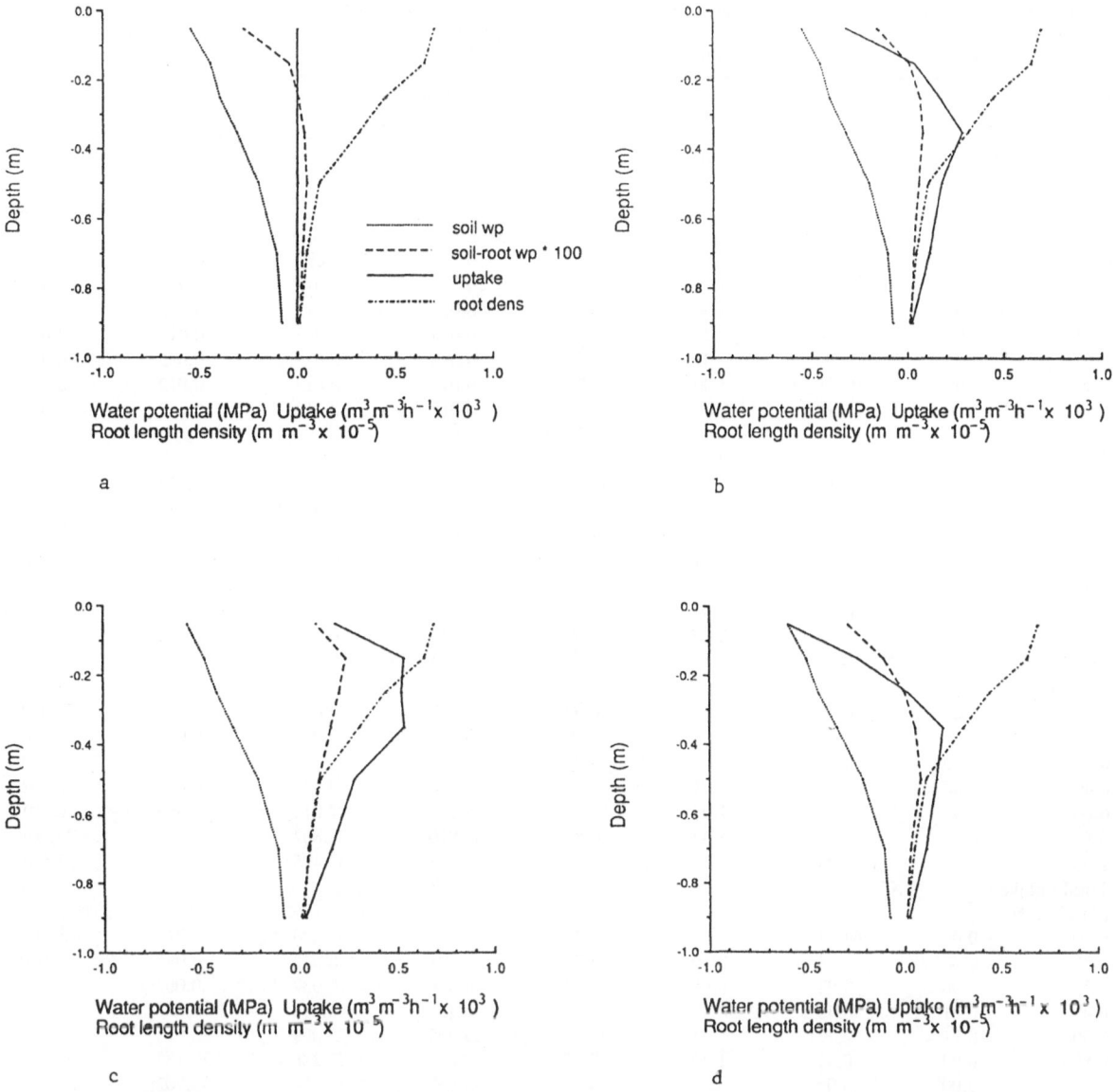

Fig. 2. Soil water potentials, soil-root water potential differences, root length densities, and water uptake simulated through the soil profile at (a) 0300 h, (b) 0900 h, (c) 1500 h, and (d) 2100 h on 4 July 1983 for the N1 treatment at Bouwing.

sive, while at 1500 h, about 3% of $F_{R,N,i}$ was by mass flow. Diurnal changes in total $U_{N,i}$ (2.25 vs. 1.71 mg m^{-2} h^{-1}) for all layers reflected declining concentrations in the uppermost layer. The simulated fertilizer treatments had little effect on NH$_4^+$, so that the dynamics of NH$_4^+$ uptake in treatment N3 (Table 3b) were similar to those in treatment N1.

NO$_3^-$ in the upper 0.40 m of the soil profile has been extracted in treatment N1 to the minimum concentration (0.030 g m^{-3} [3]) above which uptake can occur (Table 3c). Below 0.40 m, NO$_3^-$ was still available. At 0300 h, $F_{R,N,i}$ in this zone was entirely diffusive, while 30% of $F_{R,N,i}$ at 1500 h was by mass flow. The ratio of mass flow to diffusion varied with $F_{R,W}$ and L_d in each layer. Concentration gradients ($C_{Ns,i} - C_{Nr,i}$) were reduced by mass flow such that diffusion was reduced and total movement was little affected by time of day (4.92 vs 5.11 mg m^{-2} h^{-1}).

Table 3. Simulated movement through mass flow (MF) and diffusion (DF), and solution concentrations in the bulk soil (C_{Np}) and rhizosphere (C_{Nr}) of winter wheat for NH_4^+ under (a) low (N1) and (b) high (N3) and for NO_3^- under (c) low (N1) and (d) high (N3) N fertilization treatments at 0300 and 1500 h on 4 July 1983

Depth (m)	Movement ($g\,N\,m^{-3}$ soil $h^{-1} * 10^3$)				Concentration ($g\,N\,m^{-3}$ water)			
	0300 h		1500 h		0300 h		1500 h	
	MF	DF	MF	DF	C_{Np}	C_{Nr}	C_{Np}	C_{Nr}
(a) NH_4^+ N1								
0.10	0.00	21.23	0.56	15.30	0.541	0.095	0.411	0.079
0.20	0.00	0.00	0.00	0.00	0.012	0.012	0.012	0.012
0.30	0.00	0.00	0.00	0.00	0.012	0.012	0.012	0.012
0.40	0.00	0.04	0.00	0.03	0.015	0.013	0.015	0.013
0.60	0.00	0.59	0.01	0.60	0.096	0.027	0.095	0.027
0.80	0.00	0.00	0.00	0.00	0.012	0.012	0.012	0.012
1.00	0.00	0.00	0.00	0.00	0.012	0.012	0.012	0.012
Total uptake	2.25		1.71 ($g\,m^{-2}\,h^{-1} 10^3$)					
(b) NH_4^+ N3								
0.10	0.00	31.20	0.77	30.50	1.159	0.299	1.053	0.138
0.20	0.00	0.04	0.01	0.02	0.014	0.013	0.013	0.013
0.30	0.00	0.00	0.00	0.00	0.012	0.012	0.012	0.012
0.40	0.00	0.37	0.04	0.35	0.055	0.019	0.054	0.020
0.60	0.00	0.29	0.04	0.29	0.191	0.053	0.191	0.056
0.80	0.00	0.00	0.00	0.00	0.012	0.012	0.012	0.012
1.00	0.00	0.00	0.00	0.00	0.012	0.012	0.012	0.012
Total uptake	3.22		3.24 ($g\,m^{-2}\,h^{-1} * 10^3$)					
(c) NO_3^- N1								
0.10	0.00	0.00	0.00	0.00	0.030	0.030	0.030	0.030
0.20	0.00	0.00	0.00	0.00	0.030	0.030	0.030	0.030
0.30	0.00	0.00	0.00	0.00	0.030	0.030	0.030	0.030
0.40	0.00	0.00	0.01	0.25	0.030	0.030	0.042	0.032
0.60	0.00	14.61	1.03	13.89	8.179	7.014	7.587	6.501
0.80	0.01	7.93	3.42	4.89	23.816	22.862	23.628	23.061
1.00	0.00	2.06	1.57	0.63	33.297	32.495	33.313	33.081
Total uptake	4.92		5.11 ($g\,m^{-2}\,h^{-1} * 10^3$)					
(d) NO_3^- N3								
0.10	−0.09	64.14	48.91	70.39	69.857	68.626	66.716	65.309
0.20	−0.00	2.64	0.10	2.43	0.091	0.042	0.087	0.040
0.30	0.00	0.92	0.05	0.77	0.066	0.037	0.062	0.036
0.40	0.01	18.46	4.48	13.59	6.133	4.915	5.667	4.760
0.60	0.01	2.79	5.31	−2.32	25.905	25.009	25.975	26.722
0.80	0.00	0.55	1.36	−0.77	28.093	27.386	28.133	29.114
1.00	0.00	0.05	0.16	−0.10	33.669	33.054	33.702	35.019
Total uptake	9.29		14.80 ($g\,m^{-2}\,h^{-1} * 10^3$)					

Treatment N3 caused a large increase of $C_{Ns,i}$, and consequently of $F_{R,N,i}$ in the top soil layer (Table 3d), and smaller increases below. Because $C_{Nr,i}$ in the top layer (68.6 g m^{-3}) was much higher than $C_{KNr,i}$ (0.40 g m^{-3} [3]), only a small ($C_{Ns,i} - C_{Nr,i}$) was necessary to sustain high $U_{N,i}$. Diurnal changes in the temperature of the upper 0.10 m of soil between 0300 and 1500 h caused some change in $U_{N,i\,max}$ (Eq. 4) and hence in $F_{R,N,i}$. Below 0.40 m, high $C_{Ns,i}$ caused mass flow to exceed $U_{N,i}$ at 1500 h, such that negative ($C_{Ns,i} - C_{Nr,i}$) and hence negative diffusion occurred. Higher ion concentrations in the rhizosphere than in the soil solution have been observed experimentally when mass flow exceeds uptake [35]. In the absence of mass flow, however, similar uptake rates were maintained by diffusion alone in this zone.

Seasonal

Water uptake. Water uptake at the seasonal level is represented for treatment N1 in Fig. 3. Model runs began on 1 October 1982 with a stable water table at 1.50 m, and an initial volumetric water content (ϕ in $m^3\ m^{-3}$) at field capacity through the soil profile. By 7 February 1983, ϕ of the entire soil profile was estimated to be 0.43–0.45, although lower values were recorded lower in the profile. In general, both simulated and recorded ϕ remained near 0.40 until the end of May, after which depletion of water occurred at all depths. Some depletion was simulated from the 0.0–0.3 m soil zone before the end of May, such that ϕ from April through July was underestimated by about 0.03. However, the simulated rate of depletion was consistent with that recorded for this zone. Depletion from the lower soil zones was not simulated before the end of May, although rates of depletion simulated during June and July were higher than those recorded. Consequently, ϕ at 0.3–0.6 m was underestimated in later July. Seasonal trends in ϕ were similar for the other two treatments.

N uptake. Nitrogen uptake at the seasonal level is represented in Fig. 4 for the soil and Fig. 5 for the crop. Model runs were initialized on 1 October 1982 with an assumed NH_4^+ and NO_3^-

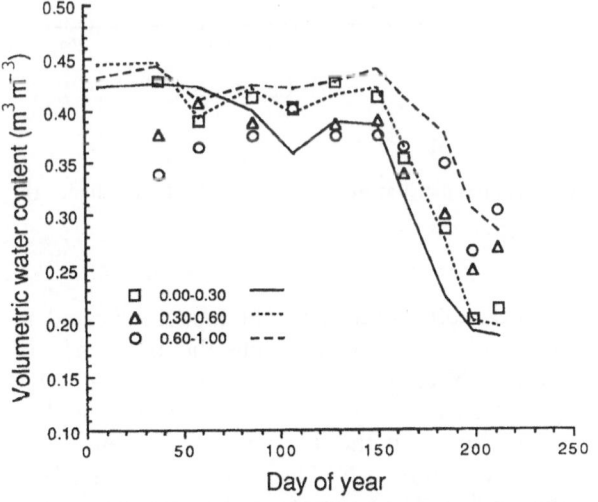

Fig. 3. Seasonal trends in volumetric water content simulated (lines) and recorded (symbols) for 0.0–0.3 m, 0.3–0.6 m, and 0.6–1.0 m at Bouwing during 1983.

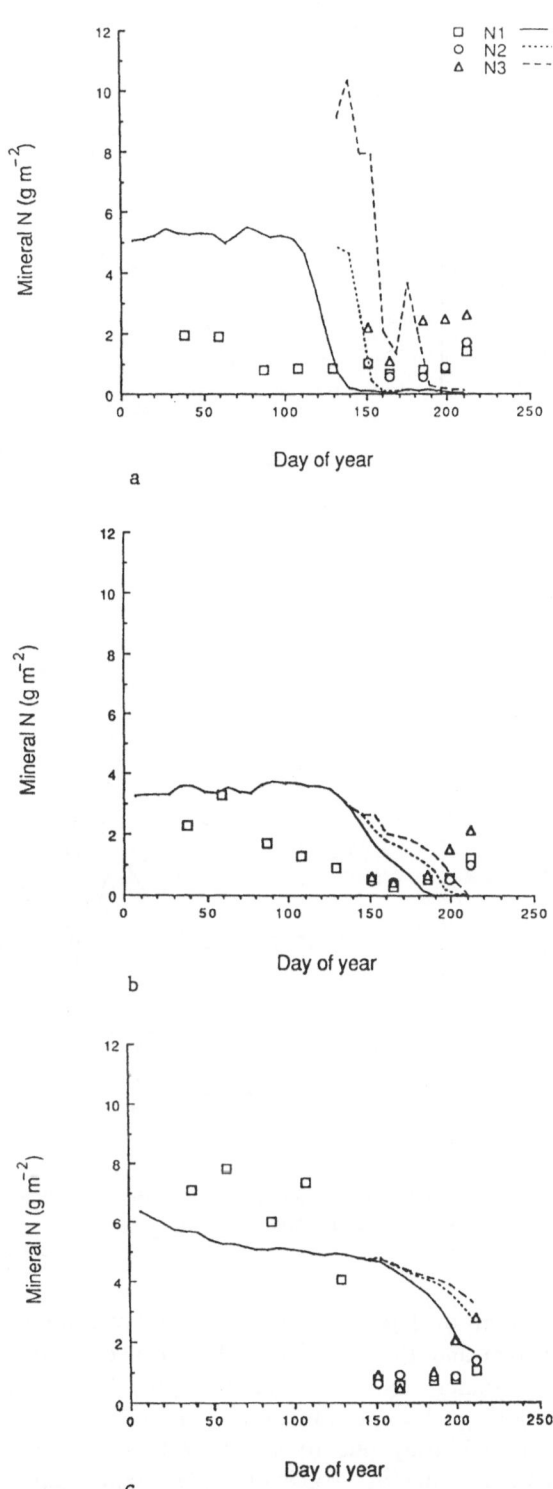

Fig. 4. Seasonal trends in mineral N content simulated (lines) and recorded (symbols) for (a) 0.0–0.3 m, (b) 0.3–0.6 m, and (c) 0.6–1.0 m at Bouwing during 1983.

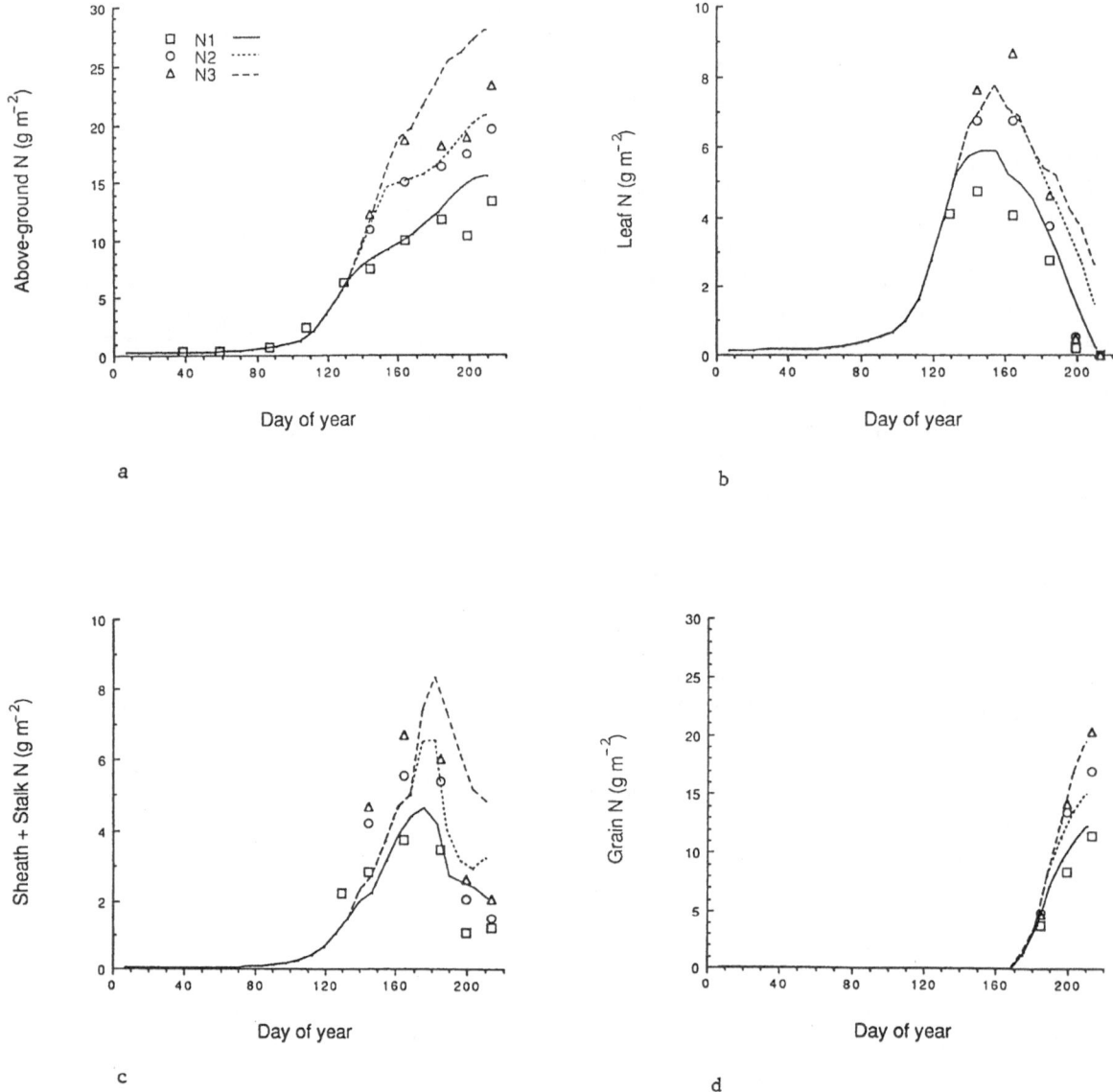

Fig. 5. Seasonal trends in mineral N uptake simulated (lines) and recorded (symbols) at Bouwing during 1983 for (a) shoots, (b) leaves, (c) stems + sheaths, and (d) grain.

content of 1 μg g^{-1} in each soil layer above 0.60 m and 15 μg g^{-1} in each soil layer below. The mineral N content in the upper 0.30 m of soil (Fig. 4a) was overestimated by the model until mid-May due to mineralization from the simulated manure application. The rapid deple-tion of mineral N simulated from this soil zone during May was not apparent in the field data, such that N reached lower values than those

recorded after mid-May. The model indicated an increase in soil N following the 6.0 g m^{-2} of mineral N applied in treatment N2 on 13 May (day 133) and following the 12.0 g m^{-2} applied on 13 May and the 4.0 g m^{-2} on 22 June (day 172) in treatment N3. However, almost none of the applied N was recovered in the field data. With the addition of 16.0 g m^{-2} of mineral N in treatment N3, cumulative denitrification simulated from 1 October 1982 to 1 August 1983 increased from 1.4 to 1.8 g m^{-2}, while cumulative net leaching increased from 5.1 to 6.0 g m^{-2} over that of treatment N1. These increases did not account for the apparent disappearance of the mineral N in the field.

Mineral N levels in the 0.3–0.6 m soil zone were overestimated by 1 to 2 g m^{-2} until mid-June (Fig. 4b). Simulated values remained between 3 and 4 g m^{-2} until mid-May, and then declined to less than 0.5 g m^{-2} by the end of June, while recorded values declined from 3 g m^{-2} to 0.5 g m^{-2} from early March to mid-June. Fertilizer treatment had only a small effect on these levels in both the simulated and recorded data. The model did not reproduce the partial recovery of mineral N levels during July.

Mineral N levels simulated in the 0.6–1.0 m soil zone were close to those recorded until the end of May (Fig. 4c). However, the model did not reproduce the rapid disappearance of mineral N from this zone during May, such that N levels in this zone were overestimated thereafter. As for the higher soil zone, the model did not reproduce the partial recovery of mineral N levels during July.

N assimilation and remobilization. The time integrals of the net N uptake rates for the shoots are shown in Fig. 5a as the cumulative N uptake by the crop over the growing season. These net rates are equivalent to the gross rates, examples of which were shown in Table 3, less the N retained in the roots and lost in senescent material. These integrals are compared with determinations of total above-ground N for the three treatments. The model accurately reproduced the rapid increase in uptake following the fertilizer applications in May and June. About 75% of the simulated cumulative N uptake at the end of the season had been taken up before anthesis (day 170), as commonly observed in field trials [21]. Uptake in the N3 treatment was overestimated during late June and July.

The time integrals of N partitioned to and withdrawn from the leaves are shown with field data for total leaf N content in Fig. 5b. The model was able to reproduce the time course of leaf N content for the three treatments, as well as the delayed withdrawal of leaf N for treat-

ments N2 and N3, which was apparent in this field trial and others [40]. However, the rate of N withdrawal during July was underestimated. Similarly, while the time course of N accumulation in the stalks was reproduced (Fig. 5c), N withdrawal from the stalks during July was underestimated, notably for treatment N3. The accumulation of N in the grain, however, was accurately simulated (Fig. 5d), indicating that late season uptake of N may have been overestimated. Such an overestimation is consistent with the overestimated depletion of mineral N in the 0 to 0.6 m soil zone during July (Fig. 4a and b), indicating that the values for $C_{Nr,i\,min}$ in Eq. 4 may be too low for field applications, or that some mineral N is not available to the root system. The simulated N harvest index was about 0.75, as recorded in this and other field trials [21].

The effect of fertilizer treatment on crop DM accumulation (Fig. 6a) was less than that on N accumulation (Fig. 5a), as apparent in both the field and simulated results. The reduced rate of DM growth simulated for treatment N1 during July was caused by reduced content of simulated leaf N (Fig. 5b) and consequent reduced CO_2 fixation. Accumulated DM for treatment N3 was overestimated during July, as was accumulated N uptake. The model reproduced seasonal trends in leaf DM (Fig. 6b), although the rate of leaf senescence during mid and late July was underestimated. Stem + sheath (Fig. 6c) and grain (Fig. 6d) growth were relatively less affected by N treatment than was leaf growth, and seasonal trends were more closely reproduced by the model. Grain DM accumulation was overestimated for treatment N3 during July.

Values between 0.05 and 0.10% have been suggested as the critical range of C_{Np} below which N may limit wheat yields [41]. Simulated C_{Np} in treatment N1 declined below this range by mid-April, and remained low for the rest of the season (Fig. 7a). Treatments N2 and N3 allowed C_{Np} to remain above this range until the end of June and July respectively. Treatment N1 caused simulated C_{Cp} to remain above those of treatments N2 and N3 (Fig. 7b), although reduced CH_2O production eventually caused C_{Cp} to decline. Higher C_{Cp} under lower N fertilization has been recorded experimentally in wheat [40].

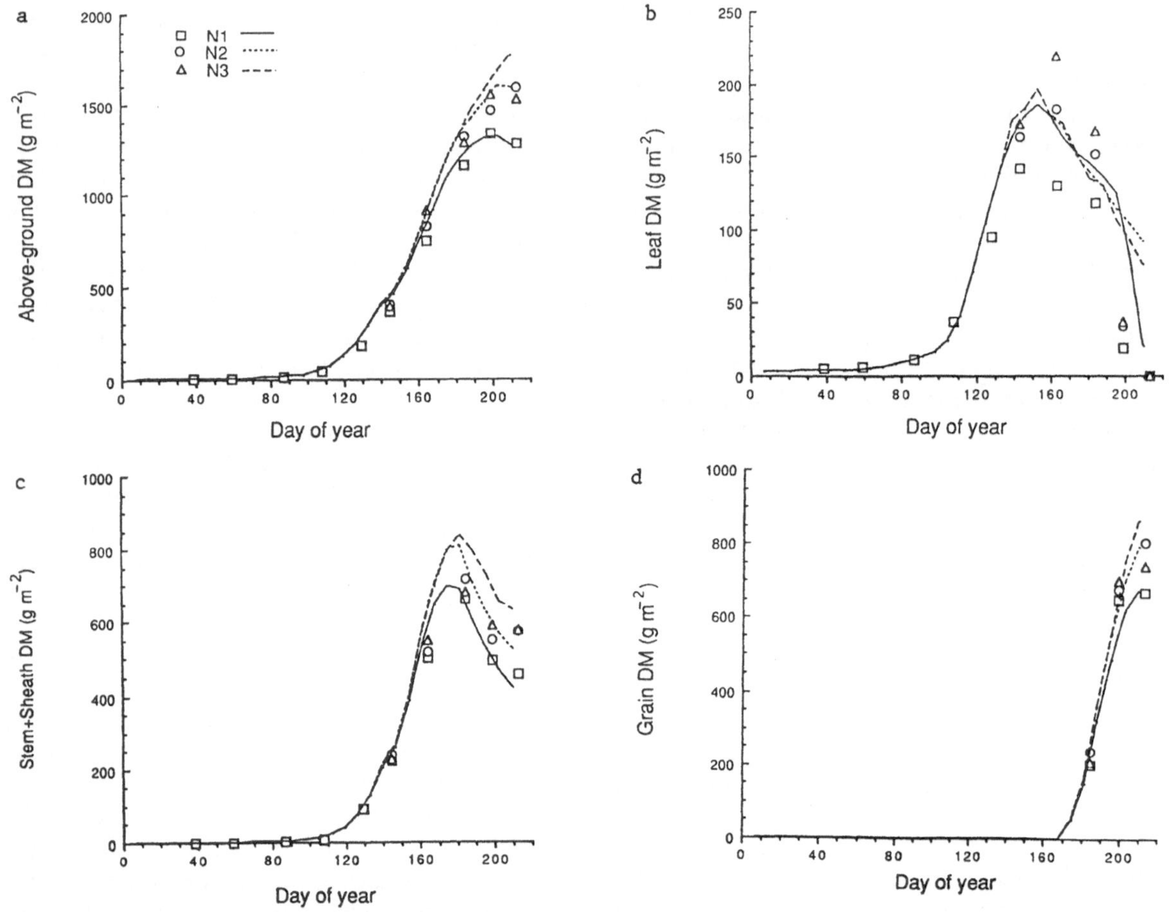

210

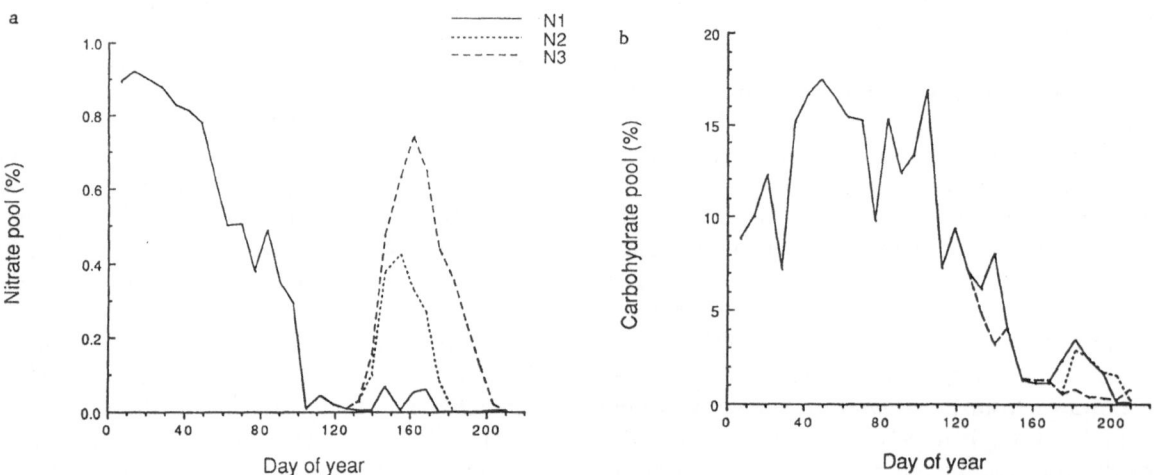

Fig. 6. Seasonal trends in DM growth simulated (lines) and recorded (symbols) at Bouwing during 1983 for (a) shoots, (b) leaves, (c) stems + sheaths, and (d) grain.

Fig. 7. Seasonal trends in simulated (a) mineral N and (b) carbohydrate stored by the crop as the primary products of uptake and fixation.

Discussion

The objective of simulation modelling is to integrate the findings of research at higher levels of spatial, temporal and biological resolution into representations of system behavior at lower levels of resolution. The first objective of this simulation study was thus to examine model behavior at the spatial and temporal resolution of the horizon and the hour, and to determine whether this behavior was consistent with that observed in detailed experiments.

The redistribution of water through the root system from wetter to drier soil zones (Fig. 2b and 2d) provides a theoretical basis for experimental observations [1], and indicates that short term patterns of water uptake by root systems were reproduced in the model. Under the conditions of this experiment, the rate of this redistribution was less than 0.1 mm h^{-1} for the entire profile and occurred when crop water potential was higher than those of the drier soil zones near the surface. This simulated rate was insufficient to replace water uptake by roots during the day, as also found experimentally [1].

The short term root dynamics reproduced in the model allowed it to simulate longer term patterns of water uptake (Fig. 3), although some evidence of excessive removal from the upper soil zones was apparent. More detailed validation of simulated water uptake is limited by uncertainty in the estimation of related phenomena (e.g. transpiration rates, unsaturated hydraulic conductivity), and the low temporal and spatial resolution at which uptake can be recorded.

The simulated short term dynamics of N uptake by root systems (Table 3) indicated that uptake was little influenced by transpiration, as changes in diffusion rates caused by altered concentration gradients partially offset changes in mass flow. The limited role of mass flow in N uptake is in agreement with theoretical findings [36]. The simulated accumulation of nutrient ions in the rhizosphere when both transpiration and nutrient concentrations are high (Table 3d) provides a theoretical basis for the accumulation of ions about the root observed in radioisotope studies [35]. The time integrals of the simulated short term uptake rates allowed the model to

reproduce seasonal trends in crop N content (Fig. 5) for the fertilizer treatments in this study. However, a longer-term validation of the uptake algorithms could not be carried out from the field data.

The cause of the divergence between simulated and recorded data for the mineral N content of the upper soil zone (Fig. 4a) is not clear. In other such fertilizer experiments, most mineral N applied to the soil in early spring was recovered in samples taken two months later [19]. However, applications in later spring were consistently not recovered in later soil samples taken from this and other sites in the field study [20]. Further study is required to identify the cause of this apparent disappearance. The extent to which the model reproduced seasonal trends in soil mineral N was limited by an apparent underestimation of both its downward movement over winter, and its uptake from the lower part of the soil profile during spring (Fig. 4c). The accurate simulation of seasonal water movement is an important prerequisite for that of seasonal N uptake.

Underlying validation of the estimated rates of water and N uptake by root systems is validation of the estimated geometry of root systems, notably the spatial distribution of L_d and r_1. Profiles of L_d simulated in this study (Fig. 2) were consistent with those recorded at the same stage of crop growth from the same site the following year [20]. Because values arising from this distribution are used in estimating water and nutrient uptake (Eqs 2 to 4), there is a real need for a better understanding of the geometry of root proliferation, and of the processes that control it. The accurate reproduction of rooting patterns through the soil profile, in conjunction with basic algorithms for the uptake and assimilation of nutrients, will allow modelling to become an effective tool in the study of nutrient dynamics in the agroecosystem.

Conclusions

This simulation study indicates that the understanding of component processes that occur at a higher level of resolution in the agroecosystem may be used to understand the integrated behavior of the agroecosystem at a lower level of resolution. It also indicates that there are some areas where our understanding of these processes limits the extent to which we may understand this behavior.

Acknowledgements

This research was partially supported by a grant from the National Science Foundation and utilized the CRAY-2 facility of the National Center for Supercomputing Applications at the University of Illinois in Urbana-Champaign.

References

1. Baker JM and van Bavel CHM (1988) Water transfer through cotton plants connecting soil regions of differing water potentials. Agron J 80: 993–997
2. Barber SA (1962) A diffusion and mass-flow concept of soil nutrient availability. Soil Sci 99: 39–49
3. Barber SA (1984) Soil Nutrient Bioavailability: a Mechanistic Approach. Wiley & Sons New York, 398 p
4. Barfield BJ, Duncan WG and Haan CT (1977) Simulating the response of corn to irrigation in humid areas. paper no 77–2005 Am Soc Agric Eng annual meeting North Carolina State Univ Raleigh NC
5. Beek J and Frissel MJ (1973) Simulation of nitrogen behavior in soils. PUDOC Wageningen Netherlands
6. Caassen N and Barber SA (1976) Simulation model for nutrient uptake from soil by a growing plant root system. Agron J 68: 961–964
7. Cushman JH (1979) An analytical solution to solute transport near root surfaces for low initial concentration. Soil Sci Soc Am J 43: 1087–1095
8. Fleisher Z, Kenig A, Ravina I and Hagin J (1987) Model of ammonia volatilization from calcareous soils. Plant and Soil 103: 205–212
9. Gardner WR (1960) Dynamic aspects of water availability to plants. Soil Sci 89: 63–73
10. Gardner WR (1983) Soil properties and efficient water use: an overview. In: Limitations to Efficient Water Use in Crop Production. Taylor HM, Jordan WR and Sinclair TR (eds) pp 45–64 Am Soc Agron
11. Gilmour JT, Clark MD and Sigua GC (1985) Estimating net nitrogen mineralization from carbon dioxide evolution. Soil Sci Soc Am J 49: 1398–1402
12. Grant RF (1989) Simulation of carbon accumulation and partitioning in maize. Agron J 81: 563–571
13. Grant RF (1989) Test of a simple biochemical model for photosynthesis of maize and soybean leaves. Agric For Meteorol 48: 59–74
14. Grant RF (1989) Simulation of maize phenology. Agron J 81: 451–457
15. Grant RF (1990) Dynamic simulation of water deficit effects upon maize yield. Agric Systems, in press
16. Grant RF, Peters DB, Larson EM and Huck MG (1989) Simulation of canopy photosynthesis in maize and soybean. Agric For Meteorol 48: 75–91
17. Grant RF, Izaurralde RC and Chanasyk DS (1990) Soil temperatures under different residue covers: simulation and experimental verification. Can J Soil Sci, in press
18. Greenwood DJ, Neeteson JJ and Draycott A (1985a) Response of potatoes to N fertilizer: dynamic model. Plant Soil 85: 185–203
19. Greenwood DJ, Neeteson JJ and Draycott A (1985b) Response of potatoes to N fertilizer: quantitative relations for components of growth. Plant Soil 85: 163–183
20. Groot JJR and Verberne ELJ (1991) Response of wheat to nitrogen fertilization, a data set to validate simulation models for nitrogen dynamics in crop and soil. Fert Res 27: 349–383
21. Herzog H (1986) Source and sink during the reproductive period in wheat, Adv Agron and Crop Sci supplment 8 Paul Parey Scientific Publishers Berlin, 104 p
22. Jones CA and Kiniry JR (1986) CERES-maize. Texas A & M University Press College Sta TX
23. Juma NG and McGill WB (1986) Decomposition and nutrient cycling in agro-ecosystems. In: Microfloral and Faunal Interactions in Natural and Agro-ecosystems. Mitchell MJ and Nakas JP (eds) pp 74–136 Nijhoff/Junk Netherlands
24. Kenig A (1987) The problem of quantifying limits to nitrogen use in relation to crop growth modeling. In Plant Growth Modeling for Resource Management. Wisiol K and Hesketh JD (eds) CRC Press Inc Boca Raton FL
25. Keulen H van (1981) Simulation of water use and herbage growth in arid regions – a reevaluation and further development of the model 'Arid Crop'. Agric Systems 6: 159–193
26. Lauchli A (1984) Mechanisms of nutrient fluxes at membranes of the root surface and their regulation in the whole plant. In: Roots, Nutrient and Water Influx, and Plant Growth. pp 1–25 Amer Soc Agron special publication no 49 Madison WI
27. Leffelaar PA and Wessel WW (1988) Denitrification in a homogeneous, closed system: experiment and simulation. Soil Sci 146: 335–349
28. Mahli SS and McGill WB (1982) Nitrification in three Alberta soils: Effect of temperature, moisture, and substrate concentration. Soil Biol Biochem 14: 393–399
29. McGill WB, Hunt HW, Woodmansee RG and Reuss JO (1981) Phoenix, a model of the dynamics of carbon and nitrogen in grassland soils. In: Terrestrial Nitrogen Cycles. Clark FE and Rosswall T (eds) Ecol Bull 33: 49–115
30. Molina JAE, Clapp CE, Shaffer MJ, Chichester FW and Larson WE (1983) NCSOIL, a model of nitrogen and carbon transformations in soil: description, calibration and behavior. Soil Sci Soc Am J 47: 85–91
31. Neeteson JJ (1989) Assessment of Fertilizer Nitrogen Requirement of Potatoes and Sugar Beet. Ph.D. Thesis Wageningen Agricultrual University Netherlands, 141 p

32. Noordwijk M van (1987) Methods for quantification of root distribution pattern and root dynamics in the field. In: Proc 20th Colloq Int Potash Insitute Bern, pp 263–281

33. Nye PH and Mariott FHC (1969) A theoretical study of the distribution of substances around roots resulting from simultaneous diffusion and mass flow. Plant and Soil 30: 459–472

34. Paul EA and Juma NG (1981) Mineralization and immobilization of soil nitrogen by microorganisms. In: Terrestrial Nitrogen Cycles. Clark FE and Rosswall T (eds) Ecol Bull 33: 179–195

35. Rendig VV and Taylor HM (1989) Principles of Soil-Plant Interrelationships. McGraw-Hill NY, 275 p

36. Seligman NG, van Keulen H and Goudriaan J (1975) An elementary model of nitrogen uptake and redistribution by annual plant speices. Oecologia 21: 243–261

37. Siddiqi MY and Glass ADM (1982) Simultaneous consideration of tissue and substrate potassium concentration in K^+ uptake kinetics: a model. Plant Physiol 69: 283–285

38. Smith OL (1982) Soil Microbiology: A Model of Nutrient Decomposition and Nutrient Cycling. CRC Press Inc Boca Raton FL

39. Spiertz JHJ (1974) Grain growth and distribution of dry matter in the wheat plant as influenced by temperature, light energy and ear size. Neth J Agric Sci 22: 207–220

40. Spiertz JHJ and Ellen J (1978) Effects of nitrogen on crop development and grain growth of winter wheat in relation to assimilation and utilization of assimilates and nutrients. Neth J Agric Sci 21: 210–231

41. Vlassak K and Verstraeten LMJ (1985) Nitrogen nutrition of winter wheat. In: Wheat Growth and Modelling. Day W and Atkin RK (eds) pp 217–236 Plenum Publishing

42. Wareing PF, Khalifa MM Treharne KJ (1968) Rate-limiting processes in photosynthesis at saturating light intensities. Nature 220: 453–455

43. Zur B and Jones JW (1981) A model for the water relations, photosynthesis an expansive growth of crops. Water Resour Res 17: 311–320

Fertilizer Research **27**: 215–231, 1991.
© 1991 *Kluwer Academic Publishers.*

Modelling water flow, nitrogen transport and root uptake including physical non-equilibrium and optimization of the root water potential

F. Lafolie
Institut National de la Recherche Agronomique, Station de Science du Sol, Domaine St Paul, B.P. 91, 84143 Montfavet, France

Key words: Modelling, water flow, nitrogen transport, nitrogen cycle, root uptake

Abstract

A model is presented for the simulation of water flow, heat flow, and nitrate and ammonium transport. Two approaches are used for modelling plant water uptake as well as for plant nitrogen uptake. Nitrogen transformations are accounted for in a very simple way. This paper focuses mainly on water flow modelling, solute transport, and water uptake. Richards' equation is used to model water flow in layered soil profiles with a great variety of boundary conditions. Solute transport is simulated with either a simple convection dispersion equation or with a two-region physical non-equilibrium model to distinguish between mobile and immobile water and solute exchange between these two regions. A macroscopic sink term is added to Richards' equation to account for plant water uptake. This term can be calculated along two different approaches, one of which is based on the concept of root water potential. The root water potential is then continuously optimized to minimize the difference between the climatic demand and the uptake rate.

Simulation results are compared with field data from the Netherlands to illustrate the degree to which the model is able to predict water flow, solute transport and plant water uptake. The root water potential optimization model seems to provide the best prediction of water distribution. In particular the shape of the profile, revealing uptake patterns, is quite well reproduced with this model. Comparison of simulated and observed water content profiles seems also to reveal the presence of preferential pathways. The comparisons show also how predicted solute distributions can be improved by using a two-region approach rather than a simple convection-dispersion model.

Introduction

It is now widely acknowledged that the environment and the soil in particular is diversely affected and disturbed by human activities. Concern for the fate of the chemicals, pesticides, herbicides, heavy metals, fertilizers, released in the soil by agriculture and other activities is growing. Among the numerous problems we are facing, a great deal of attention is being given to the protection of surface and groundwater resources, especially with respect to nitrate. Prediction of nitrate behaviour in the long term thus becomes a subject of major concern.

The economic situation requires an agriculture that produces more while at the same time minimizing inputs. This results in an increasing demand for better management of fertilization. Among fertilizers, nitrogen is the most subject to displacement and transformations. Thus, required nitrate availability to plant and nitrogen displacement in soil may conflict. Understanding and prediction of such phenomena are needed to better decide when to apply nitrogen and how much should be applied while at the same time minimizing losses and groundwater pollution.

In response to such environmental and economical preoccupations, models have been developed in the past few years to simulate nitrogen behaviour in soils. Originally, these models were designed as tools to test our understanding of the system. There is now a trend to use models as management tools, even if their capacity of prediction is still limited. While at the beginning only few phenomena were accounted for, models are actually available that treat the whole system, including plant development and soil biochemical and physical processes.

The model we are presenting in this paper focuses on physicochemical processes and deals with biological phenomena in a rather rough way. Our attention is given to water and solute transport phenomena and to water uptake by plant roots. Nitrogen transformations are treated more simply. This model is not intended for use as a management tool, but rather as a means of testing strategies and assessing risks. In the following, we present first the assumptions and equations of the model and then some comparisons with field measurements.

Model description

The model is one-dimensional and accounts for the following phenomena: water flow, nitrate and ammonium transport, heat flux, N-cycle, plant uptake of water and nitrogen. Plant growth and root system development are not simulated and must be provided. Functions or arrays of data can be used for this purpose. The model is built around the water flow and heat flux modules. Nitrogen transformations and solute transport modules use the outputs of the previous ones.

Water flow modelling

Transport equation
We assume that one-dimensional water flow obeys Darcy's law for both saturated and unsaturated conditions. Vapour phase transport and osmotic potentials are not considered. Combin-

ing with the law for mass conservation, the classical Richards' equation is derived, [13]

$$C(\psi,z)\,\frac{\partial H}{\partial t} = \frac{\partial}{\partial z}\left[K(\psi,z)\,\frac{\partial H}{\partial z}\right]$$
$$- \Gamma(\psi_r, \psi, \rho_r(z,t)), \qquad (1)$$

where t is time (h); z is distance (m), considered to be positive downwards; ψ is the soil water pressure head (m); $H = \psi - z$ is the total pressure head (m); C is the soil capillary capacity (m^{-1}) derived from the slope of the soil-water retention curve; K is the hydraulic conductivity $(m\,h^{-1})$; and Γ (h^{-1}) is a sink term accounting for root water uptake. ψ_r is the 'root water potential' and ρ_r is a function of z and t describing the time and space variations of the root density profile. The meaning and calculation of the root water potential are explained in the plant water uptake paragraph. Note that we used the total pressure head rather than the soil matrix water potential as the dependent variable.

Soil hydraulic properties, $C(\psi)$ and $K(\psi)$ are assumed to depend on z to account for possibly layered soil profiles. Soil hydraulic properties, and in particular hydraulic conductivity are known to depend on soil structure, which cannot be considered as constant if long time simulations are to be carried out. This phenomenon is certainly important in the upper tilled layer. Provision is made in the program to include, if the information is available, a time dependence of hydraulic properties. Surface sealing is also a phenomenon that can greatly influence water flow processes such as evaporation, infiltration and thus runoff or biological processes such as plant emergence. In our model, crust development is not explicitly simulated, but a thin layer with modified hydraulic properties can be added at the soil surface.

The main limitation to predicting water flow in soils with this approach stems from preferential pathway effects that cannot be accounted for with Richards' equation. When preferential pathways are present, and depending on soil humidity and rain intensity, a certain percentage water can by-pass the porous matrix and infiltrate deeply very quickly. There is no possibility in the model to account for such phenomena and quality of predictions may sometimes be altered.

Upper boundary conditions

Equation (1) must be supplemented with appropriate boundary conditions for calculating water flux and water content distributions. Upper boundary conditions in the model account for most physical phenomena occurring at the soil surface: rains, irrigations, evaporation, runoff. Two types of boundary conditions are used. Flux-type boundary conditions (Neuman-type) that specify flux, and imposed pressure boundary conditions (Dirichlet-type) that specify pressure heads at the soil surface. A flux-type boundary condition is used for rains, sprinkler irrigation, or evaporation as long as soil infiltrability or evaporation capacity are respected. This condition writes,

$$\left[-K(\psi, z)\frac{\partial H}{\partial z} \right]\Bigg|_{z=0} = Q(t) \qquad (2)$$

where Q is the intensity of the rain, irrigation or evaporation. Q is positive or negative for inward or outward flux, respectively. Dirichlet-type boundary conditions are used when the soil is flooded or when evaporation cannot occur at the rate required by climatic conditions.

In the model, the intensity and duration of rains and sprinkler irrigations are specified as inputs and a Neumann-type boundary condition (eq. 2) is used. If the applied flux exceeds the soil infiltrability, then the surface potential attains 0 and the model shifts to a Dirichlet-type condition. In this case, the model builds up a positive pressure head at the soil surface, up to a preset threshold, and then assumes that all the water is lost by runoff.

In between rains or irrigations, the model uses the potential evaporation as prescribed flux at the soil surface. During this stage, soil conditions can be such that the soil cannot evaporate at the rate required by the climate. In this case, a potential is imposed at the soil surface and evaporation occurs at a rate lower than imposed. The potential imposed at the soil surface is calculated from climatic data and along a procedure described in [20].

Potential evapotranspiration, P_{ET} (mm day^{-1}) is calculated from standard meteorological data [1]. A classical procedure is used to distribute this constraint between plant canopy and soil. In this procedure, the potential evaporation P_E (mm day^{-1}) is calculated from the leaf area index (LAI) and the potential evapotranspiration. The difference $P_{ET} - P_E$ is used as a potential transpiration, P_T for the plant. The relation we used is:

$$P_E(t) = P_{ET}(t)\exp(-a\,\mathrm{LAI}(t)) \qquad (3)$$

where a is a plant dependent parameter. It was also assumed that $P_E(t)$ was negligible for LAI values above 3. Many relations similar to eq. (3) have been proposed in the past. Neuman et al. [20], used a similar exponential relation, while Tanji et al. [36] used a power function to increase the potential transpiration P_T with increasing values of the LAI.

Lower boundary conditions

At the bottom of the soil profile, the model can handle boundary conditions to simulate an impervious layer or a water table with changing depth. These conditions correspond to a no-flux condition and a time changing pressure head, respectively. With the first condition, a water table can appear but the model does not simulate water flow towards drain tiles or losses of any kind. Hence, if water accumulates at the bottom of the profile it can only be removed by evaporation or plant uptake. In case the water table is too deep to be reasonably included in the simulation domain, a boundary is chosen somewhat below the maximum rooting depth and a free drainage condition is applied at this boundary.

Initial conditions

The initial situation is provided either in terms of water content or soil water potential at selected depths.

Root water uptake model

The program offers the possibility to use a model based on the approach suggested by Feddes et al. [9], hereafter referred as the α-model, and a model based on the optimization of the root water potential, ψ_r.

The α-model assumes that the climatic demand is evenly distributed all over the root system. The maximum rate of extraction, $S_{max} = P_T(t)/L(t)$, where $L(t)$ denotes the total length of the root system, is weighted by a water stress response function, $\alpha(\psi)$. The function $\alpha(\psi)$ is usually a piece-wise linear function, although a smooth formulation was proposed by Van Genuchten [39]. The purpose of introducing α is to account for reduced soil water availability when the soil is too wet or inversely too dry. Thus, one can reasonably assume that α depends on the crop, on soil hydraulic properties, and on root system characteristics. This model has been widely used but requires the fitting of several parameters to define the function α. Hoogland et al. [14] modified the model to account for a possible preferential water uptake close to the soil surface. The term $\Gamma(\psi_r, \psi, \rho_r(z,t))$ in Richards' equation (1) is thus defined by:

$$\Gamma(\psi_r, \psi, \rho_r(z,t)) = \alpha(\psi)\, S_{\max}. \qquad (4)$$

In the second model, it is assumed that water uptake is induced by a gradient in water potential between the root and the surrounding soil. Adjustment of the root water potential by the plant during the day and in response to long time changes in the soil water potential has been widely used [4, 9, 20, 22, 24, 44]. Recent advances in root water uptake processes seem to indicate that the concept of root water potential is meaningful and that the root water potential is regulated by chemical messages issued by the leaves [11, 19, 45, 46]. Hence, its seems interesting to use this concept and to compare it to more simple approaches. The formulation we used for Γ is:

$$\Gamma(\psi_r, \psi, \rho_r(z,t)) = \rho_r(z,t)\, K(\psi)(\psi_r - \psi) \qquad (5)$$

where ρ_r is the root density (cm of root per cm^3 of soil). Although attractive and popular because of its somewhat mechanistic bases, this model has not been widely used. One of the main reasons for that is that the root water potential cannot be easily measured. Some authors [1, 20] suggested that the root water potential varied to take up water at the rate required by the climate. In our approach we assumed that the plant adjusted continuously its root water potential so as to minimize the difference between the climatic constraint and the uptake rate. This corresponds to the following minimization problem. At any time, find ψ_r such that the following quantity be minimum:

$$\left| \int_0^Z \Gamma(\psi_r, \psi, \rho_r(z,t)) - P_T(t) \right| \qquad (6)$$

where Z is the maximum rooting depth. In addition, we assumed that the root water potential is uniform, that the root water potential cannot be lower than a prescribed $\psi_{\lim}$ value, which can be viewed as a wilting potential, and that water flow from roots to soil is not possible. Note that the only resistance to transpiration is assumed to reside in the soil. Additional resistance terms could be easily included to account for root radial and axial resistance to flow.

The assumption of uniformity of the root water potential is certainly too strong, but actually we do not have sufficient knowledge about physiological mechanisms to add rules that would allow to calculate a non-uniform root water potential. The last assumption is needed because situations occur where water flow is possible (eq. (5)) from the root into the soil due to a root water potential higher than the soil water potential at some depths. These situations are related to a contrasted water distribution in the profile, in which cases only a part of the root system works. Although the experiment reported by Baker and Van Bavel [2] seems to reveal that water flows from one part of the root system to another and possibly from the roots into the soil, the sink term is forced to zero for these cases.

Numerical solution
Richards' equation was kept in the mass conservation form (H in place of ψ) to avoid the discretization of first-order derivatives, which always lowers the quality of the solution. The right hand side of equation (1) was approximated by a 3-point centered scheme providing an approximation to the second order. The resulting system of non-linear differential equations was solved by a fully implicit scheme and a fixed-point iterative technique. This part of the numerical solution has been tested against analytical and other numerical solutions and proved to be always very accurate. Discretization schemes of higher orders for space derivatives and more sophisticated time

integration methods did not prove to bring significant improvements. The finite difference grid can have variable size increments ranging from some millimeters close to the surface to several centimeters. An automatic procedure was set up to adjust the time step to physical conditions. This procedure is based on the number of iterations required to reach the convergence in the fixed point iteration process, but also accounts for abrupt changes such as the beginning of rains, irrigations, etc. The time step classically varies between some seconds to one or two hours.

At each time step, optimization of the root water potential, when required, is achieved by an iterative algorithm searching for the ψ_r that minimizes the difference between the amount of water taken up over the time step and the climatic demand. Hence, for each time increment, Richards' equation is solved several times with successive estimates of ψ_r. Convergence is usually achieved within one or two iterations, since the initial estimate corresponding to the root water potential at the previous time step is close to the new one. This algorithm is an improvement of the tangent method used in nonlinear parameter estimation procedures.

Solute transport modelling

The various mechanisms, convection, dispersion, diffusion, adsorption, exchange between mobile and immobile water, etc. involved in solute transport are well known and have been studied at length in the past and with particular attention in the last ten years [10, 21, 23, 27, 30, 31, 32, 43]. Usually, the convection-dispersion equation (CDE) is used to model transport of non-reactive species, nitrate for example, while transport of adsorbed species, ammonium for example, is modelled with the LEA (Local Equilibrium Assumption) model, the extension of the CDE for instantaneously adsorbed species [29, 36].

Under steady state water flow, it was demonstrated [26, 27, 31], that the LEA was appropriate for modelling solute transport through weakly structured porous media only. Besides, abundant literature has illustrated the effects on solute displacement of preferential pathways and stagnant water regions, located inside or outside

the aggregates [21, 43]. Most solute transport studies in the field have been analyzed under the assumption that a steady flow existed and the LEA adequately described solute transport. Besides the pioneer work of Gaudet [10], it was not until recently, [29], that the two-region first-order physical non-equilibrium model, proposed by Van Genuchten [37] for steady state flow conditions, was coupled with Richards' equation to analyze solute transport experiments.

The program we developed in the past years offers the possibility to simulate nitrate and ammonium transport with either the LEA or the two-region first-order physical non-equilibrium model (FO).

LEA Model

In this model all the sources of dispersion are lumped into one parameter, the dispersion coefficient $D(\theta, v)$, and all the water is assumed to be active in the transport process. In addition, it is assumed that adsorption is instantaneous and described by a linear Freundlich isotherm. The adsorbed concentration S (mg(g soil)$^{-1}$) is then related to the aqueous concentration, C (mg cm^{-3}), by: $S = k C$, where k (cm^3 g^{-1}) is the partition coefficient. The partial differential equation describing solute transport writes

$$\frac{\partial(\theta + \rho_s k)C}{\partial t} = \frac{\partial}{\partial z}\left[\theta D(\theta, v)\frac{\partial C}{\partial z} - v\theta C\right]$$
$$- \Gamma_1 + \Gamma_2 - \Gamma_3 \quad (7)$$

where $\rho_s(k$ m$^{-3})$ is bulk density, t is time (h), z is depth (m), and v(m h^{-1}) is the pore water velocity defined as the ratio of the Darcy's flux q to the water content θ. Γ_1, Γ_2, Γ_3 are sink and source terms to account for plant uptake, sources and sinks of solute, respectively. D is the dispersion coefficient classically related to the pore water velocity and the diffusion coefficient by:

$$D(\theta, v) = D_m(\theta) + \lambda|v| \quad (8)$$

where $D_m(\theta)$ is the effective molecular diffusion coefficient and λ is the dispersivity for longitudinal dispersion [4]. D_m and λ may change with depth. D_m is often very small as compared to $\lambda|v|$ and hence neglected in equation (8). The linear dependence of D on v expressed in equation (8) holds only at large enough velocities. As pointed out by Sposito [32], this relation might no longer be valid at low saturation ratios. The flux of water, q, is given by the resolution of Richards' equation. Note that the retardation factor $R = \theta + \rho_s k$ reduces to θ for non-sorbed solutes.

A third-type boundary condition [41], is applied at the soil surface. This condition writes:

$$\left[\theta D(\theta, v) \frac{\partial C}{\partial z} - v\theta C \right]\Bigg|_{z=0} = Q_s(t) \qquad (9)$$

where $Q_s(t)$ is the flux of solute applied. A free drainage condition or a no flux condition is used at the bottom of the profile.

FO Model

In this model, water in the soil is partitioned into a mobile part and an immobile part. Immobile zones can be located inside the aggregates as well as outside. Division of soil water can be based on an analysis of the porosity or can be obtained from some preset threshold in the $\psi(\theta)$ or $K(\theta)$ relationships. Russo et al. [29] proposed to define a critical water content θ_c, below which no liquid flux occurs, as the water content where liquid films break down and $K \to 0$. All the water above θ_c is then mobile. Gaudet [10] showed on column experiments carried out with a sandy soil that the immobile water content is not constant but changes with total water content. He obtained a nonlinear relation that exhibited a decrease in immobile water content when global water content increased. This nonlinear relation was approximated by a linear relation: $\theta_m = a\theta + b$.

Solute in mobile water is transported by convection and dispersion, and is exchanged with immobile water under the influence of concentration gradients. Solute in immobile water regions is transported by diffusion only. In the first-order physical non-equilibrium model it is assumed that solute exchange between the two regions can be modelled by a first-order process. The set of equations is as follows

$$\frac{\partial(R_m C_m)}{\partial t} + \frac{\partial(R_{im} C_{im})}{\partial t}$$

$$= \frac{\partial}{\partial z}\left[\theta_m D(\theta_m, v) \frac{\partial C_m}{\partial z} - v\theta_m C_m \right]$$
$$- \Gamma_1 + \Gamma_2 - \Gamma_3 \qquad (10)$$

$$\frac{\partial(R_{im} C_{im})}{\partial t} = \beta(C_m \theta_m - C_{im} \theta_{im}) - \hat{\Gamma}_1 + \hat{\Gamma}_2 - \hat{\Gamma}_3 \qquad (11)$$

where subscripts $_i$ and $_{im}$ refer to variables related to mobile and immobile water, respectively. R_m and R_{im} are the retardation factors for the mobile and immobile zones, respectively. Note that partition coefficients in the Freundlich isotherms can be different for the mobile and immobile zones due to differences in mineralogy [42]. Note also that solute concentration inside immobile regions depends only on depth. Cross-sectional concentration gradients are not considered, although they exist. $C_{im}(z)$ thus represents an average concentration in immobile zones at depth z. Due to this simplification, the first order transport parameter β depends on time, soil structure, effective diffusion coefficient, flow velocity, etc. [23, 30]. The dependence on flow velocity is not very clear and in our study we assume that β is a characteristic of the porous medium only. In addition we adopted the relation proposed by Gaudet [10] and given above to relate θ_m to θ. Immobile water content is then: $\theta_{im} = \theta - \theta_m$.

Boundary conditions are required for the mobile phase equation only, and are similar to those defined for the LEA.

One equation (eq. (7)) or set of equations (eqs. (10) and (11)) is used for nitrate transport modelling and another is used for ammonium. Sink and source terms are added to the right hand side of equations (7), (10) and (11) to account for root uptake and N-cycle transformations. Modelling of nitrogen transformations and nitrogen uptake are described later in the paper.

Numerical solution of transport equations

Finite difference approximations are derived for spatial derivatives in equations (7) and (10). A particular care is given to the approximation of convective terms. It is well known that a fully upwind finite difference approximation of convective terms smoothes out spurious oscillations but introduces artificial spreading of solute fronts. A procedure was developed to control the amount of artificial spreading necessary to avoid numerical oscillations. This procedure is based on a weighted upwind approximation of first-order terms. A fully implicit time marching scheme is used. Note that transport equations are linear while they were non-linear for water flow. Hence no iteration procedure is needed.

Modelling nitrogen uptake by plants
We assumed that nitrate only is taken up by plant roots. Two models can be used in the program. One choice is to simulate nitrate uptake as a passive process resulting only from the uptake of water. In this case, the term Γ_1 is defined by $\Gamma_1 = \Gamma\,C$ where Γ is the root uptake term for water. Consequently, nitrate uptake occurs only when water uptake is active, and location of nitrate uptake in the profile is directly related to water uptake patterns. An other possibility is to use a Michaelis-Menten kinetic as proposed by Nye and Marriott [25]. In this case the flux of nitrate per unit length of root is defined by:

$$F = F_{max}\frac{C}{K + C} \qquad (12)$$

where F_{max} is the maximum uptake rate and K is the Michaelis-Menten constant. Values for K in the literature are usually of 10^{-5} to 10^{-6} M. The term Γ_1 is then simply: $\Gamma_1 = F\,\rho_r$. With this last formula, the program must be provided with F_{max}. Note that F_{max} can be obtained from either direct measurements or derived from dry matter production. Note also that nitrate uptake occurs even if water is not taken up, e.g. roots not active in water uptake, or during the night.

With this last model, nitrate availability may be reduced in some layers after uptake and if nitrification rate is too low or if nitrogen is not transported from other layers. There is no provision in the model to compensate these local deficits by increasing uptakes in other layers.

Modelling nitrogen transformations
Nitrogen cycle is modelled in a very simple way. Figure 1 presents a diagram of nitrogen transformations included in the model. One pool only is considered for organic matter. Carbon and

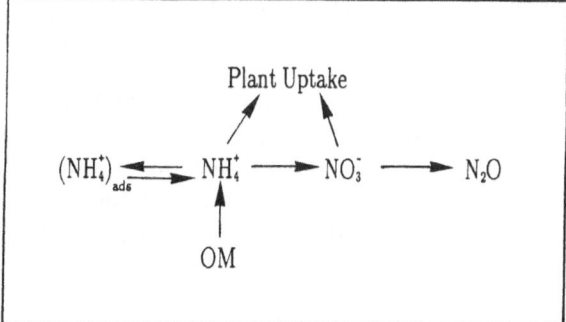

Fig. 1. Scheme of nitrogen transformations.

microbial dynamics are not modelled. Mineralization of organic matter, nitrification of ammonium, and denitrification are modelled as first order processes [5, 15, 16, 17, 28, 35, 36]. We assumed in addition that these phenomena could occur in mobile as well as in immobile water regions. Equations for, respectively, mineralization of organic matter, nitrification of ammonium, and denitrification are as follows:

$$O.M. \rightarrow NH_4$$

$$\theta\,\frac{\partial C_{NH_4}}{\partial t} = k_m f_m(\theta)g_m(T)\rho_S C_{OM} \qquad (13)$$

$$NH_4 \rightarrow NO_3 \qquad \frac{\partial C_{NO_3}}{\partial t} = k_n f_n(\theta)g_n(T)C_{NH_4} \qquad (14)$$

$$NO_3 \rightarrow N_2O \qquad \frac{\partial C_{NO_3}}{\partial t} = k_d f_d(\theta)g_d(T)C_{NO_3} \qquad (15)$$

where C_{OM}, C_{NO_3} and C_{NH_4} are nitrogen concentration for, respectively, organic matter, nitrate and ammonium, k_m, k_n, k_d are optimum first-order rate parameters for, respectively, mineralization, nitrification and denitrification and f and g response functions for, respectively, water content and soil temperature. These equations result in sink or source terms to be added at the right-hand side of solute transport equations. For the three processes above, dependence on soil temperature is classically modelled with a Q_{10} function [15, 34]. The Q_{10} function we use writes:

$$g(T) = 1.072^{(T - T_{opt})} \qquad (16)$$

where T_{opt} is the temperature at which the transformation rate is maximum. Optimum temperatures for each process may differ.

Concerning mineralization and nitrification, we assumed that an optimal water content range, $[\theta_{lopt}, \theta_{hopt}]$ existed and that the mineralization and nitrification rates were reduced for water content below or above this interval [33]. Following Johnson et al. [15], the function f writes:

$$f(\theta) = f_S + (1 - f_S) \left[\frac{\theta_S - \theta}{\theta_S - \theta_{hopt}} \right]^m$$
$$\theta_{hopt} \leq \theta \leq \theta_S \quad (17a)$$

$$f(\theta) = 1 \qquad\qquad \theta_{lopt} \leq \theta \leq \theta_{hopt}$$
$$(17b)$$

$$f(\theta) = \left[\frac{\theta - \theta_W}{\theta_{lopt} - \theta_W} \right]^m \qquad \theta_W \leq \theta \leq \theta_{lopt} \quad (17c)$$

where m is an empirical constant, θ_S is the saturated water content, θ_W is water content below which no activity is possible, and f_S is rate at soil saturation. Dependence of denitrification on soil water content is inspired from Johnson et al. [15] and Rolston et al. [28]. The response function for denitrification is:

$$f_d(\theta) = 0 . \qquad\qquad \theta \leq \theta_d \quad (18a)$$

$$f_d(\theta) = \left[\frac{\theta - \theta_d}{\theta_S - \theta_d} \right]^m \qquad \theta_d \leq \theta \leq \theta_S \quad (18b)$$

Modelling heat flux

As already mentioned, nitrogen transformations depend on temperature, but hydraulic conductivity, diffusion coefficients and root activity depend also on temperature. Modelling of heat flux in soil is thus required. The total heat flux is the sum of a diffusion flux induced by temperature gradients and a convective flux induced by water flow [6]. Assuming that the solid phase and the aqueous phase are at the same temperature, the classical transport equation writes:

$$C_h(\theta, z) \frac{\partial T}{\partial t} = \frac{\partial}{\partial z} \left[\lambda(\theta, z) \frac{\partial T}{\partial z} \right]$$
$$- C_W \frac{\partial}{\partial z} (qT) \quad (19)$$

where T (°C) is temperature, q (m h^{-1}) is the flux of water, C_h (J kg^{-1} °C^{-1}) is the volumetric heat capacity of the soil depending on mineral composition and water content, λ (W m^{-1} °C^{-1}) is the heat conductivity depending also on mineral composition and water content, and C_w is water volumetric heat capacity. Let n be the total porosity of the soil, then C_h is related to the mineral volumetric heat capacity, C_m, by: $C_h = C_m(1 - n) + C_W\theta$. The heat conductivity, λ, can be obtained by fitting the model to measurements or can be estimated from the model of De Vries [7]. Equation (19) is supplemented with top and bottom boundary conditons. At a depth large enough, one can safely assume that the soil temperature is constant. In a first approximation, soil surface temperature can be taken as equal to air temperature. Minimum and maximum air temperature can be obtained in routine from meteorological data, and a sine function can be used to reproduce temperature variation during the day. Although not very rigorous, this approach has the advantage to use easily accessible data as boundary conditions. A different approach, more rigorous, is to estimate the flux of heat into the soil from the energy balance and to use it in a flux-type boundary condition. This approach is complicated when a crop covers the soil.

Resolution of equation (19) is carried out with numerical techniques similar to those used for solute transport and water flow equations. Note that the water flow component induced by temperature gradients was not considered. Consequently, the heat transport equation and the water flow equation are not coupled.

Simulations

In the following we present simulation results and compare them to measurements in the Netherlands [12]. The model was run for the period February 14 to August 21, 1984, at the Bouwing site. The field trial consisted of a wheat crop grown with three different nitrogen dressings. Model predictions are compared with field measurements for water content and mineral nitrogen profiles at various dates.

Water flow

Hydraulic characteristics

The soil profile is split into two layers (0–40 cm and 40–100 cm) with bulk densities of 1.35 and 1.40, respectively. Hydraulic characteristics, $\psi(\theta)$ and $K(\theta)$, were provided in the form of tables, for the topsoil (0–40 cm) and the subsoil (40–60 cm), respectively. Van Genuchten [38] proposed the following closed-form expression for the $\psi(\theta)$ relationship

$$\frac{\theta - \theta_r}{\theta_S - \theta_r} = [1 + |\alpha\psi|^n]^{-m} \tag{20}$$

where θ_S is the saturated water content, θ_r, α, and n are parameters, and $m = 1 - 1/n$. θ_r is often referred to as the residual water content. This expression was fitted to the ψ–θ data. Fig. 2 presents an example of such adjustment. The parameters θ_S, θ_r, α (m^{-1}) and n were fitted to 0.494, 0.000, 0.996, 1.212 and to 0.494, 0.009, 0.711, 1.220 for the topsoil and the subsoil,

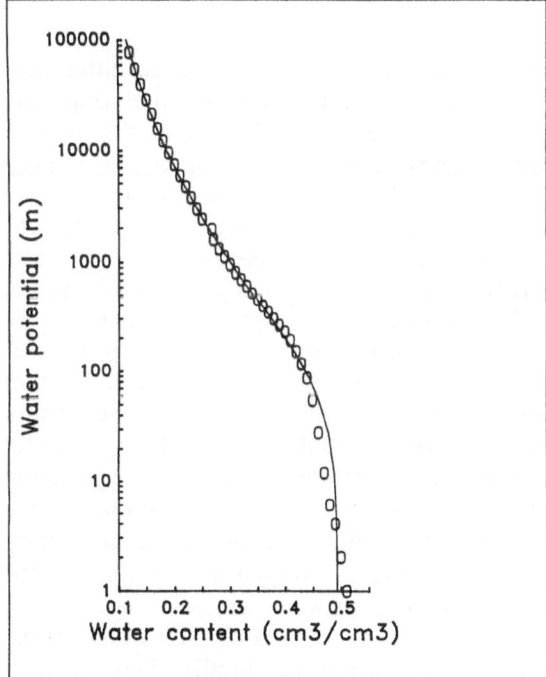

Fig. 2. Soil water potential versus volumetric water content. Open circles are measurements and the solid line is the fitted Van Genuchten analytical relationship.

respectively. In each case the fit was very good. The main discrepancies were noted close to saturation where the predicted θ_S was slightly lower than the measured one. This can probably be related to soil structure and presence of some large pores. Although available under the form of couples $(K - \theta)$, the unsaturated hydraulic conductivity was predicted from the capillary-based model of Mualem [18] later modified by Van Genuchten [38]. The saturated hydraulic conductivity was taken from the data set [12]. The predicted unsaturated hydraulic conductivity was always above the measured one. This kind of discrepancy is usual with Mualem's model and is a serious problem when hydraulic conductivity data are not available.

Initial and boundary conditions

Data for water content, N-nitrate and N-ammonium were available on February 14 for soil layers of about 20 cm thick and to a depth of 100 cm. This date was thus taken as time of start for the simulation. Initial profiles can be found in Figures 5 and 7.

The bottom boundary condition was derived from the time course of the groundwater table. This one was available for the period April 2 to August 6 only. A constant value was used from the beginning to April 2 and from August 6 to the end.

The surface boundary condition consists of rains and evaporation rates. In absence of data regarding rain intensities, we assumed they were close to 5 mm h^{-1}. Fig. 3 presents the distribution of rains during the period under study.

The daily evapotranspiration rate was calculated with the Penman formula. The following standard meteorological data were used: minimum and maximum temperature, daily total global radiation, vapor pressure at 9 a.m., and daily average wind speed. Figure 3 presents the calculated evapotranspiration. We can observe that the formula predicts values too low at the beginning of the period, and too high (up to 7 or 8 mm d^{-1}) at the end. This is probably due to the fact that we used in the formula constants fitted for France and not adequate for the Netherlands. A sine function was used to generate hourly data over the daylight period.

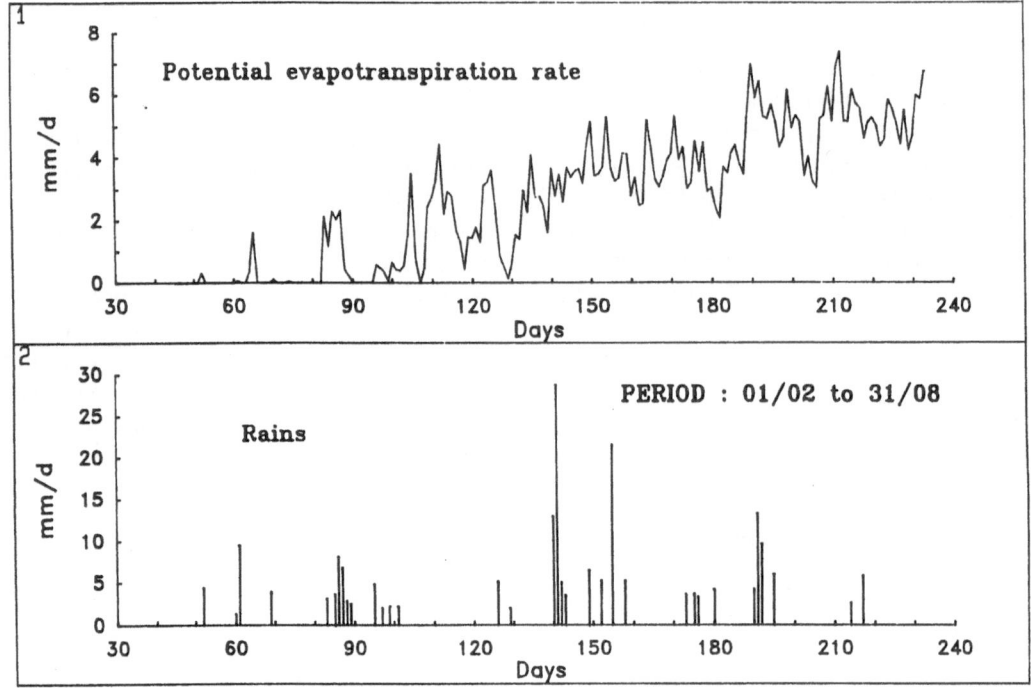

Fig. 3. (1) Potential evapotranspiration rate (mm day⁻¹) calculated from basic meteorological data. (2) Intensity of rain events (mm day⁻¹) for the simulation period.

Root water uptake

The LAI and the root density profile were updated every 24 hours. The root density profile was assumed to be steady after the last measurement. The LAI was used to calculate the potential transpiration rate as indicated previously (eq. (3)). Simulations were carried out with the two water uptake models described before. The root water optimization model does not require any parameter, while the α-model needs the weighting function α. We used a piece-wise linear function as in Belmans et al. [3]. Optimum water uptake, $(\alpha(h) = 1)$, was assumed to extend from $h = -10$ to $h = -1000$ cm, and a linear decrease was assumed from saturation down to -10 cm and from -1000 cm down to the wilting point $(h = -15000$ cm$)$.

Comparison

Figure 4 presents the evolution of observed and simulated water content profile for treatment N1 throughout the growing season. Water content profiles at early times are very poorly simulated. Differences up to 0.1 are observed between cal-culated and field data values. These differences disappear later in the season. Analyzing field data and rain sequences, it appears that a large part of early rains does not appear in a mass balance calculated between soil surface and 1 meter depth. This can only be explained by transport by preferential pathways.

Differences in water uptake models are revealed by water content profiles. Figure 5 provides a comparison of water content profiles as observed and simulated with the two root water uptake models. First it seems that the water content profiles simulated with the root water potential optimization are in better agreement with observed water content distributions than those obtained with the α-model. At later times, plant water uptake is shifted to the bottom of the profile for the root water potential optimization model, in response to a better soil water availability in that part of the profile. This phenomenon is not as marked for the α-model in which water is extracted everywhere in the profile unless soil water potential is lower than a prescribed threshold. Consequently, large differ-

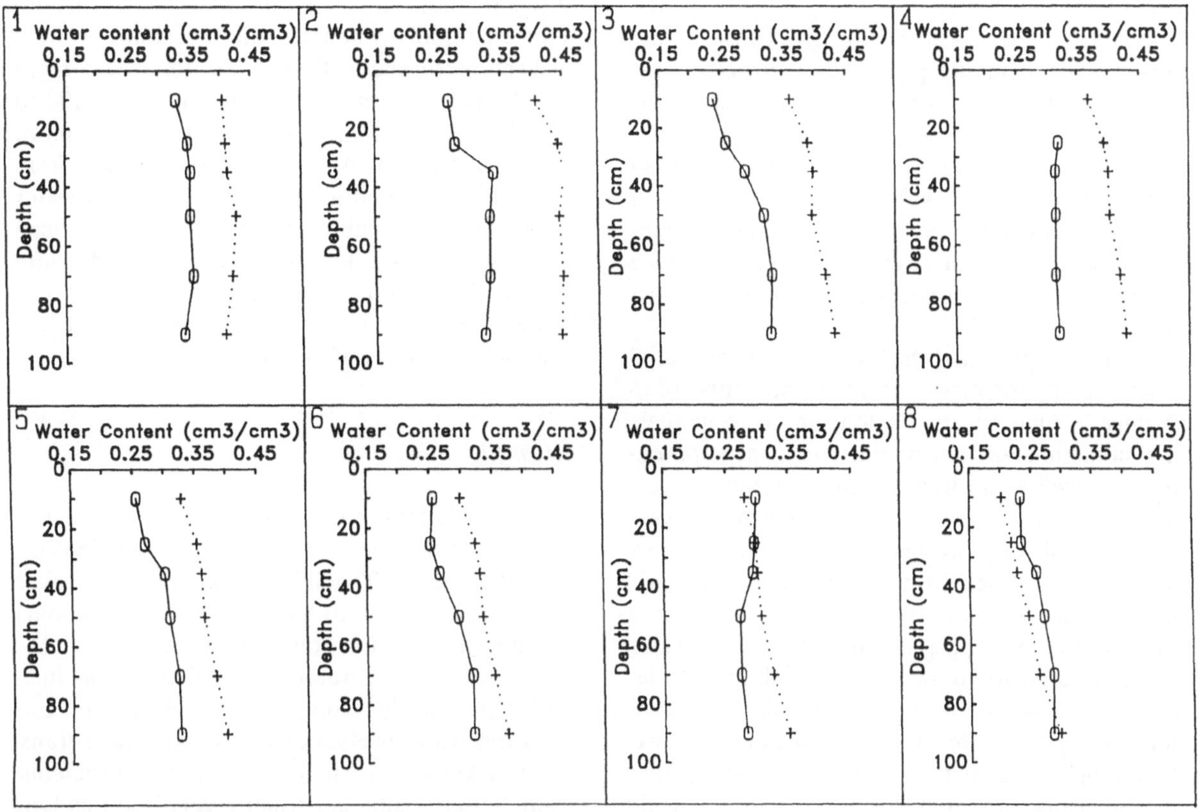

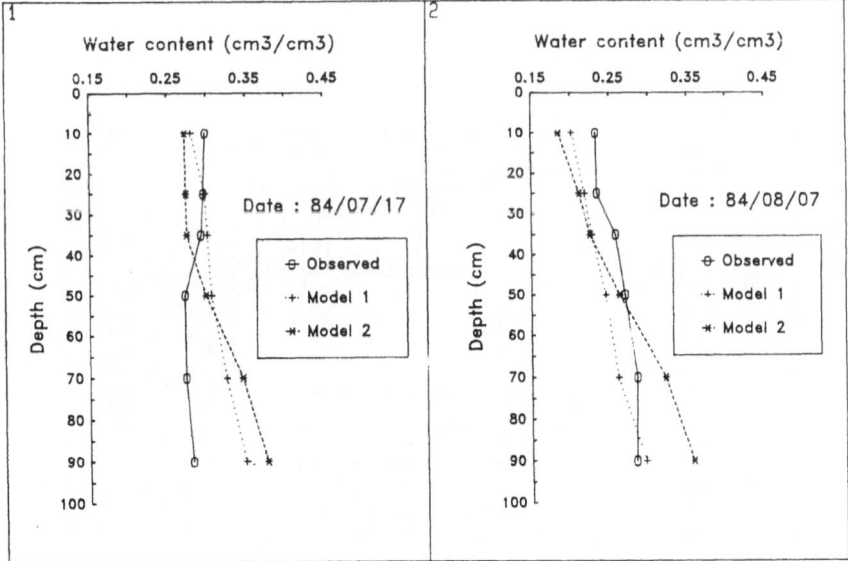

Fig. 4. Calculated and measured soil water content profiles for treatment N1 at various dates. (1): March 13; (2) April 24; (3) May 8; (4) May 28; (5) June 19; (6) July 3; (7) July 17; (8) August 7.

Fig. 5. Observed and simulated soil water content profiles at two dates close to the end of the simulation. Model 1 is the root water potential optimization model and Model 2 is the α-model.

ences in water contents can be observed in the lower part of the profile (Fig. 5). Close to the surface, the α-model predicts the lowest water contents since roots still extract water with this model. Note that the shape of the profile is quite well reproduced with the root water potential optimization option while it is absolutely not with the α-model. The predicted water content values lower than observed are explained by a probably too high climatic constraint as previously noted.

Figure 6 presents the root water potential calculated by the model for the period June 18 to August 7. One observes a long trend to increasing minimum and maximum root water potentials, as well as an increase in daily amplitudes. The long trend corresponds to a progressively reduced soil water availability, while hourly variations are induced by the variable climatic demand during the day. It is interesting to observe that in the first part of the plot, root water potential oscillations are of very weak amplitude, corresponding to a period where soil water content is high. In this range of water contents, hydraulic conductivity is high and small changes in soil water potential correspond to relatively large amounts of water. Hence, small water potential gradients between the roots and the soil suffice to move water from the soil into the roots. Due to the non-linearity of the $\psi(\theta)$ and

$K(\theta)$ relationships, the root water potential increases rapidly when a value of about -10 to -20 m is reached. From that time, day-night amplitudes increase and reach values of -120 to -140 m. A water stress is predicted at the end of July/beginning of August. This stress arises from a too high P_{ET} and a strongly reduced soil water availability. Probably, it would have not been calculated with a lower climatic demand. Note that a 23 mm rain temporarily lowers the root water potential but it starts increasing again within a few days after the rain.

Heat flux model

The parameters required for heat flux modelling were not available and were taken from plausible literature data. The initial condition was taken uniform and equal to 5°C. The bottom boundary condition was also constant and equal to 5°C. Soil surface temperature was calculated as indicated previously. Comparison could not be carried out with observations. However, heat transport is known not to be a source of problems. Simulation results are usually very good as long as we are not interested in soil temperature itself but interested in its consequences on biological processes for which modelling is much more uncertain.

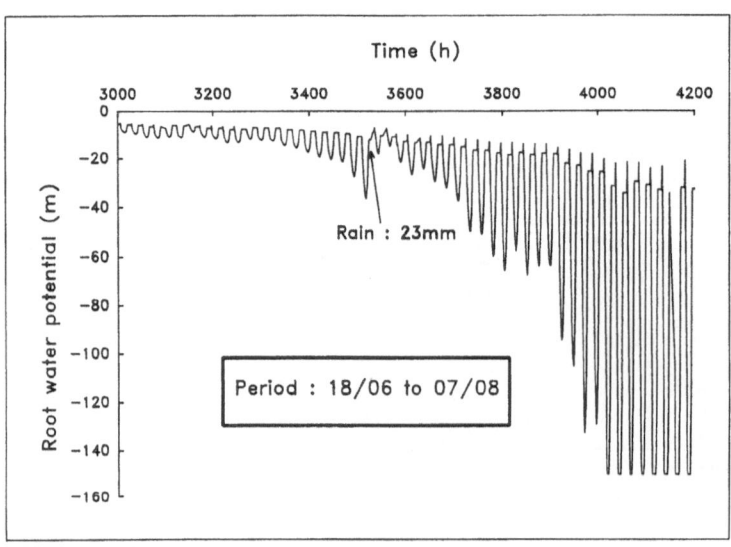

Fig. 6. Calculated root water potential for the period June 18 to August 7. An arrow indicates the date of a 23 mm rain.

Solute transport model

No indications were available about soil structure or mobile-immobile water, although the shape of the $\psi(\theta)$ relationship close to saturation denotes the presence of some structural pores. Consequently, the model was first run with the LEA option and a second time with the FO option to try to account for soil structure effects.

Transport parameters

Adsorption of ammonium was assumed to obey a linear Freundlich isotherm and a k value of $0.75 \, cm^3 \, g^{-1}$ was assumed (see eq. (7)). This value is taken from Kaluarachchi and Parker [16] for a similar material. These authors showed that the nitrate profiles are not very sensitive to k values. Hence, this arbitrary but plausible value should not have much effect on the results.

Dispersion coefficients are calculated according to equation (8). The molecular diffusion coefficient was ignored. For the LEA model, dispersivity values of $\lambda = 3 \, cm$ and $\lambda = 5 \, cm$ were arbitrarily chosen for the topsoil and the subsoil, respectively. For the two-region model, these values were set at 1 and 3, respectively. Such differences between the two models are explained by the fact that the dispersion coefficient for the LEA model accounts for all the sources of dispersion and in particular for dispersion induced by solute exchange between the two water regions. The values retained are in the range of values reported in the field [30, 31]. Dispersivity for the untilled is larger than for the tilled upper layer because of a probably wider pore size distribution.

Parameters for water partitioning and exchange rates were taken from the literature [23, 37]. Similarly to Russo et al. [29], the lower threshold for the immobile water content was set at 0.15 which corresponds to a water potential of approximately $-300 \, m$. All the water above this threshold was regarded as mobile. Values of 0.05 (h^{-1}) and 0.005 (h^{-1}) were selected for the first-order exchange rate parameter (β in eq. 11). The higher value is in the range found for soil columns made up of aggregates. The second value is lower to account for larger immobile water regions and possibly lower diffusion coefficients in response to a higher soil bulk density.

N-Cycle parameters

Mineralization and nitrification constants were taken from Johnsson et al. [15]. The mineralization rate was set at 0.036 (d^{-1}) and the nitrification rate was set to 4.8 (d^{-1}). Optimum temperature for mineralization and nitrification was set at 35°C. No mineralization was assumed below 5°C. Optimum water content for nitrification was set at $0.35 \, cm^3 \, cm^{-3}$. Denitrification and immobilization of ammonium and nitrate were not accounted for during the period simulated.

Initial and boundary conditions

The initial conditions for nitrate and ammonium are derived from the amounts measured in 20 cm thick layers and corresponding water contents and bulk densities. Fertilizations are applied with the rains closest to the dates of fertilizations. Fertilizer is assumed to dissolve instantaneously with the first rain. This constitutes an approximation that may induce errors. The surface boundary condition is a flux-type boundary condition with concentration equal to zero except for rains where the fertilizations are applied.

Comparison

Figure 7 presents the observed and simulated mineral nitrogen profiles at various dates for treatment N1. Although no attempts were made at fitting parameters, the two-region model seems to reproduce reasonably the evolution of the mineral nitrogen profile, at least up to the end of April/beginning of May. This may be fortuitous given that most transport parameters were taken from the literature and not measured. However, the rapid leaching observed with the single region model strengthens our conviction that the two-region model provides a better description of solute transport in agricultural soils and thus should preferentially be used. The best agreement is obtained during a period where water flux is important and transport mechanisms dominate the biological mechanisms. When water flux is small and most applied fertilizer has been taken up, biological mechanisms and nitrogen uptake are preponderant in nitrogen distribution. By that time, differences between the two models tend to smooth out. It is

228

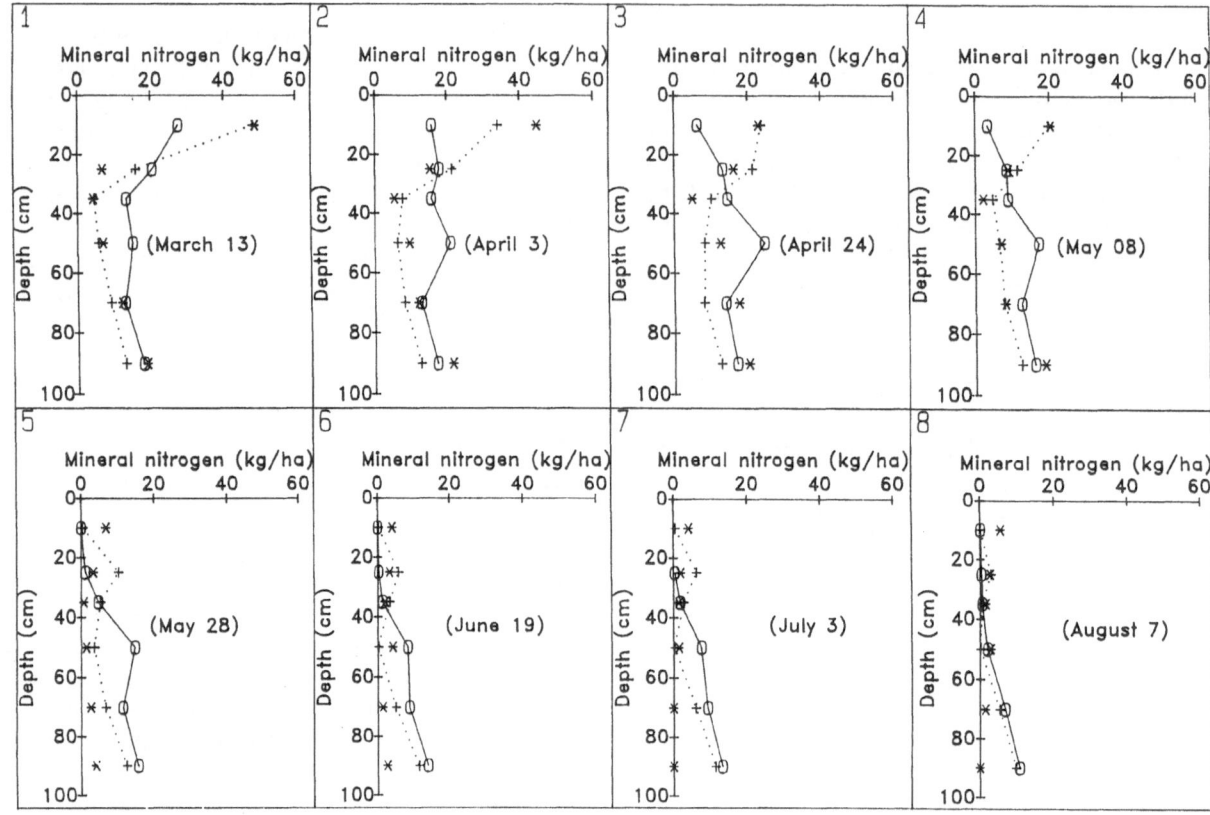

Fig. 7. Observed (∗) and simulated (O and +) mineral nitrogen profiles at different dates for treatment N1. The α-model was used to calculate root water uptake. The solid line was calculated with the LEA option (O) and the dotted line (+) with the FO model. Dates of sampling are indicated on the plots.

then more difficult to judge of the goodness of the fit between simulation and observations. However, a rapid decrease is observed in nitrogen content in the (80–100 cm) layer after May 8 which is not reproduced by the models.

The choice of the root water uptake model has only a minor effect on nitrogen distribution. Differences become noticeable from only the middle of the simulated period. This is not surprising since in the first part of the simulation, where water flux is important, the soil is very wet and the two water uptake models predict very similar uptake patterns. Because of different simulated water distributions at the end of the growing season, the choice of a root water uptake model could have significant effects on water flow for first fall or winter rains.

Figure 8 presents mineral nitrogen profiles for treatments N2 and N3. For the first and some-

are always larger than observed ones. Field data show that a large amount of fertilizer applied with the last two fertilizations disappears very quickly. Volatilization and immobilization are probably responsible for this disappearance. These two processes are not simulated and this explained partly the discrepancies between observed and predicted values. On June 19, first sampling after the last fertilization (June 6), observed profiles for N2 and N3 treatments display a clear increase near 50 cm depth that is not reproduced by the model. This bump could result from transport through preferential pathways. Amounts of nitrogen in the lower part of the profile are correctly reproduced and leaching of nitrate below 1 meter depth remains at zero from all treatments during the simulated period. However, given the large amounts of nitrate that remain in the upper part of the profile for treat-

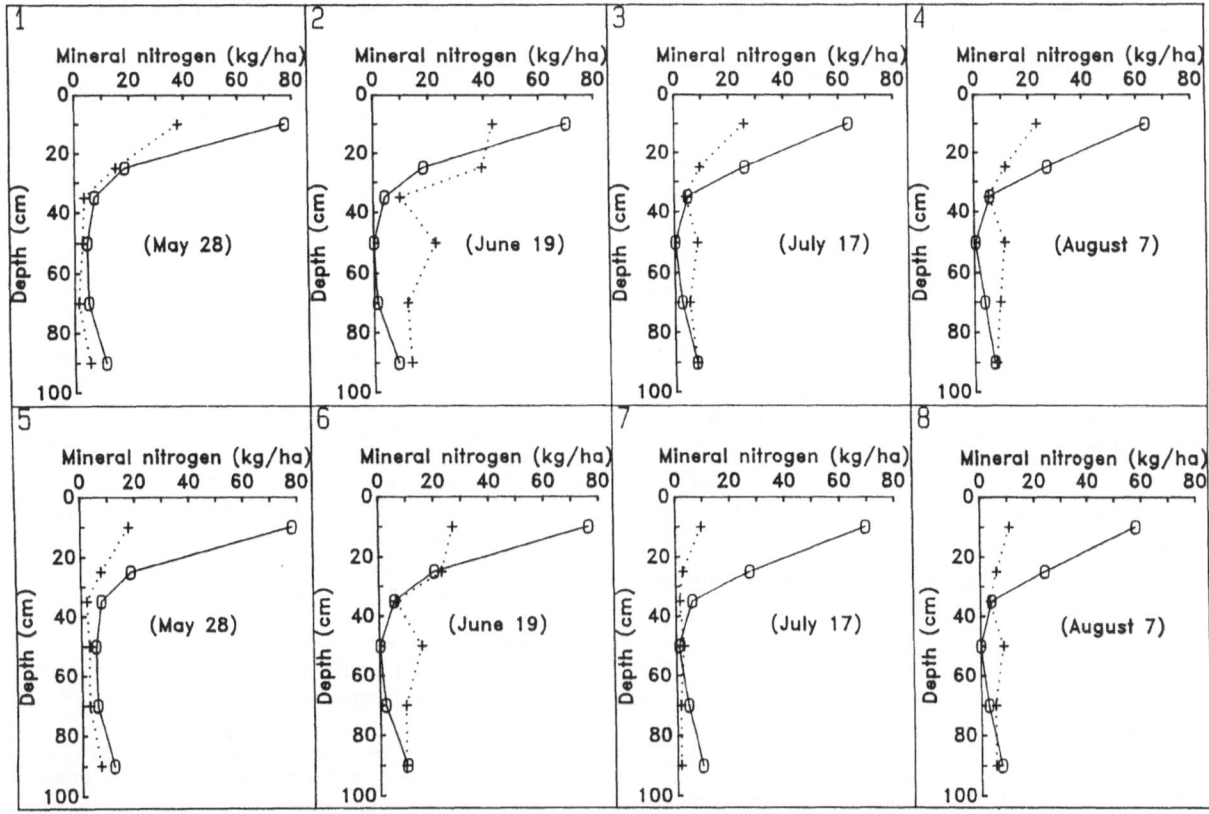

Fig. 8. Observed (+) and simulated (O) mineral nitrogen profiles for treatments N2 (1), (2), (3), (4) and N3 (5), (6), (7), (8). Simulations were carried out with the FO model. Dates of sampling are indicated on the plots.

ments N2 and N3, the model could very well predict nitrate leaching with the first fall or winter rains.

Conclusions

The model presented in this paper is preferentially oriented towards simulation of transport processes rather than biological processes. Modelling of N-transformations is done in a very simple way. A great care was taken when modelling water flow and solute transport. In particular, the model offers the possibility to use the concept of mobile-immobile water to simulate solute transport. In the past years, many solute transport studies have been carried out under steady-state flow conditions that proved the usefulness of this concept. To our knowledge, very few attempts have been made to use this ap-

proach under transient flow conditions. The results presented in this paper illustrate the large differences in nitrate leaching as predicted by the two models. Differences are much more important under more frequent water applications, such as in an irrigated field. Comparison of simulations with nitrogen data recorded for treatments N2 and N3 have illustrated some of the problems posed when modelling the nitrogen cycle. In particular, the vanishing, many times observed, of large amounts of nitrate just after fertilization remains an enigma.

The simulations have shown the effects of different plant water uptake modelling approaches on water distribution. Although the concept of root water potential requires improvements and suffers from some limitations, uniqueness for example, it seems to provide a reasonable description of water uptake patterns. An improvement in deriving macroscopic sink terms for water uptake modelling would be to account in some way for root spatial disposition and not only for root density. Also, the calculations presented have illustrated the differences from the α-model.

This exercise highlighted some of the weaknesses of mechanistic models for predicting water flow. Apart from errors in measurements, the large differences between observed and simulated water content profiles can be explained by preferential flow effects only. As well, an analysis of water content profiles for periods where the potential transpiration is negligible shows that water amounts in the profile are not increased by rains. These defaults in water balance can only be explained by preferential flows. A better prediction of water flow including macropore and soil structure effects is probably also a major challenge in modelling not only water flow but also the transport of fertilizers and chemical substances.

References

1. Ababou R (1981) Modélisation des transferts hydriques dans le sol en irrigation localisée. Thèse Université de Grenoble

2. Baker JM and Van Bavel CHM (1986) Resistance of plant roots to water loss. Agron J 78: 641–644

3. Belmans C, Wesseling JG and Feddes RA (1983) Simulation model of the water balance of a cropped soil: SWATRE. J Hydrol 63: 271–286

4. Bresler E, McNeal BL and Carter DL (1982) Saline and sodic soils. Springer-Verlag, New York, 236 p

5. Cho CM (1971) Convective transport of ammonium with nitrification in soil. Can J Soil Sci 51: 339–350

6. Chu SY, Sposito G and Jury WA (1983) The cross-coupling transport coefficient for the steady flow of heat in soil under a gradient of water content. Soil Sci Soc Am J 47: 21–25

7. De Vries DA (1963) Thermal properties of soils. In: Van Wijk WR (eds) Physics of plant environment. North-Holland Publi Comp, Amsterdam, pp 210–235

8. Feddes RA, Bresler E and Neuman SP (1974) Field test of a modified numerical model for water uptake by root systems. Water Resour Res 10: 1199–1216

9. Feddes RA, Kowalik PJ and Zaradny H (1978) Simulation of field water use and crop yield. Halsted Press, John Wiley & Sons (eds), New York, 188 p

10. Gaudet JP (1978) Transfert d'eau et de soluté dans les sols non-saturés. Mesure et simulation. Thèse Université de Grenoble

11. Gollan T, Passioura JB and Munns R (1986) Soil water status affects the stomatal conductance of fully turgid wheat and sunflower leaves. Aust J Plant Physiol 13: 459–464

12. Groot JJR and Verberne ELJ (1991) Response of wheat to nitrogen fertilization, a data set to validate simulation models for nitrogen dynamics in crop and soil. Fert Res 27: 349–383

13. Hillel D (1986) Application of soil physics. New York, Academic Press, 385 p

14. Hoogland JC, Feddes RA and Belmans C (1981) Root water uptake model depending on soil water pressure head and maximum extraction rate. Acta Hortic 119: 123–136

15. Johnsson H, Bergström L, Jansson P and Paustian K (1987) Simulated nitrogen dynamics and losses in a layered agricultural soil. Agric Ecosyst Environ 18: 333–356

16. Kaluarachchi JJ and Parker JC (1988) Finite element model of nitrogen species transformation and transport in the unsaturated zone. J Hydrol 103: 249–274

17. Misra C, Nielsen DR and Biggar JW (1974) Nitrogen transformations in soil during leaching; 1. Theoretical considerations. Soil Sci Soc Am Proc 38: 289–293

18. Mualem Y (1976) A new model for predicting the hydraulic conductivity of unsaturated porous media. Water Resour Res 12: 513–522

19. Munns R and King RW (1988) Abscisic acid is not the only stomatal inhibitor in the transpiration stream of wheat plants. Plant Physiol 88: 703–708

20. Neuman SP, Feddes RA and Bresler E (1975) Finite element analysis of two-dimensional flow in soils considering water uptake by roots: I. Theory. Soil Sci Soc Am Proc 39: 224–230

21. Nielsen DR, Van Genuchten MTh and Biggar JW (1986) Water flow and solute transport processes in the unsaturated zone. Water Resour Res 22: 89s–108s

22. Nimah MN and Hanks RJ (1973) Model for estimating soil water, plant, and atmospheric interrelations: I. Description and sensitivity. Soil Sci Soc Am Proc 37: 522–527

23. Nkedi-Kizza P, Biggar JW, Van Genuchten MTh, Wierenga PJ, Selim HM, Davidson JM and Nielsen DR (1983) Modelling tritium and chloride 36 transport through an aggregated oxisol. Water Resour Res 19: 691–700

24. Novak V (1987) Estimation of soil-water extraction patterns by roots. Agric Water Manage 12: 271–278

25. Nye PH and Marriott FHC (1969) A theoretical study of the distribution of substances around roots resulting from simultaneous diffusion and mass flow. Plant and Soil 30: 459–472

26. Passioura JB (1971) Hydrodynamic disperions in aggregated media 1. Theory. Soil Sci 111: 339–344

27. Rao PSC, Rolston DE, Jessup RE and Davidson JM (1980) Solute transport in aggregated porous media: Theoretical and experimental evaluation. Soil Sci Soc Am J 44: 1139–1146

28. Rolston DE, Rao PSC, Davidson JM and Jessup RE (1984) Simulation of denitrification losses of nitrate fertilizer applied to uncropped, cropped, and manure-amended field plots. Soil Sci 137: 270–279

29. Russo D, Jury WA and Butters GL (1989) Numerical analysis of solute transport during transient irrigation 2. The effect of immobile water. Water Resour Res 25: 2119–2127

30. Schulin R, Wierenga PJ, Flühler H and Leuenberger J (1987) Solute transport through a stony soil. Soil Sci Soc Am J 51: 36–42

31. Seyfried MS and Rao PSC (1987) Solute transport in undisturbed columns of an aggregated tropical soil: Preferential flow effects. Soil Sci Soc Am J 51: 1434–1444

32. Sposito G, Jury WA and Gupta VK (1986) Fundamental problems in the stochastic convection–dispersion model of solute transport in aquifers and field soils. Water Resour Res 22: 77–88

33. Stanford G and Epstein E (1974) Nitrogen mineralization–water relations in soils. Soil Sci Soc Am Proc 38: 103–107

34. Stanford G, Dzienia S and Vander Pol RA (1975) Effect of temperature on denitrification rate in soils. Soil Sci Soc Am Proc 39: 867–870

35. Starr JL, Broadbent FE and Nielsen DR (1974) Nitrogen transformations during continuous leaching. Soil Sci Soc Am Proc 38: 283–289

36. Tanji KK, Mehran M and Gupta SK (1980) Water and nitrogen flux in the root zone of irrigated maize. In: Frissel MJ and van Veen JA (eds) Simulation of nitrogen behaviour of soil–plant systems, PUDOC, Wageningen, pp 51–66

37. Van Genuchten MTh (1974) Mass transfer studies in sorbing porous media. Ph. D. New Mexico State University, 148 p

38. Van Genuchten MTh (1980) A closed-form equation for predicting the hydraulic conductivity of unsaturated soils. Soil Sci Soc Am J 44: 892–898

39. Van Genuchten MTh (1986) A numerical model for water and solute movement in and below the root zone. U.S. Salinity Laboratory Research Report

40. Van Genuchten MTh and Wierenga PJ (1976) Mass transfer studies in sorbing porous media 1. Analytical solutions. Soil Sci Soc Am J 40: 473–480

41. Van Genuchten MTh and Parker JC (1984) Boundary conditions for displacement experiments through short laboratory soil columns. Soil Sci Soc Am J 48: 703–708

42. Van Genuchten MTh, Tang DH and Guennelon R (1984) Some exact solution for solute transporter through soils containing large cylindrical macropore. Water Resour Res 20: 335–346

43. Van Genuchten MTh and FN Dalton (1986) Models for simulating salt movement in aggregated field soils. Geoderma 38: 165–183.

44. Whisler FD, Klute A and Millington RJ (1968) Analysis of steady-state evapotranspiration from a soil column. Soil Sci Soc Am Proc 32: 167–174

45. Zhang J, Schurr V and Davies WJ (1987) Control of stomatal behaviour by abscisic acid which apparently originates in the roots. J Exp Botany 192: 1174–1181

46. Zhang J and Davies WJ (1989) Abscisic acid produced in dehydrating roots may enable the plant to measure the water status of the soil. Plant Cell Environ 12: 73–81

Fertilizer Research **27**: 233–243, 1991.
© 1991 *Kluwer Academic Publishers*.

Simulating water and nitrogen behaviour in soils cropped with winter wheat

H. Vereecken, M. Vanclooster, M. Swerts & J. Diels
Laboratory of Land Management, Katholieke Universiteit Leuven, Kardinaal Mercierlaan 92, B3010 Belgium

Key words: Nitrogen transport, crop production, moisture retention characteristic, ammonia adsorption, simulation, organic N transformations

Abstract

The SWATNIT model [26], predicting water and nitrogen transport in cropped soils, was evaluated on experimental data of winter wheat for different N treatments. The experiments were monitored at three different locations on different soil types in the Netherlands. Crop growth was simulated using the SUCROS model [11] which was integrated in the SWATNIT model. Both water and nitrogen stress were incorporated. Except for the soil hydraulic properties, all model parameters were taken from literature. The model performance was evaluated on its capability to predict soil moisture profiles, nitrate and ammonia profiles, the time course of simulated total dry matter production and LAI; and crop N-uptake. Results for the simulations of the soil moisture profile indicate that the soil hydraulic properties did not reflect the actual physical behaviour of the soil with respect to soil moisture. Good agreement is found between the measured and simulated nitrate and ammonia profiles. The simulation of the nitrate content of the top layer at Bouwing was improved by increasing the NH_4^+-N-distribution coefficient thereby improving the simulation of the NH_4^+-N-content in this layer. Deviations between simulated and measured nitrate concentrations also occurred in the bottom layers (60–100 cm) of the soil profile. The phreatic ground water might influence the nitrate concentrations in the bottom layers. Concerning crop growth modelling, improvements are needed with respect to the partitioning of total dry matter production over the different plant organs in function of the stress, the calculation of the nitrogen stress and the total nitrogen uptake of the crop through a better estimate of the N-demand of the different plant organs.

Model description

SWATNIT [26] is a mechanistic deterministic one-dimensional transient model for water, solute and heat flow in unsaturated soils. Crop growth is described using the summary model SUCROS [11] adjusted for water and nitrogen stress.

The water flow in the unsaturated zone is described with the Richards' equation [16]:

$$\frac{\partial \Theta}{\partial t} = \frac{\partial}{\partial x}\left[K(\Theta) \frac{\partial H}{\partial x} \right] + S \qquad (1)$$

where Θ is volumetric water content in $m^3\, m^{-3}$, t the time in days, x the vertical distance coordinate in m, H the hydraulic head in m; and S the sink term in day^{-1}. Eq. 1 is solved using the SWATRER model [15]. The model uses an implicit finite difference solution technique with explicit linearization. Ritchie's model [18] is used to calculate actual soil evaporation. The sink term S is described with the Hoogland model [8]:

$$S(h, z) = \alpha(h)S_{max}(z) \qquad (2)$$

where $\alpha(h)$ is a dimensionless sink term variable,

ranging between 0 and 1 as a function of the pressure head; and $S_{max}(z)$ is the maximum water uptake at depth z (m^3 m^{-3} day^{-1}). The variation of S_{max} with depth is calculated as:

$$S_{max}(z) = a - b|z| \qquad (3)$$

where a is the maximum extraction rate at the surface (m^3 m^{-3} day^{-1}); and b is a reduction coefficient (m^3 m^{-3} day^{-1}).

Different models for the $h(\Theta)$ and $K(h)$ relation can be chosen in the model. In this paper we used the Van Genuchten equation [23] with 4 parameters to describe the moisture retention characteristic:

$$\Theta = \Theta_r + (\Theta_s - \Theta_r)/(1 + (\alpha h)^n) \qquad (4)$$

where Θ is the moisture content (m^3 m^{-3}), Θ_r the residual moisture content (m^3 m^{-3}), Θ_s the saturated moisture content (m^3 m^{-3}) and α (cm^{-1}) and n shape parameters. The Gardner equation [5] describes the $K(h)$ relationship:

$$K(h) = K_{sat}/(1 + (bh)^{ng}) \qquad (5)$$

with K_{sat} the saturated hydraulic conductivity (m day^{-1}), and b and ng parameters. The top and bottom conditions for water transport are identical to the SWATRER model. Either a Neumann or Dirichlet boundary type condition can be imposed at the bottom of the soil profile. Both boundary type conditions can also be used at the top of the soil profile.

The convection-dispersion equation is used to model the transport of solutes in the unsaturated zone:

$$\frac{\partial(c_{t,i})}{\partial t} = \frac{\partial}{\partial x}\left[D(\Theta, q)\Theta\frac{\partial c_i}{\partial x}\right] - \frac{\partial(qc_i)}{\partial x} + \phi_i - U_i \qquad (6)$$

where c_i is the concentration of either NH_4^+ ($i = 1$) or NO_3^- ($i = 2$) in the soil solution (mg 1^{-1}), $c_{t,i}$ the total solute concentration (mg 1^{-1}), q the water flux (m day^{-1}), $D(\Theta, q)$ the apparent diffusion coefficient (m^2 day^{-1}), ϕ_i the rate of transformation of the different species (mg 1^{-1} day^{-1}); and U_i the rate of uptake of the different species by plants (mg 1^{-1} day^{-1}). The apparent diffusion coefficient $D(\Theta, q)$ is defined as:

$$D(\Theta, q) = \lambda|v| + D_0 a\,e^{-b\Theta} \qquad (7)$$

with λ the dispersivity (m), v the pore velocity (m s^{-1}), D_0 the diffusion coefficient in a pure liquid phase (m^2 day^{-1}); and a, b empirical constants. In the case of NH_4^+, the solute is distributed between a sorbed and solution phase [27], according:

$$c_{t,1} = \rho c_s + \Theta c_1 \qquad (8)$$

where ρ is the bulk density (kg dm^{-3}), c_1 the concentration in the liquid (mg 1^{-1}), c_s the sorbed concentration (mg kg^{-1}), Θ the moisture content (m^3 m^{-3}) and $c_{t,1}$ the total concentration. Sorption is assumed to be instantaneous and reversible, described by:

$$c_s = K_e c_1 \qquad (9)$$

where K_e is the distribution coefficient (dm^3 kg^{-1}). Introducing Eq. (8) and (9) in Eq. (6) for NH_4^+ results in:

$$\frac{\partial c_{t,1}}{\partial t}(\Theta + \rho K_e) = \frac{\partial}{\partial x}\left[D(\Theta, q)\Theta\frac{\partial c_1}{\partial x}\right]$$
$$- \frac{\partial(qc_1)}{\partial x} + \phi_i - U_i \qquad (10)$$

This equation is solved using a Crank-Nicolson finite difference scheme with an approximation of the second derivative in order to reduce numerical dispersion [2]. The flux q and moisture content Θ are obtained from the solution of the Richards' equation. The sink term ϕ_i describes the transformation of the NH_4^+ or NO_3^- species in the soil. Mathematically, the sink term for NH_4^+ (ϕ_1) and NO_3^- (ϕ_2) can be written as:

$$\phi_1 = ((-k_n - k_v)c_1 + \xi_1)\Theta \qquad (11)$$

$$\phi_2 = (k_n c_1 - k_d c_2 + \xi_2)\Theta \qquad (12)$$

where k_n, k_v, k_d represent, respectively, the first order kinetic rate coefficients for nitrification, volatilization and denitrification (day^{-1}), c_1 and c_2 represent the concentration of NH_4^+ and NO_3^- in the soil solution (mg 1^{-1}), Θ the moisture content (m^3 m^{-3}) and ξ_1 and ξ_2 the net production or immobilization (mg 1^{-1} day^{-1}) of respectively NH_4^+ and NO_3^- from organic N-transformation processes.

First-order reaction type equations are used to model these processes. Inorganic N-transformations considered are nitrification, denitrification, volatilization and ion exchange of NH_4^+–N. Organic N-transformations (mineralization and immobilization) are described by three organic matter pools [10]. A fast cycling pool of organic matter-microbial biomass complex receiving fresh organic matter (soil litter pool) and a slow cycling pool of stabilized decomposition products (soil humus pool) is distinguished. A third pool is considered receiving slurry and/or manure derived faeces (soil manure pool). The model assumes that the nitrogen demand for internal cycling of carbon and humus formation is governed by a constant C/N ratio of decomposer biomass and humification products (r_0) the synthesis efficiency (f_e) and the humification fraction (f_h). The actual C/N ratio of soil litter is assumed to determine the release of nitrogen. The C/N ratio of humified organic matter is equal for both microbial biomass and metabolites. For more details on the equations the reader is referred to [10, 26].

The system of first-order differential equations, describing these organic transformation processes is solved using Euler's integration technique. The mineralized or immobilized amounts of NH_4^+ or NO_3^- enter the sink term of either Eq. (11) or (12). Because the above described transformation processes of nitrogen in the soil are not only influenced by the moisture content, but also by the soil temperature, we included the simulation of heat flow in soil. The heat flow equation incorporated in the model is:

$$\rho C_h \frac{\partial T}{\partial t} = \frac{\partial}{\partial x}\left[K_h(\Theta) \frac{\partial T}{\partial x} \right] \qquad (13)$$

where ρ is the wet bulk density (kg dm^{-3}), C_h the heat capacity of the soil (J m^{-3} $°C^{-1}$), $K_h(\Theta)$ the thermal conductivity (J $m^{-1}s^{-1}$ $°C^{-1}$); and T the temperature (°C).

The volumetric heat capacity and the thermal conductivity are calculated as proposed by De Vries [3]. Eq. (13) is solved using a finite difference technique [22]. Upper boundary conditions are formulated as presented by Kirkham and Powers [12]:

$$T = T_a + \gamma \sin(2\pi t)/p \qquad (14)$$

where T is the soil surface temperature (°C), T_a the average measured soil surface temperature (°C), γ the amplitude of the sine wave (°C), p the time period needed to complete one wave cycle (days), and t the time in days. The bottom boundary condition is either assumed constant or having a zero flux condition.

Crop growth is calculated on daily basis using the summary model SUCROS [11] which was fully integrated in the SWATNIT model. The SUCROS model simulates the time course of the dry matter production in function of air temperature and irradiation. The dry matter production is divided into roots, leaves, stems and storage organs according to partitioning factors which are functions of the phenological stage of the crop. For the specific purpose of this study, the model was modified. The phenological stage, DVS, is calculated according to an approach proposed by Groot [6] which is specific for winter wheat. The gross CO_2 assimilation rate ($DTGA$, kg CO_2 ha^{-1} day^{-1}), which is the basis for the calculation of the dry matter production is reduced for water and nitrogen stress. The influence of water stress on $DTGA$ is defined by:

$$WRED = T_a/T_p \qquad (15)$$

where T_a is the actual transpiration (m day^{-1}) and T_p is the potential crop transpiration (m day^{-1}). T_a is calculated by the water transport submodel.

Nitrogen stress [21] is calculated as:

$$NRED = (ANCL - LNCL)/(MNCL - LNCL) \qquad (16)$$

where $MNCL$ is the minimum N concentration in the leaves for unrestricted growth (kg kg^{-1}), $LNCL$ the irreversibly incorporated N in the leaves (kg kg^{-1}) and $ANCL$ the actual N concentration in the leaves (kg kg^{-1}). Values of $NRED$ can vary between 0 and 1. The nitrogen demand of the crop is calculated according to the PAPRAN model [21]. Maximum levels of N in the vegetative organs are functions of the development stage of the crop. Under limited supply of N the actual amount taken up by the crop is partitioned among the different organs (stems, leaves, roots and seeds) in proportion to their relative demands. The mechanism of nitrogen uptake by the roots is described by a macroscopic type of uptake model [9, 14]. Distinction is made between convective and diffusive uptake.

The root distribution pattern (length-density) is calculated following an approach used in the CERES-WHEAT model [19]. In this model, the total growth of roots is determined by the biomass translocated to the root system. Increase in root biomass is translated to root length increase using a specific root weight factor $(m \, kg^{-1})$. This increase in root biomass is distributed over the different soil layers using a root preference factor which is a function of the moisture content of the soil layers. The rooting depth is calculated from the daily accumulation of thermal time corrected with a soil water deficit factor for photosynthesis and transpiration.

Materials and methods

The SWATNIT model was evaluated on experimental data for winter wheat obtained at three different locations in the Netherlands (Bouwing (Wageningen), Eest and PAGV (Lelystad)) during the growing season 1983–1984 [7]. The growing periods are from 27/10/1983 to 07/08/1984 at the Bouwing, from 21/10/1983 to 09/08/1984 at the Eest, and from 21/10/1983 to 07/08/1984 at PAGV. At each location, three different mineral nitrogen treatments were applied as shown in Table 1. Daily weather data for the growing period (rainfall, minimum and max-

Table 1. Different nitrogen treatments at the Bouwing, Eest and PAGV in kg ha^{-1}

	N1	N2	N3
Bouwing	70/–/–	70/60/40	70/120/40
Eest	50/60/–	50/60/40	50/60/40
PAGV	80/–/–	80/60/40	80/120/40

imum temperature, wind speed, global radiation and vapour pressure) were obtained from meteorological stations nearby the experimental sites. The potential evapotranspiration was calculated following an approach by Groot [8]. The parameters for the calculation of the actual soil evaporation were estimated from texture [20]. Every week the groundwater table depth was determined at Bouwing and Eest starting from April up to harvest. At PAGV, the measurements already started in March and continued until the end of July. Because groundwater table depths are needed to quantify the bottom boundary condition from the start of the simulation (sowing), a constant groundwater table depth was assumed equal to the first measurement. The moisture retention characteristic $h(\Theta)$, the hydraulic conductivity $K(h)$ and the bulk density were determined at three different depths at PAGV and Eest [7]. The soil hydraulic properties for the Bouwing were taken from the Staringreeks [27]. The Van Genuchten [23] and Gardner [5] equations were fitted to the $h(\Theta)$ and $k(h)$ data respectively. Parameter values are shown in Table 2. Parameters for the root water uptake model were set equal to: $h_0(-10 \, cm)$, $h_1(-100 \, cm)$, $h_2(-500 \, cm)$, $h_3(-500 \, cm)$, $h_4(-16000 \, cm)$, a (0.03 day^{-1}) and b (2.10^{-5}). The α reduction factor is 0 at h_0, 1 at h_1, h_2 and h_3, 0 at h_4 and varies linearly between h_0 and h_1 and between h_4 and h_2 or h_1.

Table 2. Parameters of the Van Genuchten equation and the Gardner equation for the moisture retention characteristic and the hydraulic conductivity respectively. K_{sat} (mm day^{-1})

	Θ_r	Θ_s	α	n	K_{sat}	b	ng
Bouwing							
Layer 0–40 cm	0.012	0.533	0.0005	0.413	76.55	5.85	1.37
40–120 cm	0.012	0.534	0.0004	0.415	56.28	7.14	1.30
Eest							
PAGV							
Layer 0–25 cm	0.065	0.419	0.0012	1.02	2135.58	3.04	1.35
25–40 cm	0.093	0.486	0.0016	1.12	270.90	0.29	1.67
40–100 cm	0.118	0.527	0.0015	1.08	2656.80	2.37	1.22

The first order reaction rate constants for nitrification and denitrification were set equal to 0.01 day^{-1}. Higher nitrification rate constants (1.8 day^{-1}) were assumed to occur in the top 20 cm. Decomposition rates for the litter (0.035 day^{-1}) and humus pool (7.10^{-5}) were taken from literature [10]. The synthesis efficiency constant f_e was set equal to 0.2, the humification fraction f_h to 0.5 and the C/N ratio of microorganisms and humified products to 10 [10]. The C/N ratio of litter equals 77 [16]. Humus was assumed to have a C/N of 10. The solute transport parameters D_0 (120 mm^2 day^{-1}), a (0.001), b (1) and λ (17 mm) were taken from [27].

Different literature references were used to obtain the input data for the SUCROS model. Distribution patterns between shoot-root growth and among shoot organs are functions of the development stage [6, 13]. The relation between death rate of leaves and development stage is obtained from [15]. Photosynthesis and maximum assimilation rate are functions of the temperature [6]. The LAI is calculated from the leave weight using a constant leaf area ratio of 0.002 kg ha^{-1}. The conversion of root mass to root length was done with a specific root length factor of 75 m g^{-1}. The maximum assimilation rate was set equal to 40 kg CO$_2$ ha^{-1} leaf^{-1} hr^{-1}. A Q$_{10}$ of 2 was applied to calculate the temperature effect on maintenance respiration. A light reflection coefficient of 0.8 was used. The maximum nitrogen concentration in the leaves is a function of the crop development stage [6].

Results and discussion

The SWATNIT model was applied at the three different sites for the different N treatments. The model performance was compared with measured moisture content profiles, measured NO$_3^-$-N and NH$_4^+$-N concentration profiles and crop growth data. Initial simulations for the Bouwing site using treatment N3 gave too high moisture content profiles for all soil layers. Assuming that the soil moisture profile is in equilibrium with the groundwater table during the winter, new shape parameters were calculated approximating the measured moisture contents. The α parameter was changed to 0.005 for the top layer and to

0.004 for the bottom layer. The n parameter was estimated with pedotransfer functions [25] and was changed to 0.613. Using the same estimation technique, the residual moisture content was set equal to 0.211. Simulation results with above corrections are shown in Fig. 1. Results for the other layers were identical. Although there is a tendency to follow the temporal moisture content variations, the simulated moisture content values are systematically too high except for the period from July up to harvest. Changing the shape parameters of the $h(\Theta)$ relation did not improve the simulation results. Probably the hydraulic conductivity relation is not appropriate for the given soil. Results for the PAGV site are shown in Fig. 2. The moisture content in every

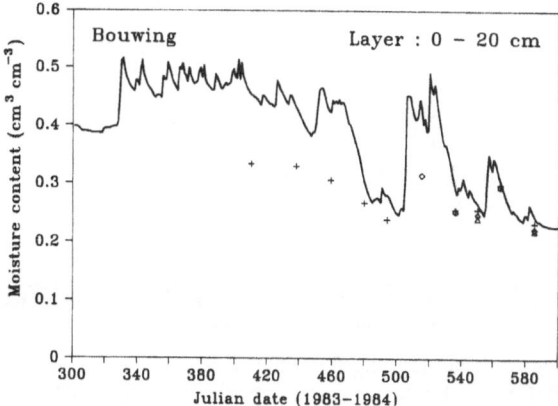

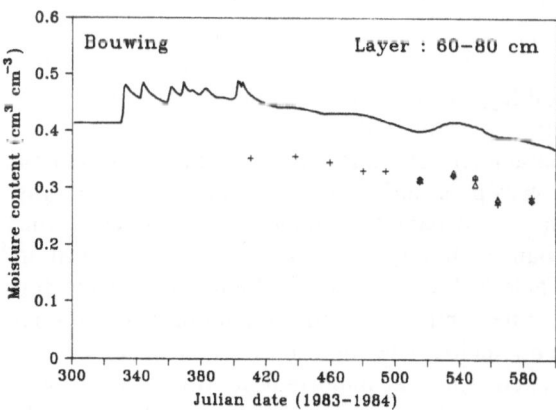

Fig. 1. Simulated and measured moisture contents at the Bouwing for the N1 treatment, 0–20 cm (top) and 60–80 cm (bottom). Plus signs indicate the mean measured values, squares and triangles represent the moisture contents from the N2 and N3 treatment.

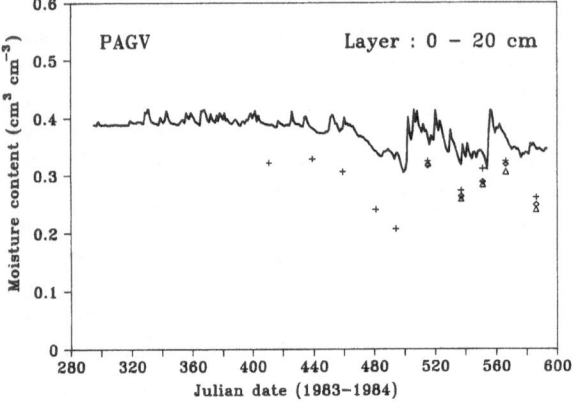

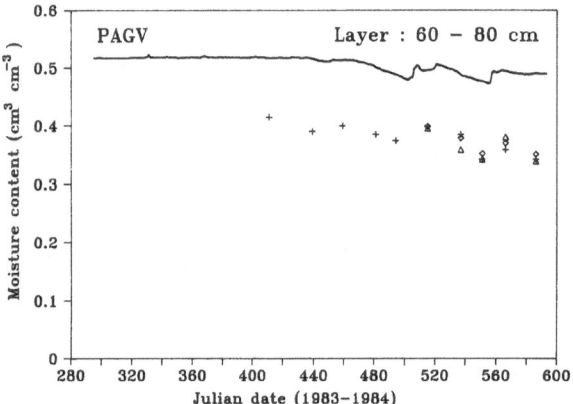

Fig. 2. Simulated and measured moisture contents at the PAGV site for the N1 treatment, 0–20 cm (top) and 60–80 cm (bottom). Plus signs indicate the mean measured values, squares and triangles represent the moisture contents from the N2 and N3 treatment.

Table 3. The simulated components of the water balances for the different experimental sites (mm)

	Bouwing	Eest	PAGV	PAGV*
T_p	263.0	269.7	263.3	263.8
E_p	149.6	155.6	156.2	155.7
T_a	262.8	269.5	263.1	263.7
E_a	136.6	144.8	144.2	143.9
R	539.5	584.3	584.3	584.3
C	69.9	225.4	115.2	170.5
D	313.9	434.4	328.2	475.0
Δ	−103.9	−39.0	−36.2	−128.0

Legend: T_p = Potential Transpiration, E_p = Potential Evaporation, T_a = Actual Transpiration, E_a = Actual Evaporation, R = Rainfall, C = Capillary rise, D = Drainage, Δ = Change in moisture content, * = with adjusted α parameters.

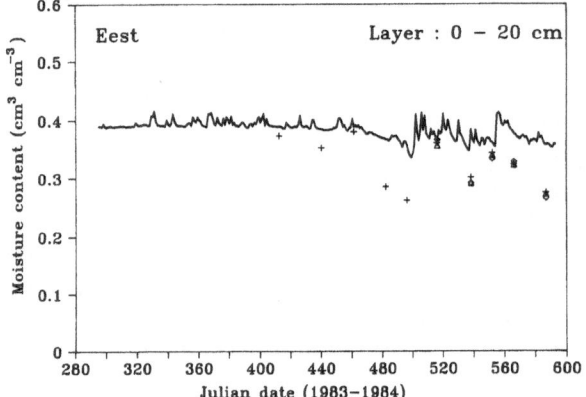

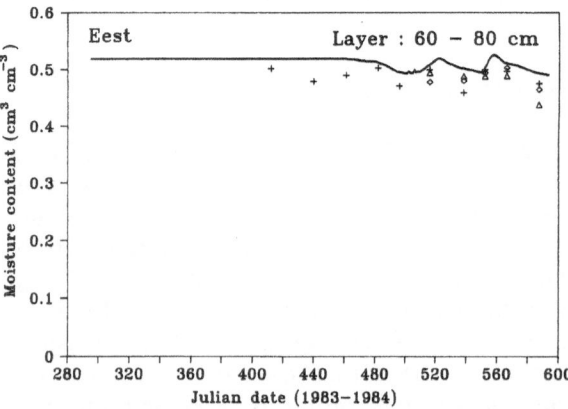

Fig. 3. Simulated and measured moisture contents at the Eest site for the N1 treatment, 0–20 cm (top) and 60–80 cm (bottom). Plus signs indicate the mean measured values, squares and triangles represent the moisture contents from the N2 and N3 treatment.

soil layer is overestimated but the temporal variation is simulated quite well. Correcting the α parameter (10 times higher) shifted the curves downwards and gave good agreement with the measured data (not shown). The effect of this change on the soil water balance is shown in Table 3. More water has drained out of the soil profile compared to the simulation with original parameters. The ratio T_a/T_p is however not affected which is important for the crop growth simulation. In cases with no groundwater table present, the sensitivity of this ratio to changes in soil hydraulic properties might be more important. In free draining loess soils e.g., the T_a/T_p ratio is more influenced by changes in hydraulic parameters than by changes in the parameters of the root water uptake model [24].

Results of the simulation for the Eest are illustrated in Fig. 3. Again no good agreement is found between measured and simulated moisture contents although the temporal variations are simulated well. Only for the deeper layers, there is a very good agreement. Also here, incorrect hydraulic properties for the top layers might explain the difference between simulated and measured data.

To evaluate the SWATNIT performance with respect to nitrogen transport and transformations, the simulated nitrogen concentrations (NO_3^--N, NH_4^+-N or total mineral N) are compared with measured concentration profiles at different depths. Simulations were done for all treatments and all sites but only the N1 treatment will be discussed in detail. The other two treatments contain too few measurements for a detailed analysis. For the top layer (0–20 cm) the simulated nitrate concentrations are in good agreement with the measurements and follow the temporal variation. Only the simulated nitrate concentrations in the top layer of the Bouwing site overestimate the measured data clearly (Fig. 4). This is caused by the too low K_e value which does not correspond to the high clay and organic matter content of this soil. Increasing its value to 1.5 clearly improved both the nitrate and ammonia profiles without increasing the nitrogen stress. A K_e value of 3.0 gives the best agreement with the measured nitrate concentrations

but gives too high NH_4^+-N concentrations (Fig. 5) and increases nitrogen stress. Judging from the other simulation results for the top layers (Eest and PAGV) a value of 1.5 seems acceptable. Very good agreements with measurements are obtained for the deeper layers of all sites for the N1 treatment (Fig. 6). Underestimation of the nitrate amounts occurs in the layers between 60–100 cm for the N1 treatments (Fig. 7). For all N1 treatments high concentrations of nitrate are measured during winter and spring at these depths. This is not caused by matrix flow, otherwise nitrate peaks should be clearly visible at different depths with time. It is also unlikely that by-pass flow would occur to such a depth in winter time. Probably these amounts are induced

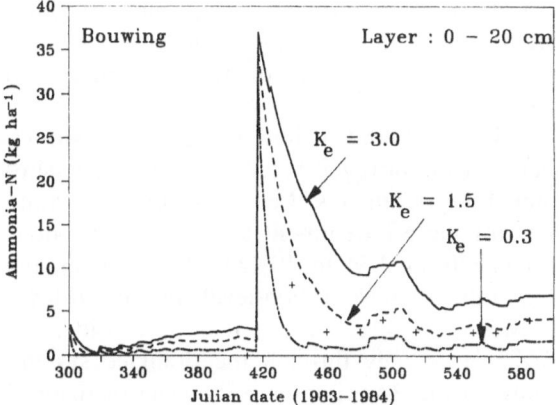

Fig. 5. Simulated and measured amounts of NH_4^+-N for the top layer at the Bouwing. N1 treatment. Three simulation runs are shown for three different K_e values.

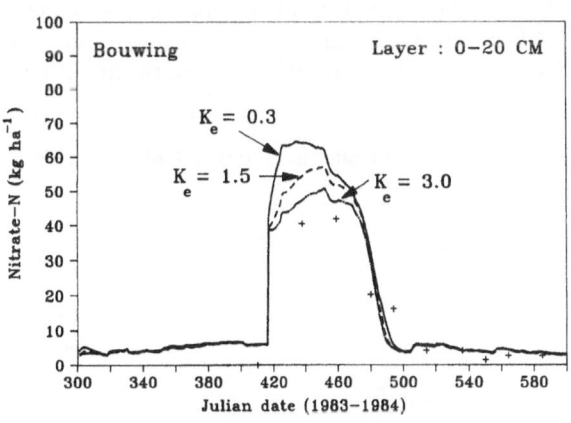

Fig. 4. Simulated and measured amounts of NO_3^--N for the top layer at the Bouwing, N1 treatment. Three simulation runs are shown for three different K_e values.

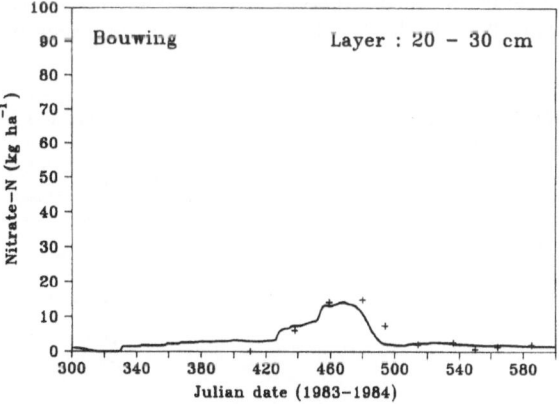

Fig. 6. Simulated and measured amounts of NO_3^--N for the layer between 20 and 30 cm, the Bouwing, N1 treatment.

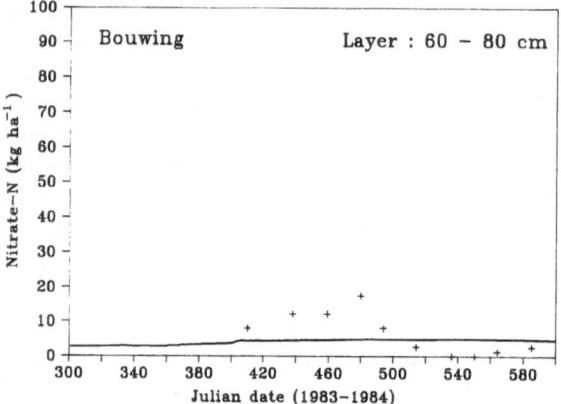

Fig. 7. Simulated and measured amount of NO_3^--N for the layer between 60 and 80 cm, the Bouwing, N1 treatment.

Table 4. Different components of the nitrogen balance in kg ha^{-1} for the three different treatments at the Bouwing 1983–1984 at a depth of 150 cm below soil surface

	N1	N2	N3
N-leaching	10.12	17.24	22.53
plant uptake	154.76	266.23	284.4
denitrification	4.50	7.28	11.57
min./immob.	69.45	69.50	69.50
N-fertilizer	70.00	170.00	230.00
dry deposition	28.10	28.10	28.10
wet deposition	14.85	14.85	14.85
change	13.55	7.69	23.83
balance error	0.53	−1.11	0.12

Table 5. Different components of the nitrogen balance in kg ha^{-1} for the three different treatments at the Eest 1983–84 at a depth of 150 cm below soil surface

	N1	N2	N3
N-leaching	4.71	7.28	4.47
plant uptake	191.62	232.16	230.55
denitrification	14.37	14.14	13.89
min./immob.	68.8	69.96	68.67
N-fertilizer	110.0	150.0	150.0
dry deposition	28.76	28.76	28.76
wet deposition	16.46	16.46	16.46
change	13.74	11.14	17.04
balance error	−0.42	0.46	−1.06

by the high phreatic water table occurring in this period. In general, the NH_4^+-N-profiles are simulated quite well. In many cases, model evaluation is obscured due to the low measured amounts.

Tables 4, 5 and 6 show the different components of the nitrogen balance for the three sites and three treatments. Leaching losses are quite low for the N1 treatment at the three locations ranging from 2.34 to 10.12 kg ha^{-1}. Given the high nitrate concentration levels in the soil layer between 80–100 cm, losses may be considerably higher especially through the drainage system. Only for the Bouwing there is a clear increase of nitrate leaching with increasing fertilizer doses. The net mineralization amounts are comparable for the Eest and Bouwing site, ranging from 68.8 to 69.96 kg ha^{-1}. The amount of mineralized organic-N at the PAGV varies between 48.12 and 48.37 kg ha^{-1}. Thus simulated differences in

mineralization amounts between treatments are very small.

Results for the crop growth modelling are shown in Tables 7 and 8 and Figs. 8 and 9. These are obtained with one set of crop growth parameters applied to all treatments and locations. The main objective of this crop growth modelling effort is the estimation of the time course of the LAI development and N-plant uptake for the different fertilizer treatments. Both quantities

Table 6. Different compounds of the nitrogen balance in kg ha^{-1} for the three different treatments at PAGV 1983–84 at a depth of 150 cm below soil surface

	N1	N2	N3
N-leaching	2.34	2.34	2.37
plant uptake	143.8	239.91	295.13
denitrification	15.06	16.32	19.27
min./immob.	48.12	48.21	48.37
N-fertilizer	80.00	180.00	240.00
dry deposition	28.20	28.20	28.20
wet deposition	16.13	16.13	16.13
change	14.15	12.76	14.70
balance error	−2.9	1.21	1.23

Table 7. Simulated and measured total dry matter production (TDM), grain yield (GY), and aboveground dry matter production (VEG) in kg ha^{-1} for different N-treatments at the different sites

Bouwing	N1		N2		N3	
	meas.	sim.	meas.	sim.	meas.	sim.
TDM	15162.4	12483.1	16230.5	16529.4	17111.4	16529.4
GY	7580.5	4228.8	7553.8	8180.5	8168.2	8180.5
VEG	7581.9	8254.3	8676.7	8348.9	8943.2	8348.9
Eest						
TDM	16308.8	14594.2	16926.0	16194.2	16622.6	16105.2
GY	7434.8	6190.0	7616.5	7717.5	7520.7	7664.6
VEG	8834.0	8404.2	9309.5	8476.7	9109.9	8440.6
PAGV						
TDM	15901.5	13718.9	17158.8	16334.5	17133.0	16652.6
GY	7363.3	4461.5	8028.3	8052.7	7692.2	8307.6
VEG	8538.2	8256.4	9130.5	8381.8	9440.8	8445.0

Table 8. Simulated and measured nitrogen uptake in the aboveground plant parts for the different treatments

	N1		N2		N3	
	meas.	sim.	meas.	sim.	meas.	sim.
Bouwing	194.7	115.1	263.9	237.7	275.5	257.7
Eest	168.7	167.6	204.5	206.2	224.6	204.7
PAGV	183.1	124.0	191.9	213.6	220.0	226.2

are needed to calculate and evaluate the soil water and nitrogen balance. For all sites and treatments, the simulated time course of the LAI development follows the temporal variation in the measurements. From Fig. 11 it can be concluded that there is a tendency to overestimate the LAI on the N1 treatments in the period from early February to the end of April. Probably the partitioning of the dry matter production between the root/shoot and/or among leaves and stems is not correct.

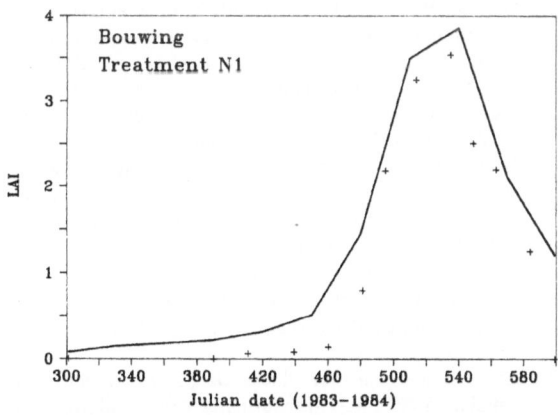

Fig. 8. Simulated and measured LAI values, the Bouwing, N1 treatment.

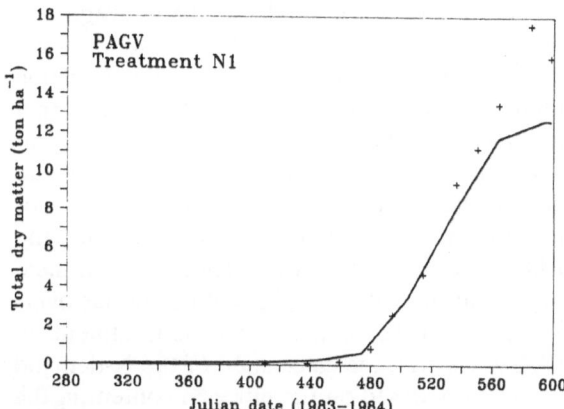

Fig. 9. Simulated and measured total dry matter production (TDM), PAGV, N1 treatment.

Simulation of the time course of total dry matter production (TDM) is fairly well simulated for most treatments as is illustrated by Fig. 9. At the Bouwing, the time course of TDM production is slightly overestimated for all treatments. The simulated final TDM production is however lower than the measured ones for all treatments and sites except for the N3 treatment at the Bouwing site. Especially the grain yield of the N1 treatments is severely underestimated. This is caused by the occurrence of nitrogen stress during the final crop growth stage. Clearly the method of N stress calculation needs refinement.

It is however necessary in the evaluation of these values to incorporate the measurement error variability caused by spatial variability and measurement errors. Mean squared deviations, derived from the pure error, obtained from experiments on winter wheat at Helecine (Belgium) show values varying between 900 to 1900 kg ha^{-1} for dry matter production of vegetative plant organs. The MSD value for grain yield ranges between 400 to 600 kg ha^{-1} (Comité Toegepaste Bodemkunde, unpublished). This indicates that the prediction of the grain yields for the N1 treatments is not very good but that for the other treatments they are quite acceptable. Predicted values for the dry matter production of vegetative plant organs are well within range for all treatments. For a proper evaluation it is therefore necessary to have an idea of the variability which is inherent to these measured values.

The total nitrogen uptake of the aboveground plant organs is shown in Table 8. Good agreement is found between measured and simulated total nitrogen uptake for the three different treatments at the Eest. Large differences occur between the measured and simulated uptake values for the N1 treatments of the other sites. For the N2 and N3 treatments of the Bouwing and PAGV the differences are smaller between measured and simulated uptake, with a serious overestimation of the uptake for the N3 treatments at PAGV. Clearly, the calculation of the N-demand of the crop, based on the nitrogen content in the leaves is not accurate enough. More detailed information with respect to the nitrogen demand of the different plant organs is needed to simulate more accurately the total nitrogen uptake.

Conclusions

The SWATNIT model was used to simulate the nitrogen and water transport in soils cropped with winter wheat. Combined with the SUCROS model, crop growth was predicted. Experimental data were obtained from three different sites in the Netherlands with different nitrogen applications.

To improve the simulation of the moisture content profiles a more correct description of the soil hydraulic properties is needed. Good agreement was found for the NO_3^--N and NH_4^+-N concentration profiles for the soil layers up to 80 cm depth. Adjustment of the NH_4^+-N distribution coefficient K_e for the Bouwing site was

necessary to avoid too high nitrate concentrations in the top layer, thereby improving the result of the simulated NH_4^+-N concentrations. This higher K_e value accounts for the high clay and organic matter content. NO_3^--N contents were underestimated for the layer between 80 to 100 cm depth. Probably the presence of the phreatic groundwater influences the NO_3^--N contents. Crop growth was predicted using one set of model parameters. The time course of the total dry matter production and LAI development is simulated quite well taken into account the variability of the measurements. Improvements can probably be obtained by a better partitioning of the total dry matter production between the different plant organs. Comparison between simulated and measured nitrogen uptake shows that the calculation of the nitrogen demand of the crop needs to be improved.

References

1. Belmans C, Wesseling JG and Feddes RA (1983) Simulation model of a water balance of a cropped soil: SWATRE. J Hydrol 63: 271–286
2. Bresler E (1973) Simultaneous transport of solute and water under transient unsaturated flow conditions. Water Resour Res 9: 975–986
3. De Vries DA (1975) Heat transfer in soils. In: De Vries DA and Afgan NH (eds) Heat and mass transfer in the biosphere, John Wiley and Sons, New York, pp 5–28
4. Dierckx J, Belmans C and Pauwels P (1986) SWATRER, a computer package for modelling the field water balance. Reference manual, February 1986, pp 114. Laboratory of Soil and Water Management, Leuven, Belgium
5. Gardner WR (1958) Some steady state solutions of the unsaturated moisture flow equation with application to evaporation from a water table. Soil Sci 85: 228–232
6. Groot JJR (1987) Simulation of nitrogen balance in a system of winter wheat and soil. Simulation report CABO-TT nr. 13, pp 69. Wageningen
7. Groot JRR and Verberne ELJ (1991) Response of wheat to nitrogen fertilization, a data set to validate simulation models for nitrogen dynamics in crop and soil. Fert Res 27: 349–383
8. Hoogland J, Belmans C and Feddes RA (1981) Root water uptake model depending on soil water pressure head and maximum extraction rate. Acta Hortic 119: 123–136
9. Huwe B and Van der Ploeg RR (1988) Modelle zur Simulation des Stickstoffhaushaltes von Standorten mit unterschiedlicher landwirtschaftlicher Nutzung, pp 213. Inst für Wasserbau, Universität Stuttgart, Stuttgart

10. Johnsson H, Bergström L, Jansson PE and Paustian K (1987) Simulated nitrogen dynamics and losses in a layered agricultural soil. Agric. Ecosyst. Environ 18: 333–356

11. Keulen H van, Penning de Vries FWT and Drees EM (1982) A summary model for crop growth. In: Penning de Vries FWT and van Laar HH (eds), Simulation of crop growth and crop production, pp 87–97. Pudoc, Wageningen

12. Kirkham D and Powers WL (1972) Advanced Soil Physics. John Wiley and Sons, New York

13. Marletto V and Van Keulen H (1984) Winter wheat experiments in the Netherlands and Italy analysed by the SUCROS model. Simulation report CABO-TT nr. 3, pp 61, Wageningen

14. McIsaac G, Martin D and Watts D (1985) Users guide to NITWAT – a nitrogen and water management model. Agricultural Engng. Department, University of Nebraska, Lincoln, Nebraska

15. Penning de Vries FWT and Van Laar HH (1982) Simulation of crop growth and crop production, pp 307. Pudoc, Wageningen

16. Reddy KR, Khabel R and Overcash MR (1980) Carbon transformations in the land areas receiving organic wastes in relation to non point source pollution. A conceptual model. J Environ Qual 9: 434–442

17. Richards LA (1931) Capillary conduction of liquids in porous mediums. Physics 1: 318–333

18. Ritchie JT (1972) A model for predicting evaporation from a row crop with incomplete cover. Water Resour Res 8: 1204–1213

19. Ritchie JT and Otter S (1984) Description and performance of CERES-WHEAT: A user-oriented wheat yield model, pp 37. USDA-ARS, Grassland, Soil and Water Research Laboratory, Temple, Texas, USA 76503

20. Ritchie JT and Crum J (1988) Converting soil survey characterization data into IBSNAT crop model input. In: Bouma J and Bregt A (eds) Land qualities in space and time, pp 352. Pudoc, Wageningen

21. Seligman HG and Van Keulen H (1981) PAPRAN: a simulation model of annual pasture production limited by rainfall and nitrogen. In: Frissel HM and Van Veen JA (eds) Simulation of nitrogen behaviour of soil-plant systems, pp 192–221. Pudoc, Wageningen

22. Tillotson WR, Robbins WC, Wagenet RC and Hanks RJ (1980) Soil water, solute and plant growth simulation, pp 53. Bulletin 502 Utah State Agr Exp Stn Logan, Utah

23. Van Genuchten MTh (1980) A closed-form equation for predicting the hydraulic conductivity of unsaturated soils. Soil Sci Soc Am J 44: 892–898

24. Vereecken H, Diels J and Feyen J (1988) Validation of the water balance model SWATRER. In: Wierenga PS and Bachelet D (eds) pp 545. Validation of flow and transport models for the unsaturated zone. Conference Proceedings: May 23–26. Ruidoso. New Mexico. Research Report 88–SS–04 Department of Agronomy and Horticulture, New Mexico State University, Las Cruces, N.M.

25. Vereecken H, Maes J, Feyen J and Darius P (1989) Estimating the soil moisture retention characteristic from texture, bulk density and carbon content. Soil Sci 6: 389–403

26. Vereecken H, Vanclooster M and Swerts M (1990) A model for the estimation of nitrogen leaching with regional applicability. In: Merckx R, Vereecken H and Vlassak K (eds) Fertilization and the Environment, pp 250–263. Leuven University Press, Leuven, Belgium

27. Wagenet RJ and Hutson JL (1987) LEACHM, a process-based model of water and solute movement, transformations, plant uptake and chemical reactions in the unsaturated zone. Continuum, Water Resources Institute, vol 2, pp 148. Department of Agronomy, Cornell University, Ithaca, New York

28. Wösten, JHM, Bannink MH and Beuving J (1986) Waterretentie- en doorlatenheidskarakteristieken van boven- en ondergrond in Nederland: de Starringreeks. ICW-rapport 18, Stiboka-rapport 1932, pp 73. Wageningen

Fertilizer Research **27**: 245–259, 1991.
© 1991 *Kluwer Academic Publishers.*

245

Simulation of nitrogen dynamics and biomass production in winter wheat using the Danish simulation model DAISY

S. Hansen, H.E. Jensen, N.E. Nielsen & H. Svendsen
Section of Soil and Water and Plant Nutrition, Department of Agricultural Sciences, The Royal Veterinary and Agricultural University, Thorvaldsensvej 40, DK-1871 Frederiksberg C, Copenhagen, Denmark

Key words: Simulation model, crop production, soil water dynamics, soil nitrogen dynamics, winter wheat

Abstract

A dynamic simulation model for the soil plant system is described. The model includes a number of main modules, viz., a hydrological model including a submodel for soil water dynamics, a soil temperature model, a soil nitrogen model including a submodel for soil organic matter dynamics, and a crop model including a submodel for nitrogen uptake. The soil part of the model has a one-dimensional vertical structure. The soil profile is divided into layers on the basis of physical and chemical soil characteristics. The simulation model was used to simulate soil nitrogen dynamics and biomass production in winter wheat grown at two locations at various levels of nitrogen fertilization. The simulated results were compared to experimental data including concentration of inorganic nitrogen in soil, crop yield, and nitrogen accumulated in the aboveground part of the crop. Based on this validation it is concluded that the overall performance of the model is satisfactory although some minor adjustments of the model may prove to be necessary.

Introduction

Due to the fact that losses of nitrogen from agricultural arable land to the aquatic environment have increased in many areas, in particular during the recent four decades, great concern has arisen as to how an economically and environmentally sustained agricultural crop production can be developed. This includes sustained crop yield and crop quality, sustained natural resources for crop production, high resource use efficiency in crop production and limitation of the impact of crop production on environmental quality.

The present Danish simulation model DAISY which is described in detail elsewhere [12] was developed to enable simulation of crop production, soil water dynamics and nitrogen dynamics in crop production at various agricultural man-

agement practices and strategies. Thus a model has been developed to facilitate application at field level as a management tool as well as at higher level, e.g. regionally, as part of a model system for administrative purposes.

The objective of the present study was to validate the model by comparing simulated values with experimental data from extensive field experiments [10] with various nitrogen application to winter wheat on three locations in the Netherlands.

Model description

The model comprises a number of main modules, viz., a hydrological model including a submodel for soil water dynamics, a soil temperature model, a soil nitrogen model including a

246

submodel for soil organic matter dynamics, and a crop model including a submodel for nitrogen uptake. The soil part of the model has a one-dimensional vertical structure. The soil profile is divided into layers on the basis of physical and chemical soil characteristics. A relational diagram of the overall model is shown in Fig. 1.

The hydrological model

The hydrological processes considered in the model include snow accumulation and melting, interception of precipitation by the crop canopy, evaporation from crop and soil surfaces, infiltration, water uptake by plant roots, transpiration and vertical movement of water in the soil profile. In the model snow melting is influenced by incidental radiation and soil and air tempera-

tures. Interception is determined by precipitation and by the crop canopy. The soil water dynamics is modelled by Richards' equation [23]:

$$C_w \frac{\partial \psi}{\partial t} = \frac{\partial}{\partial z} \left[K_w \frac{\partial \psi}{\partial z} - K_W \right] - S \qquad (1)$$

where C_w is specific water capacity $(m^3 m^{-3} (m\,H_2O)^{-1})$, ψ is soil water pressure potential $(m\,H_2O)$, K_w is hydraulic conductivity $(m^2 (m\,H_2O)^{-1} s^{-1})$, S is water uptake by plant roots $(m^{-3} m^{-3} s^{-1})$, t is time (s) and z is soil depth (m). The vertical flow rate q $(m^3 m^{-2} s^{-1})$ is given by the Darcy equation:

$$q = -K_w \frac{\partial \psi}{\partial z} + K_w \qquad (2)$$

Modelling of the water uptake by plant roots is based on steady state radial flow to the root

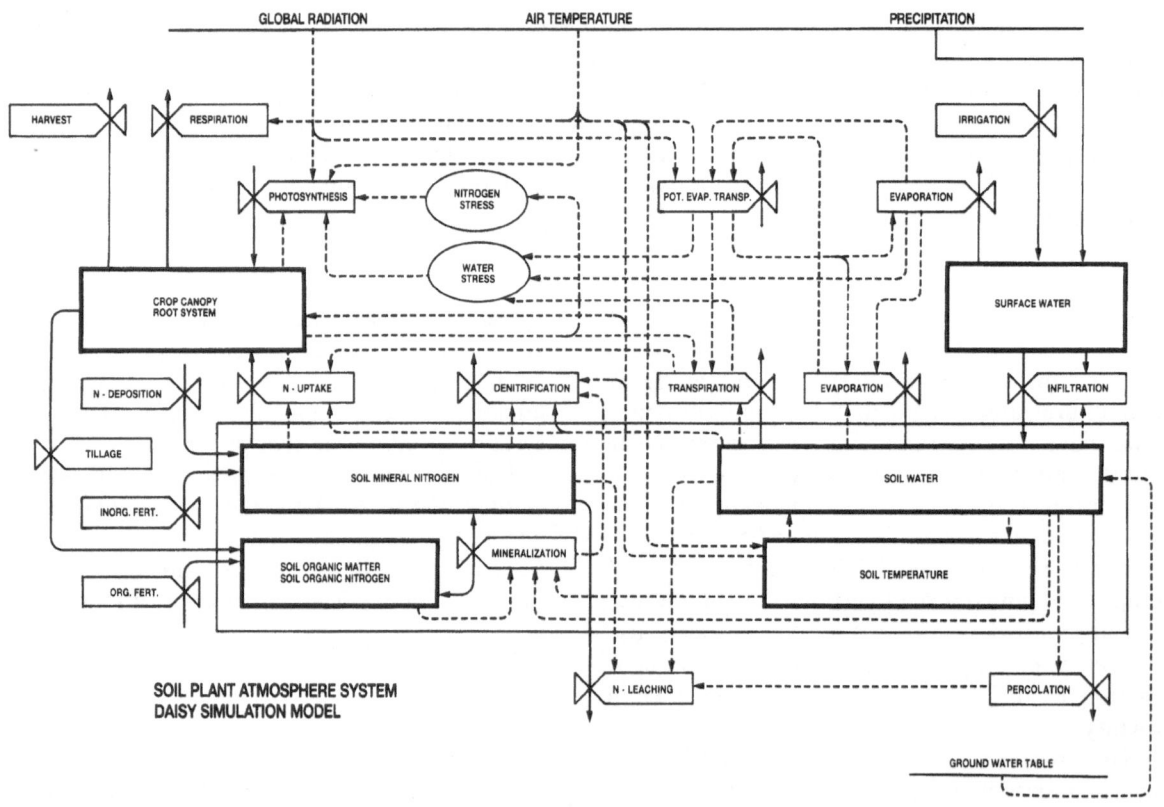

Fig. 1. Relational diagram of the DAISY model. Global radiation, air temperature and precipitation are driving variables. Rectangles denote system state variables, valve symbols denote processes while ovals denote auxiliary variables. Solid lines denote flows of matter while broken lines denote flows of information.

surfaces:

$$S = L \frac{\theta_r}{\theta_s} \frac{\int_0^{\psi_s} K_w \, d\psi - \int_0^{\psi_r} K_w \, d\psi}{-\frac{1}{2} \ln(r_r^2 \pi L)} \quad (3)$$

where L is root density (m m^{-3}), θ_r is soil water content at the root surface (m^3 m^{-3}), θ_s is soil water content at saturation, ψ_s and ψ_r is soil water pressure potential (m H$_2$O) of the bulk soil and at the root surface, respectively, and r_r is root radius (m).

The potential evapotranspiration constitutes the upper limit for the considered evaporation and transpiration processes. In the present study the potential evapotranspiration was calculated by using a modified Makkink equation [11, 16].

The soil temperature model

The soil temperature is modelled by solving an extended heat flow equation which includes the effect of frost and thaw processes:

$$\frac{\partial(C_h T_s)}{\partial t} - L_f \rho_i \frac{\partial X_i}{\partial t} = \frac{\partial}{\partial z} \left(K_h \frac{\partial T_s}{\partial z} \right)$$
$$- \rho_w c_w \frac{\partial(T_s q)}{\partial z} \quad (4)$$

where C_h is volumetric heat capacity of soil (J m^{-3} °C^{-1}), T_s is soil temperature (°C), L_f is latent heat of freezing (J kg^{-1}), ρ_i is density of ice (kg m^{-3}), X_i is volumetric fraction of ice (m^3 m^{-3}), K_h is thermal conductivity (W m^{-1} °C^{-1}) calculated according to De Vries [5], ρ_w is density of water (kg m^{-3}) and c_w is the specific heat capacity of water (J kg^{-1} °C^{-1}). The freezing process induces water flow in the soil as ice formation, and is assumed to take place in the large soil pores extracting water from small pores resulting in water flow towards the freezing zone [19].

The soil nitrogen model

The transformation and transport processes considered in the model include net mineralization of nitrogen, nitrification, denitrification, nitrogen uptake by plants and nitrogen leaching from the root zone. It appears that before mineralization can take place the soil organic matter has to be degraded and dissolved in the soil solution. These processes may be considered as the steps determining the turnover rate of soil organic matter [20]. The net mineralization is thus governed by the turnover of organic matter in the soil which conceptually is divided into three main pools, viz., dead native soil organic matter (SOM), microbial biomass (BOM) and added organic matter (AOM). Each main pool of organic matter is subdivided into two or three subpools (Fig. 2) each one being characterized by a particular carbon to nitrogen ratio and by a particular turnover time. Thus dead native soil organic matter (SOM) is subdivided into three

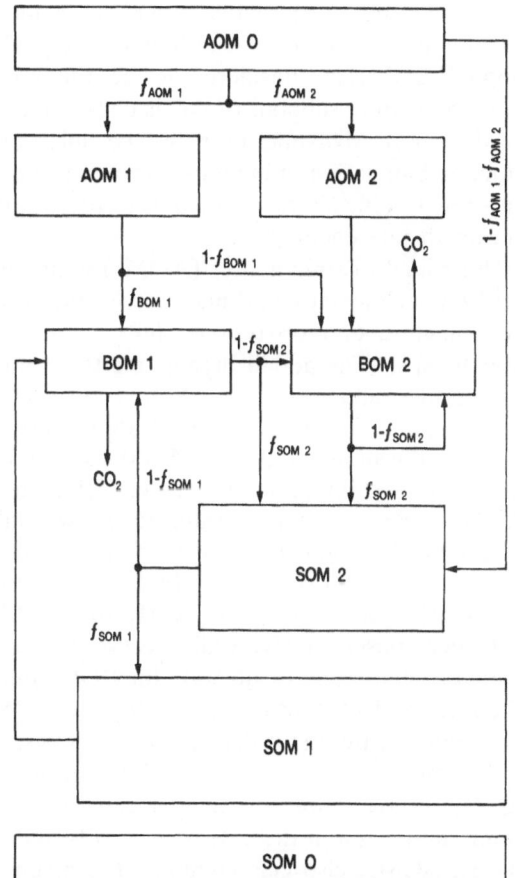

Fig. 2. The DAISY submodel of soil organic matter. Pools and subpools (1 and 2) of organic matter and related partitioning coefficients (f). AOM: added organic matter, BOM: microbial biomass, SOM: native soil organic matter.

248

subpools designated SOM_0, SOM_1 and SOM_2, respectively. The subpool SOM_0 can be neglected as it consists of almost inert organic matter. The rates of decomposition of SOM_1 and SOM_2 are simulated by first-order reaction kinetics. The subpool SOM_1 is assumed to consist of chemically stabilized organic matter, i.e. compounds with a chemical structure that implies resistance to biological attack, while the subpool SOM_2 is assumed to consist of organic matter, part of which is physically stabilized, i.e. compounds which are protected against biological attack by adsorption to soil colloids or entrapment within soil aggregates [14, 32, 33].

The microbial biomass in the soil which usually accounts for less than 3% of the organic carbon [6] is subdivided into two subpools designated BOM_1 and BOM_2, respectively, in order to obtain a relatively stable as well as a more dynamic part of the microbial biomass [33]. The subpool BOM_1 is considered to be the more stable part while subpool BOM_2 is considered to be the more dynamic part of the microbial biomass. Simulation of biomass turnover is based on growth efficiency, maintenance respiration and death rate coefficients.

The added organic matter (AOM_0) is organic fertilizer such as farmyard manure, slurry, green crop manure, or crop residues left in the field after harvest. The added organic matter is allocated to the subpools AOM_1, AOM_2 and SOM_2. The subpool AOM_1 is assumed to consist of mainly cell wall material while AOM_2 is assumed to consist of mainly water extractable cell material. The organic matter allocated to the soil subpool SOM_2 is assumed to consist of lignin mainly and other resistant compounds [31]. The subpool AOM_1 is substrate for both BOM_1 and BOM_2, and decomposes slowly, while AOM_2 which is easily decomposable is substrate for BOM_2 only. The rates of decomposition of AOM_1 and AOM_2 are simulated by first-order reaction kinetics.

The considered abiotic factors influencing the carbon turnover are soil temperature and soil water status, and in the case of subpools SOM_1, SOM_2, BOM_1, also clay content. It is assumed that no interaction exists between the effect of the various abiotic factors and that the combined effect is multiplicative. The abiotic functions adopted which were derived from various

sources in literature [1, 2, 4, 18, 21, 27, 28, 29, 30, 31] are shown in Fig. 3.

Referring to Fig. 2 a carbon balance for each pool of organic matter can be established resulting in an expression for the rate of change of carbon in each particular pool (dC_i/dt). As each pool of organic matter is characterized by a particular carbon to nitrogen ratio $[C/N]_i$, an overall organic nitrogen balance can be established resulting in an equation for net mineralization ξ_m of nitrogen:

$$\xi_m = -\sum_i [C/N]_i^{-1} \frac{dC_i}{dt} \qquad (5)$$

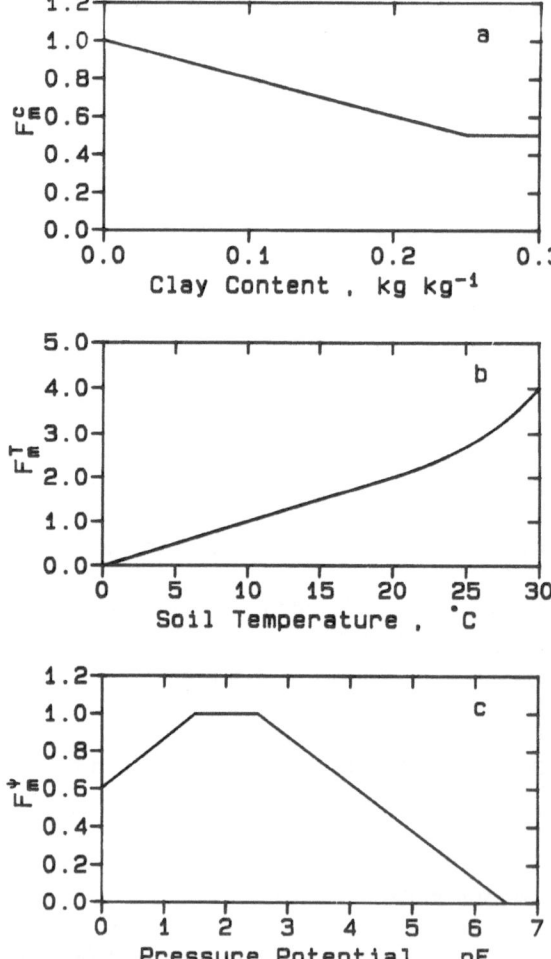

Fig. 3. Abiotic functions for adjustment of decomposition rate coefficients to clay content (a), soil temperature (b), and soil water pressure potential (c).

If net mineralization is negative, i.e. net immobilization occurs, it is assumed that NH_4^+ is utilized in preference to NO_3^- by the microbial biomass.

In well aerated arable soils at relatively high water content $(1.5 < pF < 2.5)$, pH in the range 4–8, and soil temperature higher than 5°C, microbial activity is usually limited by the availability of organic carbon and most NH_4^+ is biologically oxidized to NO_3^- as rapidly as it is formed by the process of mineralization. Under such conditions nitrification can be expressed by first-order reaction kinetics.

The considered abiotic factors influencing the nitrification are soil temperature and soil water status. It is assumed that no interaction exists between the effect of soil temperature and soil water status, respectively, and that the combined effect is multiplicative. The abiotic functions adopted which were derived from various sources in literature [7, 18, 22, 25] are shown in Fig. 4. As the O_2 concentration in soil solution usually is correlated with temperature and soil water content, the effect of O_2 concentration on the nitrification rates is implicitly included in the combined effect of soil temperature and soil water content.

Denitrification is modelled by defining a potential denitrification rate, i.e. the denitrification rate under complete anaerobic conditions. The potential denitrification rate ξ_d^* (kg N $m^{-3} s^{-1}$) is assumed to be related to soil temperature and CO_2 evolution rate [15]:

$$\xi_d^* = F_d^T(T)\alpha_d^*\xi_{CO_2} \tag{6}$$

where $F_d^T(T)$ is a soil temperature function identical to $F_m^T(T)$ (Fig. 3), α_d^* is an empirical constant and ξ_{CO_2} is CO_2 evolution rate (kg C $m^{-3} s^{-1}$). Under partly anaerobic conditions the potential denitrification rate is reduced according to the oxygen status of the soil expressed as a function of soil water content adopted from Rolston et al. [24], Fig. 5. The actual denitrification rate is either determined as the reduced potential denitrification rate or as the rate at which nitrate in soil becomes available for denitrification.

The nitrogen uptake model is based on the concept of a potential nitrogen demand simulated by the crop model, and the availability of nitrogen in the soil for plant uptake, i.e. the rate at which nitrogen can be made available at the root surfaces. The transport of nitrogen from the bulk soil to the root surfaces is based on a number of assumptions. Each root exploits an average effective volume of soil which is assumed to be a cylinder around each root. The radius of this cylinder is assumed to correspond to the average half distance between the roots. It is assumed that nitrogen is transferred to the root

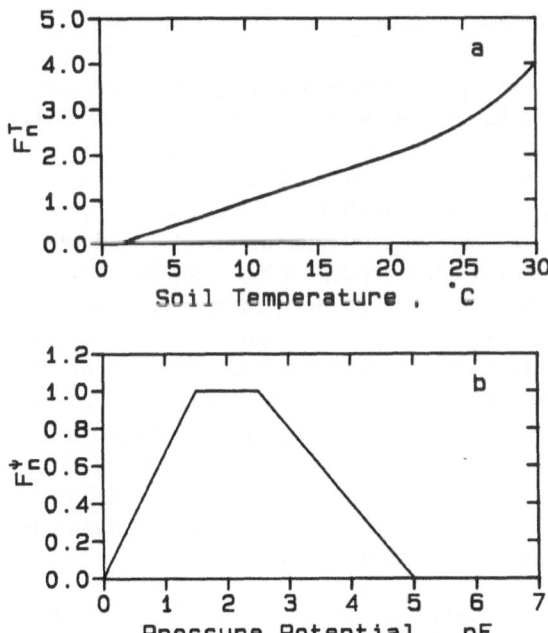

Fig. 4. Abiotic functions for adjustment of nitrification rate coefficients to soil temperature (a) and soil water pressure potential (b).

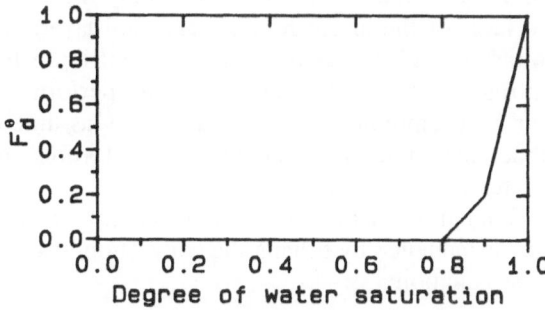

Fig. 5. Soil water content function for adjustment of denitrification rate to the degree of water saturation.

surface by mass flow and diffusion and that the concentration – distance profile develops in time in a stepwise manner, and at each timestep it approximates to a steady state profile [3]. In the present model it is assumed that nitrogen uptake equals the nitrogen flux towards the root surface. Based on these assumptions the flux of nitrogen I towards the root surface is calculated as:

$$I = \begin{cases} 2\pi Db(\bar{C} - C_r)\left[\dfrac{\beta^2 \ln \beta}{\beta^2 - 1} - \dfrac{1}{2}\right]^{-1} & \alpha = 0 \\[2ex] q_r \dfrac{(\beta^2 - 1)\bar{C} - C_r \ln \beta^2}{(\beta^2 - 1) - \ln \beta^2} & \alpha = 2 \\[2ex] q_r \dfrac{(\beta^2 - 1)(1 - \frac{1}{2}\alpha)\bar{C} - (\beta^{2-\alpha} - 1)C_r}{(\beta^2 - 1)(1 - \frac{1}{2}\alpha) - (\beta^{2-\alpha} - 1)} & \text{else} \end{cases}$$
$$(7)$$

$$\alpha = \frac{q_r}{2\pi Db}$$

$$\beta = (r_r^2 \pi L)^{-1/2}$$

where D is the diffusion coefficient ($m^2\,s^{-1}$), $\bar{C}$ is the average nitrogen concentration in solution ($kg\,m^{-3}$), C_r is the nitrogen concentration at root surface ($kg\,m^{-3}$), b is buffer power of soil, L is the root length density and q_r is the radial water flow rate ($m^3\,m^{-1}\,s^{-1}$). If the uptake is limited by the availability of nitrogen then C_r is equal to zero and hence the root acts as a zero sink. In this case total uptake of nitrogen is calculated by integrating I over the entire root system. In the case of ample nitrogen supply the total nitrogen uptake is determined by the potential nitrogen demand. Then total uptake is distributed over the entire root zone by assuming a common value of C_r to exist along the root surfaces of the entire root system. Soil layers in which $\bar{C} < C_r$ are assumed not to contribute to nitrogen uptake. The calculations are performed for both ammonium and nitrate. It is assumed that ammonium is taken up by plant roots in preference to nitrate.

Simulation of the vertical movement of nitrogen is modelled by solving the convection-dispersion equation:

$$\frac{\partial (A + \theta C)}{\partial t} = \frac{\partial}{\partial z}\left[\theta D^* \frac{\partial C}{\partial z} - qC\right] + \Phi \qquad (8)$$

where A is adsorbed nitrogen ($kg\,m^{-3}$), D^* is the dispersion coefficient ($m^2\,s^{-1}$), C is the bulk concentration of nitrogen in solution ($kg\,m^{-3}$) and Φ is a sink-source term ($kg\,m^{-3}\,s^{-1}$). The convection-dispersion equation is solved for ammonium as well as for nitrate. In the case of ammonium the relation between adsorbed and dissolved ammonium is described by an adsorption-desorption isotherm derived from Schouwenberg and Schuffelen [26], while in the case of nitrate adsorption is considered insignificant. The source-sink term for ammonium and nitrate $\Phi_{NH_4^+}$ and $\Phi_{NO_3^-}$, respectively, in the convection-dispersion equation integrates the transformation and transfer processes of ammonium and nitrate, respectively:

$$\Phi_{NH_4^+} = \xi_m + \xi_{f,NH_4^+} - \xi_{u,NH_4^+} - \xi_n \qquad (9)$$

$$\Phi_{NO_3^-} = \xi_n + \xi_{f,NO_3^-} - \xi_{u,NO_3^-} - \xi_d \qquad (10)$$

where ξ_m, ξ_n and ξ_d are net mineralization, nitrification and denitrification ($kg\,m^{-3}\,s^{-1}$), respectively, while ξ_{f,NH_4^+}, ξ_{f,NO_3^-}, and ξ_{u,NH_4^+}, ξ_{u,NO_3^-} are fertilizer input and root uptake of NH_4^+ and NO_3^- ($kg\,m^{-3}\,s^{-1}$), respectively. It is noted that $\bar{C}$ in Eq. (7) corresponds to C in Eq. (8).

The crop model

In the present model a crop is considered to consist of two parts, viz., shoot and root. The shoot is characterized by dry matter content and leaf area index, while the root system is characterized by dry matter content, rooting depth and root density. The crop model is based on the thermal unit concept which implies that crop development from emergence to harvest can be described in terms of a temperature sum. Plant emergence and leaf area index L_{ai} ($m^2\,m^{-2}$) at the early stage of crop canopy development are simulated solely as functions of air temperature sum while leaf area index at later stages of crop canopy development is simulated as of function of both air temperature sum and accumulated amounts of shoot dry matter:

$$L_{ai} = \begin{cases} 14\,W_s & \sum T_a \leq 450 \\[2ex] 14\left[1 - \dfrac{1.8(\sum T_a - 450)}{(\sum T_a - 450) + 1000}\right]W_s & \sum T_a > 450 \end{cases}$$
$$(11)$$

where W_s is the accumulated amount of shoot dry matter $(\mathrm{kg\,m^{-2}})$ and $\Sigma\, T_a$ is the air temperature sum calculated from emergence (°C days).

Simulation of crop dry matter production is based on calculation of daily gross canopy photosynthesis, partitioning of assimilates between shoot and root, and respiration of shoot and root, respectively. The calculation of gross canopy photosynthesis is based on the assumptions that gross leaf photosynthesis can be described by a single light response curve [9] and that the light distribution within the crop canopy can be described by Beer's law. The daily gross canopy photosynthesis $F_{g,1}$ $(\mathrm{kg\,CO_2\,m^{-2}\,d^{-1}})$ is calculated as:

$$F_{g,1} = \frac{1}{f}\,\varepsilon S_a \qquad (12)$$

$$\varepsilon = 0.15$$
$$-\left[0.158 - 0.094\,\frac{L_{ai}}{L_{ai}+3}\right]\frac{S_a}{S_a + 7.8\cdot 10^6} \qquad (13)$$

$$S_a = (1 - A_v)(1 - \exp(-KL_{ai}))\alpha_v S_i \qquad (14)$$

where f is the energy content of carbohydrates $(\mathrm{J\,kg^{-1}})$, ε is radiation conversion efficiency coefficient, S_a is absorbed photosynthetically active radiation $(\mathrm{J\,m^{-2}\,d^{-1}})$, A_v is the reflection coefficient of crop canopy, K is the extinction coefficient, S_i is global radiation $(\mathrm{J\,m^{-2}\,d^{-1}})$ and α_v is the fraction of photosynthetically active radiation in the global radiation. Corresponding values of ε, L_{ai} and S_a were obtained by applying the basic assumptions on crop canopies with various values of L_{ai} for days with various radiation patterns, and integrating the gross leaf photosynthesis over space and time. Equation (13) is the result of a parameterization of the relation between ε, L_{ai} and S_a.

The partitioning of assimilates between shoot and root is considered as a function of temperature sum. The function of assimilates allocated to shoot γ_s is estimated as:

$$\gamma_s = \begin{cases} 0.4 + 0.5\,\dfrac{\Sigma\, T_a}{700} & \Sigma\, T_a < 700 \\[2mm] 0.9 & \Sigma\, T_a \geq 700 \end{cases} \qquad (15)$$

The fraction of assimilates allocated to root γ_r is then equal to $1 - \gamma_s$.

Respiration is assumed to include growth respiration as well as a temperature dependent maintenance respiration [17]. Thus the daily growth of shoot ΔW_s $(\mathrm{kg\,DM\,d^{-1}})$ is calculated as:

$$\Delta W_s = Y_s(\gamma_s F_g - r_{m,s}W_s) \qquad (16)$$

where Y_s is the assimilate conversion coefficient $(\mathrm{kg\,DM\,(kg\,CH_2O)^{-1}})$ and $r_{m,s}$ is the temperature dependent $(Q_{10} = 2)$ maintenance respiration coefficient $(\mathrm{kg\,CH_2O\,(kg\,DM)^{-1}\,d^{-1}})$. Calculation of daily root growth is performed by analogy to the corresponding calculation of daily shoot growth.

Root penetration is assumed to take place if a daily net root growth ΔW_r occurs, if the soil temperature is above 4°C, and if the actual rooting depth d_r (m) is less than the maximum rooting depth allowed in the particular soil considered. Daily root penetration Δd_r (m) is calculated as [13]:

$$\Delta d_r = 0.25(T_s - 4.0) \qquad (17)$$

where T_s is the soil temperature (°C) at the actual rooting depth d_r.

The distribution of root density L in the soil profile is described in accordance with Gerwitz and Page [8]. In order to establish the root density distribution it is assumed that the total root lengh is proportional to the accumulated amount of root dry matter and that the root density at the rooting depth d_r is $0.01\ \mathrm{cm\,cm^{-3}}$.

The gross canopy photosynthesis may be limited due to water and/or nitrogen deficiency. The gross canopy photosynthesis under conditions of water deficiency $F_{g,2}$ $(\mathrm{kg\,CO_2\,m^{-2}\,d^{-1}})$ is calculated as:

$$F_{g,2} = F_{g,1}\,\frac{E_{a,c}}{E_{p,c}} \qquad (18)$$

where $E_{a,c}$ is actual crop evapotranspiration $(\mathrm{mm\,d^{-1}})$ and $E_{p,c}$ is potential crop evapotranspiration $(\mathrm{mm\,d^{-1}})$. The gross canopy photosynthesis under conditions of nitrogen deficiency $F_{g,3}$ $(\mathrm{kg\,CO_2\,m^{-2}\,d^{-1}})$ is calculated as:

$$F_{g,3} = F_{g,2} \frac{N_c - N_c^o}{N_c^a - N_c^o} \qquad (19)$$

where N_c, N_c^o and N_c^a (kg N m^{-2}) is the amount of nitrogen in the crop at the actual nitrogen supply, at extremely low nitrogen supply, and at just ample nitrogen supply, respectively. The daily potential nitrogen demand ΔN_u^p (kg N m^{-2}) is calculated as:

$$\Delta N_u^p = N_c^p - N_c \qquad (20)$$

where N_c^p (kg N m^{-2}) is the potential amount of nitrogen in the crop. The nitrogen concentrations corresponding to N_c^p, N_c^a, and N_c^o, respectively, assumed to be functions of temperature sum are illustrated in Fig. 6 for the two parts of the crop, viz., shoot and root.

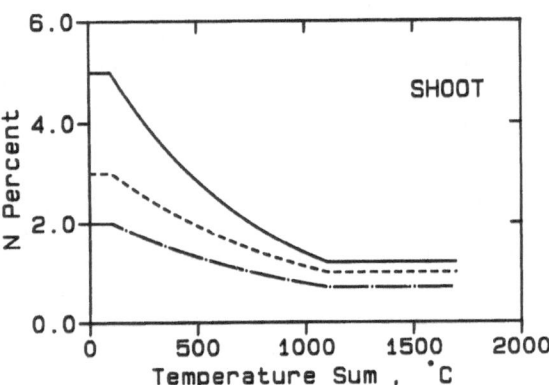

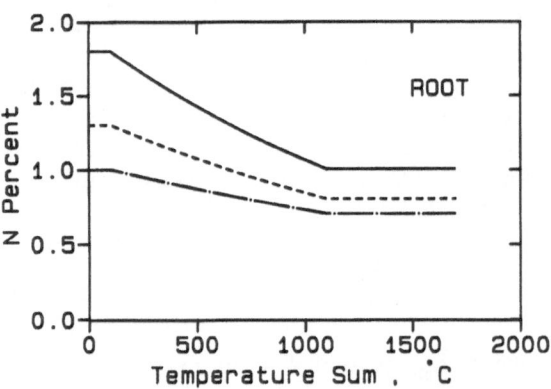

Fig. 6. Nitrogen concentrations in wheat as function of temperature sum at potential nitrogen content (solid line), content at just ample nitrogen supply (dashed line), and content at extremely low nitrogen supply (dash dotted line).

At harvest the crop nitrogen is distributed between root and shoot by using the following equation:

$$C_r = 0.9\sqrt{C_s} \qquad (21)$$

where C_r and C_s are nitrogen concentration in per cent in root and shoot dry matter, respectively. The parts of the crop which are not removed at harvest, e.g. stubble and root, are allocated to the appropriate pools of organic matter in the soil.

The experimental data set

The experimental data used in the present validation have been described in detail by Groot and Verberne [10]. The data were obtained in field experiments in 1982/1983 and 1983/1984 with winter wheat grown at various nitrogen treatments on three locations, the Bouwing, the Eest and PAGV in the Netherlands (Table 1). The objective of these experiments was to provide a data set to validate simulation models for nitrogen dynamics in crop and soil.

Simulation results and discussion

Initialization of the model

The required hydraulic properties of the soil in terms of soil water characteristics and unsaturated hydraulic conductivity were obtained from the experimental data set. The thermal properties of the soil were calculated according to De Vries [5]. The carbon content of the soil provided in the experimental data set was used in the initialization of the soil nitrogen model. The carbon content of the subsoil (below 30 cm) was assumed to be one third of the content in the topsoil (0–30 cm). Important parameters used in the soil nitrogen model are listed in Table 2. Important parameters used in the crop model are listed in Table 3. The soil water profiles and the soil nitrogen profiles, respectively, were initialized by using data obtained from the experimental data set.

Table 1. Outline of experimental treatments available for model evaluations

Location	1982–1983				1983–1984			
	Total–N* kg N ha^{-1}	N-applications kg N ha^{-1}			Total–N* kg N ha^{-1}	N-applications kg N ha^{-1}		
		Feb	May	June		Feb	May	June
The Bouwing	N1 = 115	0	0	0	N1 = 100	70	0	0
	N2 = 175	0	60	0	N2 = 200	70	60	40
	N3 = 275	0	120	40	N3 = 260	70	120	40
The Eest	N1 = 94	0	0	0	N1 = 136	50	60	0
	N2 = 154	0	60	0	N2 = 186	50	60	40
	N3 = 254	0	120	40	N3 = 186	50	60	40
PAGV	N1 = 120	80	0	0	N1 = 140	80	0	0
	N2 = 200	60	80	0	N2 = 240	80	60	40
	N3 = 300	60	140	40	N3 = 300	80	120	40

*) Incl. the content of $NH_4^+ + NO_3^-$ in the soil profile to 1 m depth in February.

Table 2. Parameters and initial values used in the soil nitrogen model

	Dimension	Units
Carbon turnover model		
Decomposition rate coefficient[a] of SOM$_1$	day^{-1}	$2.7 \cdot 10^{-6}$
Decomposition rate coefficient[a] of SOM$_2$	day^{-1}	$1.4 \cdot 10^{-1}$
Decomposition rate coefficient[a] of AOM$_1$ (root material)	day^{-1}	$7.0 \cdot 10^{-3}$
Decomposition rate coefficient[a] of AOM$_2$ (root material)	day^{-1}	$7.0 \cdot 10^{-2}$
Death rate coefficient[a] of BOM$_1$	day^{-1}	$1.0 \cdot 10^{-3}$
Death rate coefficient[a] of BOM$_2$	day^{-1}	$1.0 \cdot 10^{-2}$
Maintenance coefficient[a] of BOM$_1$	day^{-1}	$1.0 \cdot 10^{-2}$
Maintenance coefficient[a] of BOM$_2$	day^{-1}	$1.0 \cdot 10^{-2}$
Substrate utilization efficiency of BOM$_1$		0.60
Substrate utilization efficiency of BOM$_2$		0.60
Partitioning constant f_{SOM_1}		0.10
Partitioning constant f_{SOM_2}		0.40
Partitioning constant f_{BOM_1}		0.50
Initial carbon content		
Fraction of soil C allocated to SOM$_1$		0.793
Fraction of soil C allocated to SOM$_2$		0.200
Fraction of soil C allocated to BOM$_1$		0.004
Fraction of soil C allocated to BOM$_2$		0.003
Amount of C allocated to AOM$_1$ (root material)	t ha^{-1}	0.36
Amount of C allocated to AOM$_2$ (root material)	t ha^{-1}	0
Net nitrogen mineralization (ammonification)		
C/N ratio in pool SOM$_1$		
C/N ratio in pool SOM$_2$		
C/N ratio in pool BOM$_1$		6
C/N ratio in pool BOM$_2$		10
C/N ratio in pool AOM$_1$		100
Nitrification		
Turnover rate coefficient[a]	day^{-1}	0.1
Denitrification		
Empirical constant, α_d^*, in Eq. (6)	g Gas-N/g CO$_2$-C	0.1
Nitrogen movement in soil		
Longitudinal dispersivity	cm	4

[a] At standard conditions, i.e. 10°C, optimum water content and no clay content.

Table 3. Crop model parameters

	Units	Value
Extinction coefficient, K	–	0.6
Fraction of PAR in global radiation, α_v	–	0.48
Reflection coefficient of crop canopy, A_v	–	0.06
Energy content of carbohydrates, f	MJ/kg CH_2O	15.7
Assimilate conversion efficiency, Y_s shoot	kg DM/kg CH_2O	0.72
root	kg DM/kg CH_2O	0.54
Maintenance respiration coefficient (20°C), $r_{m,s}$		
Shoot, $\Sigma\, T_a \leqslant 800$	kg CH_2O/kg DM/day	0.015
Shoot, $\Sigma\, T_a > 800$	kg CH_2O/kg DM/day	0.008
Root, $\Sigma\, T_a \leqslant 800$	kg CH_2O/kg DM/day	0.065
Root, $\Sigma\, T_a > 800$	kg CH_2O/kg DM/day	0.015
Specific root length	m/kg	10^5
Root radius, r_r in Eq. (3)	mm	0.1

Driving variables

The driving variables required to run the model, i.e. daily values of global radiation, air temperature and precipitation were supplied with the experimental data set [10]. In the case of a shallow groundwater table the level of the groundwater table itself is required as a driving variable in order to obtain a correct simulation of the soil water and soil nitrogen dynamics. In the present experimental data set information on the groundwater level was available for 1984 only.

Soil water and nitrogen dynamics

The experimental data on soil water characteristics and unsaturated hydraulic conductivity and the soil water content at the Bouwing location indicate that by-pass flow dominated the water flow in the unsaturated soil. Consequently the simulated soil water dynamics proved to be unsatisfactory as by-pass water flow is not included in the model. For that reason soil nitrogen dynamics were not simulated for this location. The Eest and PAGV locations are polder soils with a shallow groundwater table which affects soil water dynamics and consequently also soil nitrogen dynamics considerably. However, information on groundwater table fluctuations was available only for the 1983–1984 experiments. For that reason biomass production and soil nitrogen dynamics were simulated for the 1983–1984 experiments on the Eest and PAGV only.

Simulated water balances for the root zone covering the period from the date of the first

fertilizer application to the date of harvest are shown in Table 4. In the beginning of the period a positive percolation occurred at both locations while later in the period upward water movement took place in particular at PAGV. In Table 4 the net results are shown which indicate that only limited nitrogen leaching from the root zone is to be expected in the period considered.

Simulated balances of inorganic nitrogen for the root zone covering the period from the date of the first fertilizer application to the date of harvest are shown in Table 5. It appears that the

Table 4. Simulated water balance for the period February 17 to August 21 and February 17 to August 22 at the Eest and the PAGV location, respectively, 1984

	Eest	PAGV
Precipitation, mm	283	283
Evapotranspiration, mm	353	357
Percolation, mm	11	−27
Change in storage, mm	−81	−47

Table 5. Simulated inorganic nitrogen balance (kg N ha^{-1}) for the period February 17 to August 22 and February 17 to August 21 for the Eest and the PAGV locations, respectively, 1984

	Eest		PAGV		
	N1	N3	N1	N2	N3
Fertilization	110	150	80	180	240
Atmospheric deposition	14	14	14	14	14
Mineralization	51	52	35	35	35
Denitrification	17	20	8	11	11
Plant uptake	179	217	151	241	255
Leaching	2	2	−3	−3	−3
Change in storage	−23	−23	−27	−20	−26

net flux of nitrogen across the lower boundary of the root zone was insignificant for both locations. Furthermore it appears that the levels of mineralization and denitrification are higher at Eest as compared to PAGV. This can be attributed partly to a higher initial carbon content and partly to more moist conditions in the root zone at Eest as compared to PAGV.

The observed and simulated nitrate concentrations in soil during the considered period are shown in Fig. 7 and Fig. 8 for the Eest and the PAGV location, respectively. The simulated profiles of nitrate concentrations show characteristic peak values occurring in the upper soil layer as a consequence of fertilizer application. The concentration peaks are recognized to deeper soil

layers with some phase deplacement, and the peak values are smaller and less distinct. This general pattern of the simulated profiles of nitrate concentrations indicates that some downward movement of nitrate has taken place during the growing season. Qualitatively and quantitatively the agreement between observed and simulated values is satisfactory with a few exceptions. In the case of Eest treatment N3, Fig. 7, the simulated disappearance of nitrate in the top layer after the second fertilizer application is too quick as compared to the experimental results. In the case of PAGV treatment N3, Fig. 8, the model overestimates the soil mineral nitrogen content after the second fertilizer application.

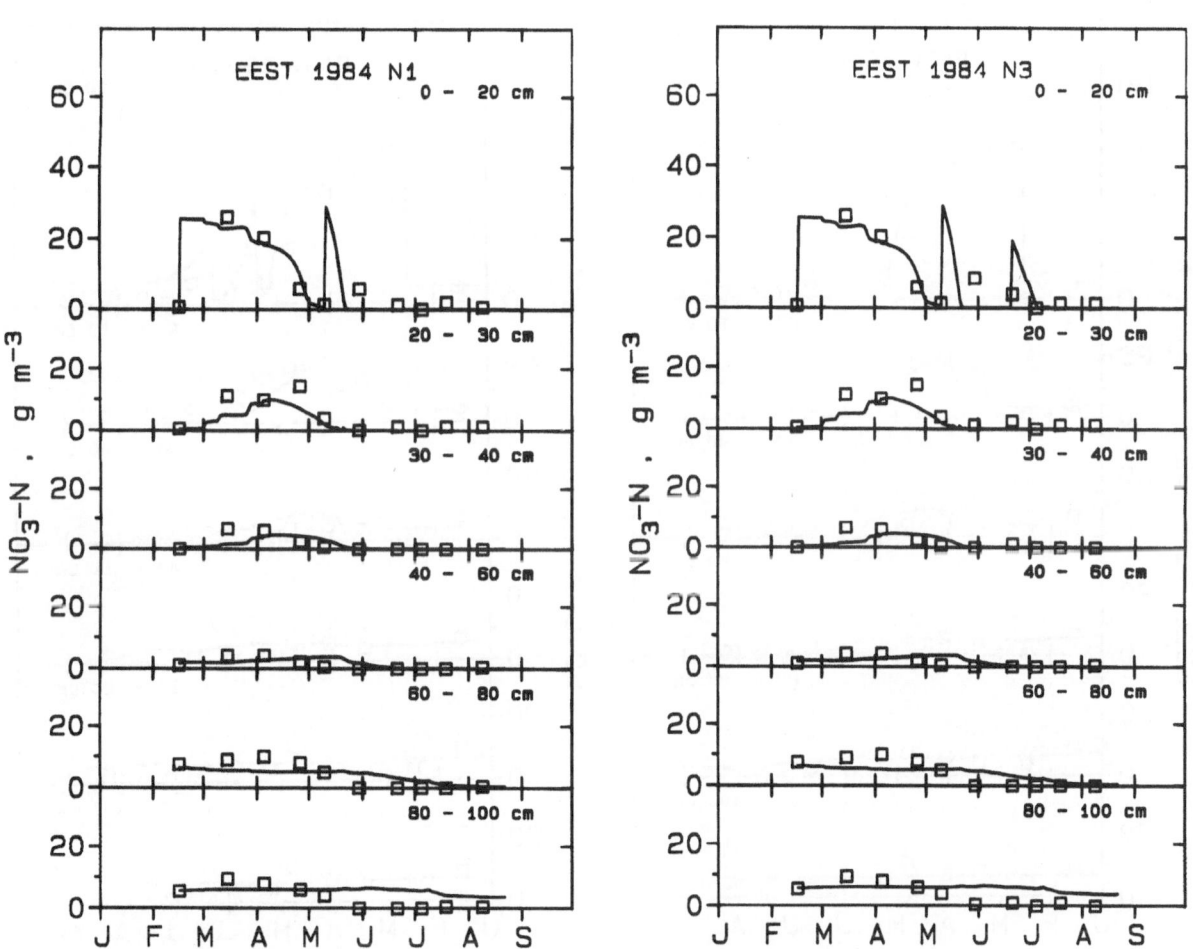

Fig. 7. Simulated and measured concentrations of nitrate during the period February–August in various soil depths for treatment N1 and N3, the East location, 1984.

256

Crop yield and nitrogen uptake

The observed and simulated course of biomass production and nitrogen uptake in the shoot part of the crop during the growing season is shown in Fig. 9 and Fig. 10 for the Eest and the PAGV location, respectively. Regarding biomass production the agreement is quite satisfactory during the first part of the growing season while later in the season the model tends to overestimate the biomass production. This is also seen in the scatter diagram, Fig. 11. However, it should be realized that the present crop model does not take into account any possible detrimental effects other than those caused by soil water or soil nitrogen deficiency.

The observed and simulated course of nitrogen uptake accumulated in the shoot part of the crop during the growing season is shown in Fig. 9 and Fig. 10 for the Eest and PAGV location, respectively. It appears that the model tends to slightly underestimate the nitrogen uptake accumulated in the shoot during the first part of the growing season. This is also seen in the scatter diagram, Fig. 11.

However, in general the agreement between simulated and observed nitrogen uptake is considered satisfactory except for the PAGV treatment N2 in which case the model clearly overestimated the nitrogen uptake although the agreement between the observed and simulated soil nitrogen profile was satisfactory. On the

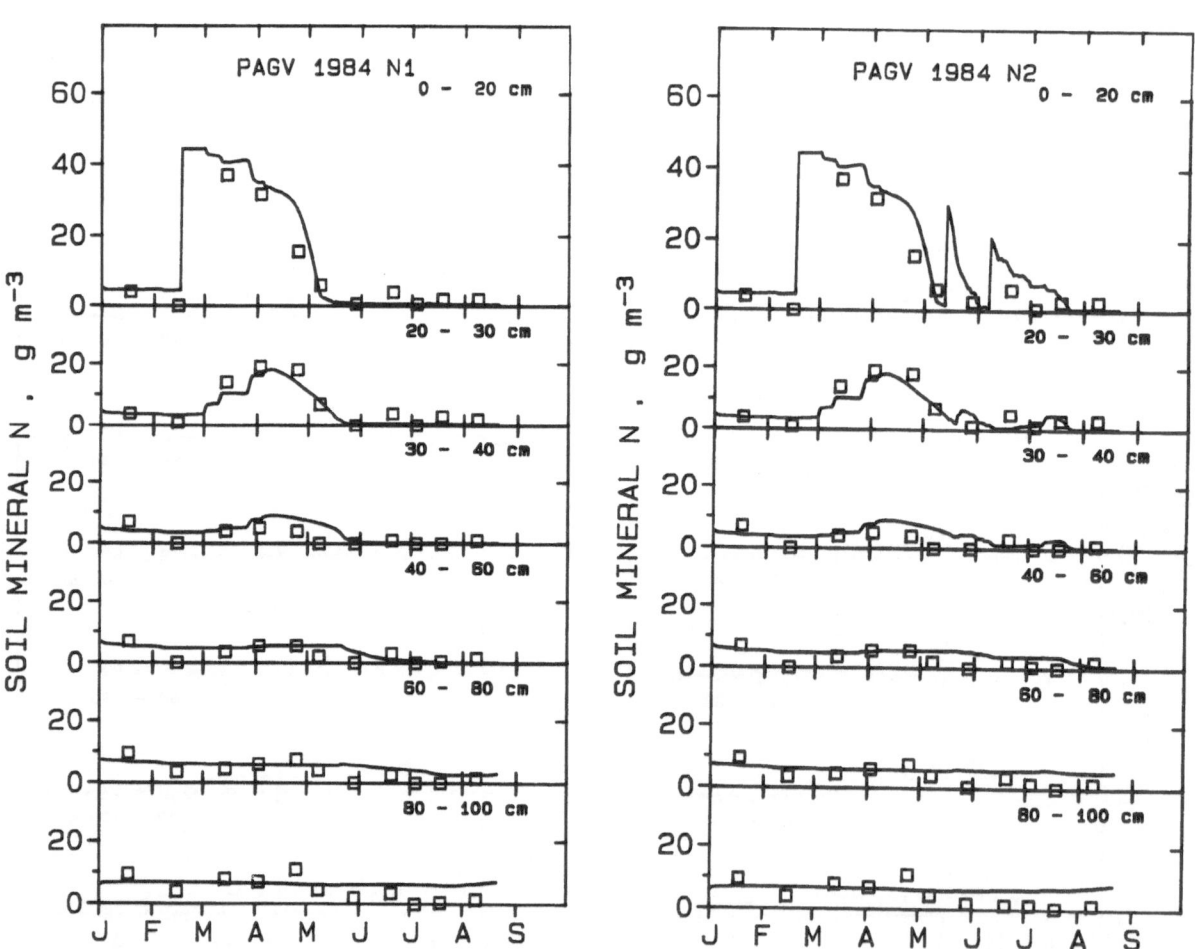

Fig. 8. Simulated and measured concentrations of mineral nitrogen during the period February–August in various soil depths for treatment N1, N2, and N3, the PAGV location, 1984.

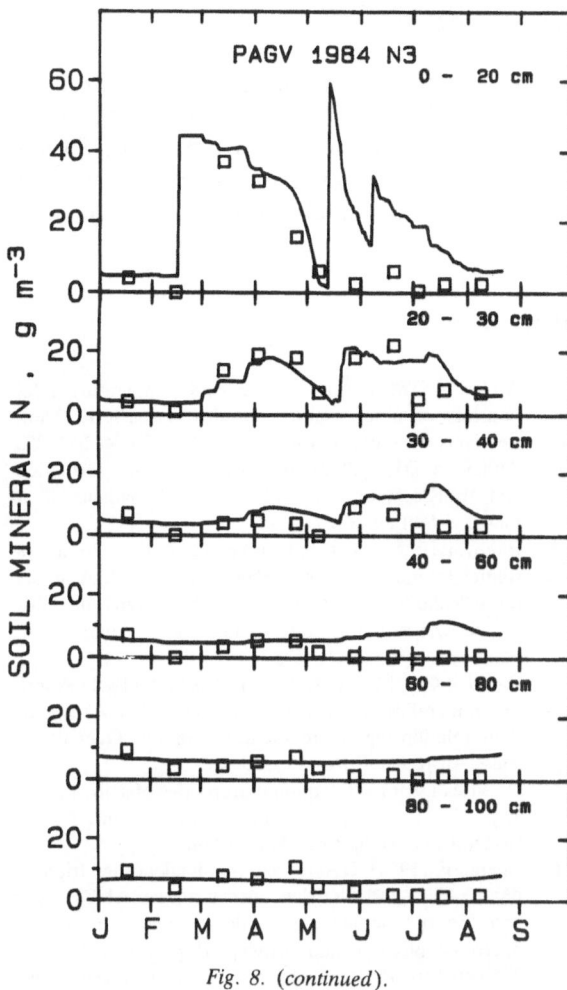

Fig. 8. (continued).

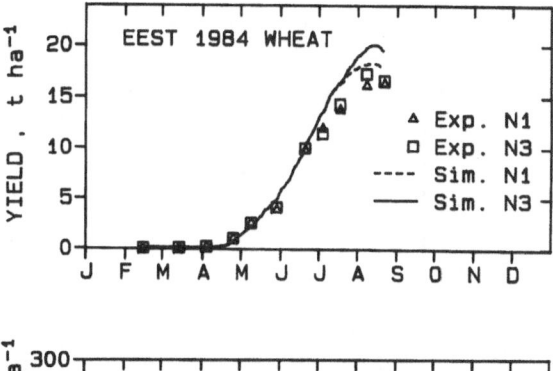

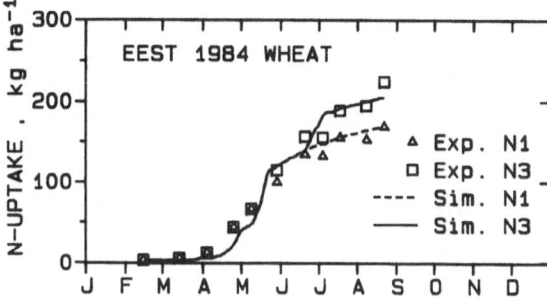

Fig. 9. Simulated and observed course of biomass production and nitrogen uptake in the aboveground part of the crop during the period February–August for treatment N1, and N3, the Eest location, 1984.

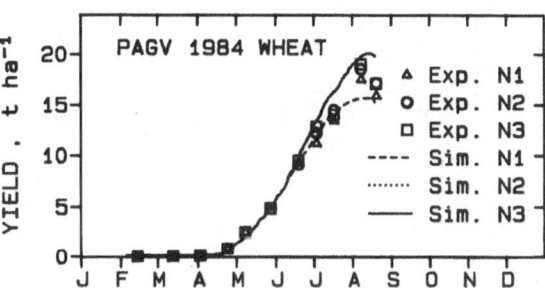

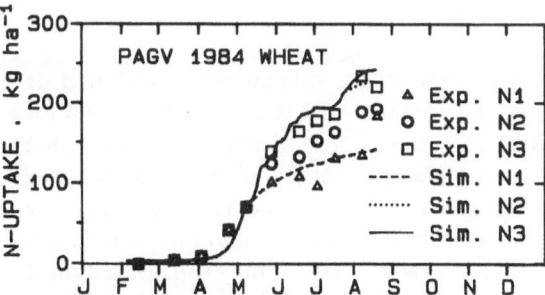

Fig. 10. Simulated and observed course of biomass production and nitrogen uptake in the aboveground part of the crop during the period February–August for treatment N1, N2, and N3, the PAGV location, 1984.

other hand in the case of PAGV treatment N3 the agreement between the simulated and observed nitrogen uptake was satisfactory while the model overestimated the soil mineral nitrogen content after the second fertilizer application. For PAGV treatment N2 and N3 the model apparently overestimates the amount of nitrogen present in the soil plant system. In the case of PAGV treatment N1 satisfactory agreement between observed and simulated results was satisfactory for nitrogen uptake as well as for the content of soil mineral nitrogen. Realizing the leaching of nitrogen to be insignificant the model probably underestimates the denitrification in the case of the heavily fertilized PAGV treatment N2 and treatment N3 as mineralization of nitrogen usually is unaffected by fertilizer nitrogen application.

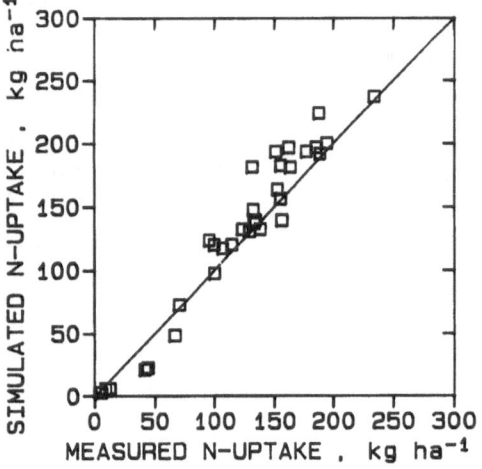

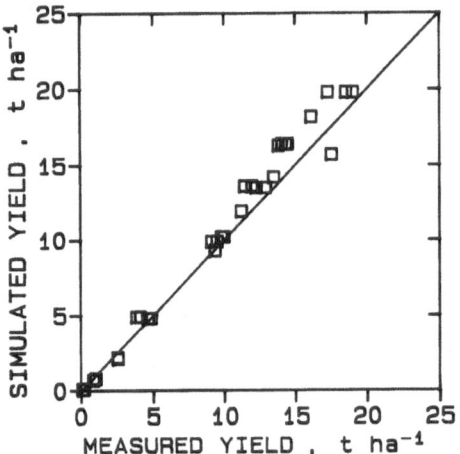

Fig. 11. Relation between simulated and observed biomass production (bottom) and nitrogen uptake (top) for all treatments in 1984 on the Eest and PAGV location.

Conclusion

The present model validation has included data from experiments with wheat grown at two locations at various rates of application of mineral fertilizer. The simulated and experimental results compared have included concentration of nitrate nitrogen in soil, crop yield and nitrogen accumulated in the shoot part of the crop. Based on this validation it is concluded that the overall performance of the model is satisfactory although some minor adjustments of the model may prove to be necessary.

Acknowledgement

We are indebted to the Institute for Soil Fertility Research, Haren, The Netherlands, for inviting us to validate our simulation model DAISY on the excellent Dutch experimental data set on soil nitrogen dynamics and wheat crop production.

References

1. Addiscott TM (1983) Kinetics and temperature relationships of mineralization and nitrification in Rothamsted soils with different histories. J Soil Sci 34: 343–353
2. Anderson DW (1979) Processes of humus formation and transformation in soils of the Canadian Great Plains. J Soil Sci 30: 77–84
3. Baldwin JP, Nye PH and Tinker PB (1973) Uptake of solutes by multiple root system from soil. III. A model for calculating the solute uptake by a randomly dispersed root system developing in a finite volume of soil. Plant and Soil 38: 621–635
4. Cambell CA, Meyers RJK and Weier KL (1981) Potential mineralizable nitrogen, decomposition rates and their relationships to temperature for five Queensland soils. Austr J Soil Res 19: 323–332
5. De Vries DA (1963) Thermal properties of soils. *In* Van Wijk WR (ed) Physics of Plant Environment. North-Holland Publishing Co., Amsterdam
6. Eiland F (1985) Determination of adenosine triphosphate (ATP) and adenylate energy charge (AEC) in soil and use of adenine nucleotides as measures of soil microbial biomass and activity. Report No. S 1777, Tidsskrift for Planteavls Specialserie, Copenhagen 1985
7. Flowers H and O'Callaghan JR (1983) Nitrification in soils incubated with pig slurry or ammonium sulphate. Soil Biol Biochem 15: 337–342
8. Gerwitz S and Page ER (1974) An empirical mathematical model to describe plant root systems. J Appl Ecol 11: 773–781
9. Goudriaan J and Van Laar HH (1978) Calculation of daily totals of the gross CO_2 assimilation of leaf canopies. Neth J Agric Sci 26: 373–382
10. Groot JJR and Verberne ELJ (1991) Response of wheat to nitrogen fertilization, a data set to validate simulation models for nitrogen dynamics in crop and soil. Fert Res 27: 349–383
11. Hansen S (1984) Estimation of potential and actual evapotranspiration. Nordic Hydrol 15: 205–212
12. Hansen S, Jensen HE, Nielsen NE and Svendsen H (1990) DAISY: A Soil Plant System Model. Danish simulation model for transformation and transport of energy and matter in the soil plant atmosphere system, 369 pp. The National Agency for Environmental Protection, Copenhagen
13. Jacobsen BF (1976) Jord, rodvækst og stofoptagelse. In: Simuleret planteproduktion. Hydroteknisk Lab-

oratorium. Den Kgl Veterinær- og Landbohøjskole, København

14. Jenkinson DS and Rayner JH (1977) The turnover of soil organic matter in some of the Rothamsted classical experiments. Soil Sci 123: 298–305

15. Lind AM (1980) Denitrification in the root zone. Tidsskr Planteavl 84: 101–110

16. Makkink GF (1957) Ekzameno de la formulo de Penman. Neth J Agric Sci 5: 290–305

17. McCree KJ (1970) An equation for the rate of respiration of white clover plants grown under controlled conditions. *In* Setlik I (ed) Prediction and Measurement of Photosynthetic Productivity, pp 221–229. Proc. IBP/PP Technical Meeting. Trebon 1969, PUDOC, Wageningen

18. Miller RD and Johnson DD (1964) The effect of soil moisture tension on carbon dioxide evolution, nitrification and nitrogen mineralization. Soil Sci Soc Amer Proc 24: 644–647

19. Miller RD (1980) Freezing phenomena in soils. *In* Hillel D (ed) Application of Soil Physics. Academic Press, New York

20. Nielsen NE, Schjørring JK and Jensen HE (1988) Efficiency of fertilizer nitrogen uptake by spring barley. *In* Jenkinson DS and Smith KA (eds) Nitrogen Efficiency in Agricultural Soils, pp 62–72. Elsevier Applied Science, London

21. Orchard WA and Cook FJ (1983) Relationship between soil respiration and soil moisture. Soil Biol Biochem 15: 447–453

22. Reichman GA, Grunes DL and Viets FG (1966) Effect of soil moisture on ammonification and nitrification in two northern plain soils. Soil Sci Soc Amer Proc 30: 363–366

23. Richard LA (1931) Capillary conductivity of liquids in porous mediums. Physics 1: 318–333

24. Rolston DE, Rao PSC, Davidson JM and Jessup RE (1984) Simulation of denitrification losses of nitrate fertilizer applied to uncropped, cropped and manure-amended field plots. Soil Science 13: 170–279

25. Sabey BR (1969) Influence of soil moisture tension on nitrate accumulation in soils. Soil Sci Soc Amer Proc 33: 263–266

26. Schouwenburg JCh van and Schuffelen AC (1963) Potassium-exchange behaviour of an illite. Neth J Agric Sci 11: 13–22

27. Stanford G, Frere MH and Schwaninger DH (1973) Temperature coefficient of soil nitrogen mineralization. Soil Science 115: 321–323

28. Stanford, G and Epstein E (1974) Nitrogen mineralization – water relations in soils. Soil Sci Soc Amer Proc 38: 103–107

29. Stott DE, Elliot LF, Papendick RI and Cambell GS (1986) Low temperature or low water potential effects on microbial decomposition of wheat residue. Soil Biol Biochem 18: 577–582

30. Sørensen LH (1975) The influence of clay on the rate of decay of amino acid metabolites synthesized in soils during decomposition of cellulose. Soil Biol Biochem 7: 171–177

31. Veen JA van and Paul EA (1981) Organic carbon dynamics in grassland soils. 1 Background information and computer simulation. Can J Soil Sci 61: 185–201

32. Veen JA van, Ladd JN and Frissel MJ (1984) Modelling C and N turnover through the microbial biomass in soil. Plant and Soil 76: 257–274

33. Veen JA van, Ladd JN and Amato M (1985) The turnover of carbon nitrogen through the microbial biomass in a sandy loam and a clay soil incubated with C^{14}-Glucose and ^{15}N-$(NH_4)_2$ SO_4 under different moisture regimes. Soil Biol Biochem 17: 747–756

Fertilizer Research **27**: 261–272, 1991.
© 1991 *Kluwer Academic Publishers.*

Simulation of the nitrogen balance in the soil and a winter wheat crop

J.J.R. Groot & P. de Willigen
Institute for Soil Fertility Research, P.O. Box 30003, 9750 RA Haren (Gn.), The Netherlands

Key words: Crop growth, immobilization, mineralization, modelling, nitrogen uptake

Abstract

A simulation model for winter wheat growth, crop nitrogen dynamics and soil nitrogen supply was tested against experimental data. When simulations of dry matter production agreed with measurements, nitrogen uptake was simulated accurately. The total amount of soil mineral nitrogen as well as the distribution of mineral nitrogen over the various soil layers were generally simulated well, except for experiments in which fertilizer was applied late in spring. In these experiments, applied nitrogen 'disappeared' because it could not be accounted for by the model. Some explanations for this disappearance are briefly discussed.

Introduction

Various models for the behaviour of nitrogen in the soil have been developed [7], and models to simulate crop nitrogen uptake and the effects of crop nitrogen status on crop growth are available [7, 26]. Most of these models were developed as research tools, and cannot be readily applied for management purposes, as they require too many site-specific input parameters that cannot easily be obtained by simple field measurements.

The purpose of the present study is to examine to what extent cereal growth, nitrogen (N) uptake and soil N dynamics can be simulated for a single field, by using input parameters that can be obtained from simple field measurements. The model was developed on the basis of existing theory and submodels for crop growth, N distribution in the crop and water and N dynamics in the soil, and was parameterized and tested against a set of experimental data from the Netherlands [11].

Model structure

The model comprises three submodels (Fig. 1) which simulate soil N dynamics, soil moisture dynamics, and crop growth and N uptake, respectively.

Soil moisture submodel

The soil is treated as a multilayered system with a finite number of compartments of variable thickness. Changes in moisture status of a compartment are calculated from the combined effect of water infiltration, extraction due to soil surface evaporation, water extraction by the roots and downward movement of water through the compartments.

Infiltration

When precipitation occurs, the first compartment is filled with water until field capacity (corresponding to a -100 cm pressure head) is reached, and excess water entering the compartment drains to the next compartment. This procedure is repeated for deeper compartments as long as drainage occurs.

Evaporation and transpiration

Potential soil surface evaporation and potential crop transpiration are calculated according to a

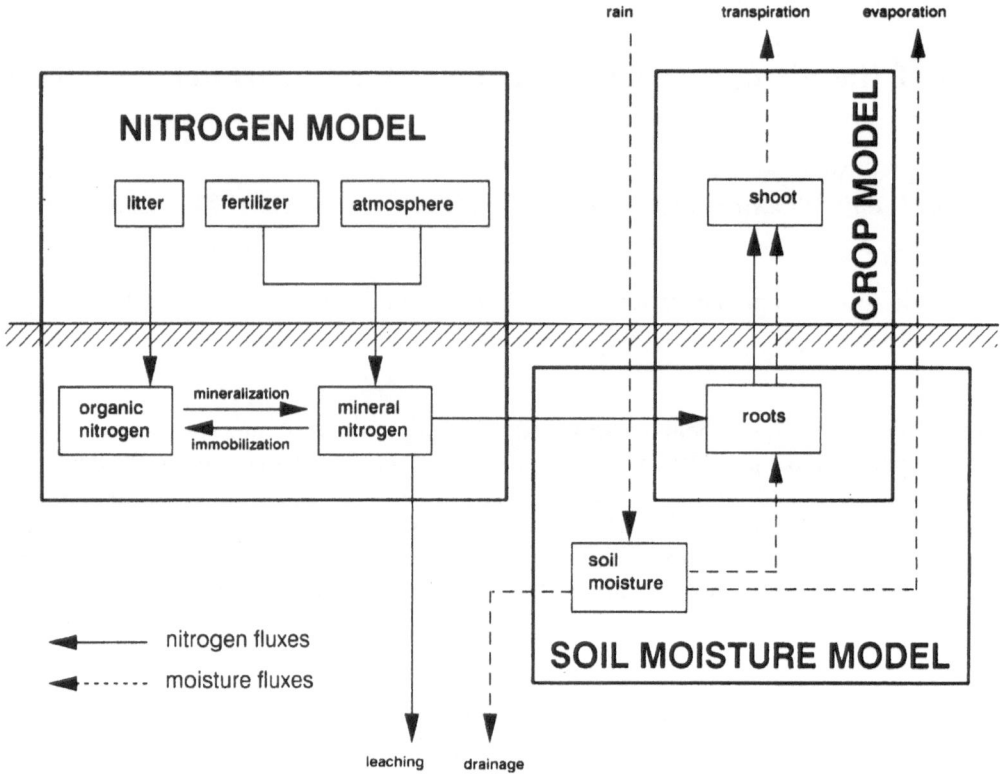

Fig. 1. Schematic representation of the model structure.

modified Penman approach [16], described in detail by Groot [9]. Actual soil surface evaporation is related to the soil moisture content of the upper soil compartment according to Van Keulen & Seligman [26], and moisture lost by evaporation is distributed over various soil compartments according to an exponential decay curve.

The actual rate of transpiration is related to the leaf area index of the crop, and to the soil moisture content of the rooted soil compartments according to Van Keulen & Seligman [26]. When actual transpiration is smaller than potential transpiration, gross canopy assimilation is reduced proportionally.

Soil nitrogen submodel

Nitrogen availability depends on the balance between N input through fertilizer application, decomposition of old organic matter (humus) and crop residues, and N output through crop uptake and transport out of the rooting zone. Denitrification and volatilization are not considered.

Mineralization – immobilization

Most models for soil organic matter dynamics distinguish between a microbial biomass pool and several soil organic matter pools, each representing material of different stability with regard to decomposition [12, 25, 28]. Generally, transformations are described by first-order kinetics and result in growth of biomass, which itself is subject to decomposition. These models have proven to be useful for long-term simulations (decades, e.g. [12, 25]), but they are hard to initialize, because good methods for partitioning soil organic matter among different pools are lacking. Moreover, parameterization is rather speculative; for each of the transformations a decomposition rate constant and a yield efficiency factor is required.

In our model, organic matter dynamics is restricted to the upper 30 cm, representing the plough layer. The only pools considered are crop residues and native soil organic matter.

Decomposition of crop residues

In the model, the rate of substrate decomposition ($C_{d,p}$, kg(C)ha^{-1} d^{-1}) is proportional to the amount of substrate (C_c, kg ha^{-1}):

$$C_{d,p} = dC/dt = k_d C_c f_{t,m} \qquad (1)$$

in which k_d is the first-order decomposition rate constant (d^{-1}) and in which $f_{t,m}$ accounts for the combined effect of soil temperature and soil moisture content according to Verbruggen [29]. The response surface given in [29] is preferred to approaches in which the effects of temperature and moisture content are considered to be multiplicative. Nitrogen fluxes are assumed to be proportional to the carbon fluxes. The rate of N release ($N_{d,p}$, kg(N)ha^{-1}d^{-1})) thus depends on the C/N ratio of the substrate (CN_s):

$$N_{d,p} = dN/dt = C_{d,p}/CN_s \qquad (2)$$

A fraction ϵ_b of the decomposed substrate is used for biomass growth, $(1 - \epsilon_b)$ is used for respiration. Thus, N required for optimum growth of biomass ($N_{\text{req},b}$, kg(N)ha^{-1}d^{-1}) is given by:

$$N_{\text{req},b} = \epsilon_b(C_{d,p}/CN_b) \qquad (3)$$

in which CN_b is the C/N-ratio of microbial biomass. When CN_s exceeds (CN_b/ϵ_b), decomposition is N-limited and net immobilization occurs. In this situation $N_{\text{req},b}$ is covered by mineral N (N_i, kg(N) ha^{-1}). When the amount of soil mineral N is not sufficient to cover $N_{\text{req},b}$, potential decomposition rates are reduced:

$$N_d = N_{d,p}(N_i/N_{\text{req},b}) \qquad (4)$$

$$C_d = C_{d,p}(N_i/N_{\text{req},b}) \qquad (5)$$

In the model nitrification is neglected, and mineralization directly yields mineral N as nitrate.

Mineralization of native soil organic matter

Only a small proportion of soil organic matter is mineralized each year (1–2% [14]), and for simulations over a single growing period, changes in the amount of soil organic matter (OM, kg ha^{-1}) may be neglected. According to Johnsson et al. [13], mineralization ($N_{d,o}$, kg(N) ha^{-1}d^{-1}) is calculated as

$$N_{d,o} = OM k_{om} f_{t,m} CN_{om} \qquad (6)$$

in which k_{om} is the specific mineralization rate constant (d^{-1}), and ($f_{t,m}$) is the combined effect of soil moisture content and temperature [29]. The C/N-ratio of soil organic matter (CN_{om}) generally ranges from 8–15 [28]; the C/N ratio of biomass (CN_b) and the growth efficiency of biomass ϵ_b are in the order of 10 and 0.33, respectively, and thus net immobilization is not expected to occur.

Nitrogen transport

Downward N transport occurs by downward water flow as calculated in the soil moisture submodel. Following the approach proposed by Burns [3], water and N entering a soil compartment are mixed with water and N already present in the soil compartment. The resulting concentration is subject to drainage.

Nitrogen uptake

In the model, different horizontal layers in the root zone are distinguished, each with its own root density and N concentration. The basic assumption of the uptake model is that the uptake rate is governed by demand as long as the transport rate of nitrogen from bulk soil to the root surface exceeds the required rate given by plant demand. The highest transport rate occurs when the concentration at the root surface is zero, the root behaving as a so-called zero sink. The zero-sink uptake rate can be shown to be proportional to the average concentration in the soil [6], where the proportionality constant depends on root density, the rate of flow of water towards the root and the diffusion coefficient of nitrogen in the soil, which in turn depends on the soil moisture content [2].

Uptake is calculated by iteration. First the nitrogen demand is divided by the total root length to obtain the required uptake per unit root length. Multiplying this by the root length in a given layer yields the required uptake from that layer. If the potential uptake rate (zero-sink uptake rate) exceeds the required uptake rate, uptake from this layer equals the required uptake. If the potential uptake is lower than the required uptake rate, uptake from this layer equals the potential uptake rate.

Total uptake by the root system is the sum of the amounts taken up from the individual layers. If the uptake in each layer can proceed at the required rate, total uptake equals nitrogen demand and no iteration is required. If not, it is assumed, as indeed is often found, that roots in favorable positions can compensate for roots in less favorable positions [5]. If total uptake is lower than the nitrogen demand, the model checks whether uptake from these layers (that is where the potential uptake was higher than the required uptake rate), can be raised sufficiently to meet the total demand. This calculation procedure implies that roots growing under favorable conditions will compensate as much as possible for roots growing under less favorable conditions. It is thus assumed that information about the necessary behavior of the plant, as far as uptake is concerned, is instantaneously available throughout the entire root system.

In field experiments with winter wheat, N uptake declines after anthesis, but the mechanism for this decline is poorly understood. In the model this decline is mimicked by reducing the potential rate of uptake by a factor which represents the total root weight relative to the root weight at anthesis.

Crop submodel

Crop growth

Crop growth is calculated with the SUCROS model [23], but the model was slightly modified to account for the effects of N on crop growth. Gross canopy photosynthesis is calculated as a function of leaf area index, radiation distribution within the canopy, and the photosynthesis-light response curve of individual leaves. Maintenance

requirements for various plant organs, calculated as a function of their weight and chemical composition according to Penning de Vries [17, 18], are subtracted from daily gross assimilation. Partitioning of remaining assimilates to leaves, stems, roots, reserves and grains is varied with the stage of crop development according to fixed empirical functions [9]. The rate of crop development is a funtion of ambient air temperature, but is modified to account for effects of vernalization and photoperiod [20]. After anthesis no vegetative growth occurs (see e.g. [9]), and assimilates are either stored in the stem as water-soluble carbohydrates (sink-limited grain growth) or allocated to the grains (source-limited grain growth). The sink strength of grains is characterized by a variety-specific potential rate of carbohydrate accumulation for single grains, which is a function of the ambient air temperature [21], multiplied by the total grain number. This number can be empirically related to total above ground dry matter at anthesis [30].

Assimilates allocated to various plant organs are converted into structural plant material, taking into account the energy required for conversion (growth respiration) as a function of protein content of the growing material [18]. In the present model, chemical composition is considered only in terms of proteins and carbohydrates. Leaf area index is calculated by multiplication of leaf weight by an average specific leaf area $(m^2 kg^{-1})$ derived from experimental data [11].

Both the rate of photosynthesis and the rate of maintenance respiration increase with rising N content of vegetative plant parts [26]. Note that this was not the case in the original SUCROS model.

The potential rate of root extension was set at a value of 18 mm d^{-1} [8]. This rate was reduced to account for the effects of soil moisture content according to Stapper [24] and temperature of the soil compartment in which root extension occurs [1, 4]. Once rooting depth and root weight are known, the root length distribution pattern is calculated. It is assumed that root length density decreases exponentially with increasing depth. The shape factor for the exponential decrease was fitted to the measurements of root length density in the data set [11].

Nitrogen uptake and distribution

Crop N demand is based on the concept of N deficiency. If the N content of a given plant part is below its maximum value corresponding to the current stage of development, a sink for N exists. Maximum values of N content were assessed from N_3-treatments in the data set [11], that is those receiving the highest rate of N fertilizer. Actual N uptake proceeds at a maximum rate until crop demand is satisfied.

Once taken up, N is distributed over stems, leaves and roots in proportion to the relative demands of these organs. Nitrogen taken up after anthesis is assumed to be reduced and stored first in vegetative tissue. Subsequently it may be translocated to grains, which lowers both the N content and photosynthetic capacity of vegetative tissue. All N is available for translocation except the N which is incorporated in structural cell material. The time constant for N translocation is assumed to be equal to that for protein decomposition, i.e. in the order of 10 days [19].

The N requirement of grains is characterized by a variety-specific rate of N accumulation in single grains, which is a function of ambient air temperature [22], multiplied by the total grain number.

N translocation results in senescence of vegetative plant parts, and the rate of senescence is proportional to the rate of N depletion.

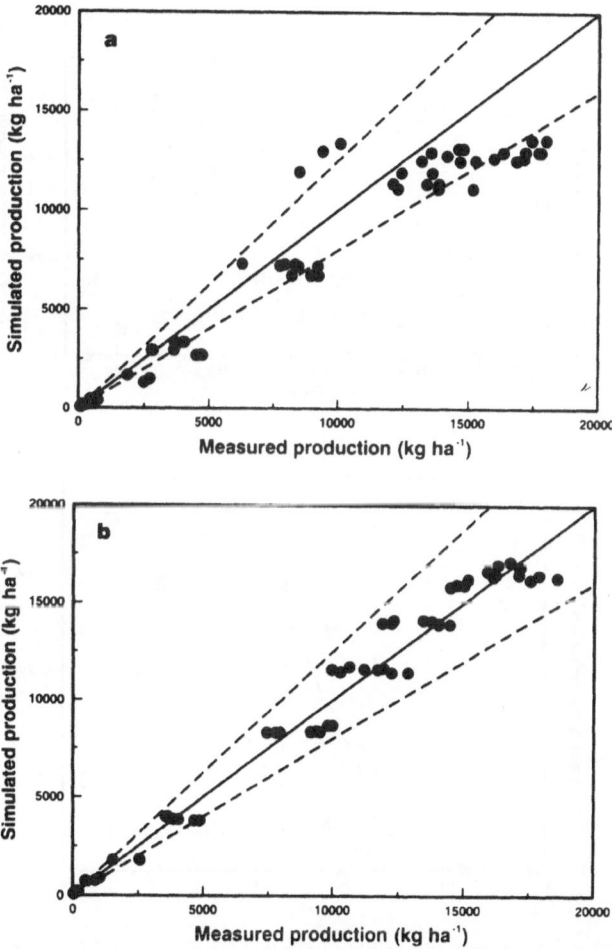

Fig. 2. Comparison of measured and simulated total dry matter production for all locations and N treatments included in the data set [11] for 1983 (a) and for 1984 (b). Broken lines indicate a 20% deviation from the 1:1 line (solid line).

Simulations

Simulations were done for winter wheat experiments in 1982/1983 and 1983/1984 [11]. The experiments were done on three different locations (the Bouwing, the Eest and PAGV), and each experiment comprised three N fertilizer treatments.

Dry matter production, leaf area development and N uptake

In 1983, for each of the locations and for all N treatments, dry matter production was generally strongly underestimated by the model (Fig. 2a), while in 1984 simulations were in agreement with the measurements (Fig. 2b).

The major reason for poor results in 1983 was that the model underestimated leaf area development, resulting in too low a dry matter production (example: Fig. 3a). In 1984, the leaf area index was calculated accurately, resulting in a proper estimate of dry matter production (example: Fig. 3b). N uptake was also too low in 1983 because N demand is proportional to dry matter production (Fig. 4a). In 1984, proper simulation of dry matter production resulted in a more accurate simulation of N uptake (Fig. 4b).

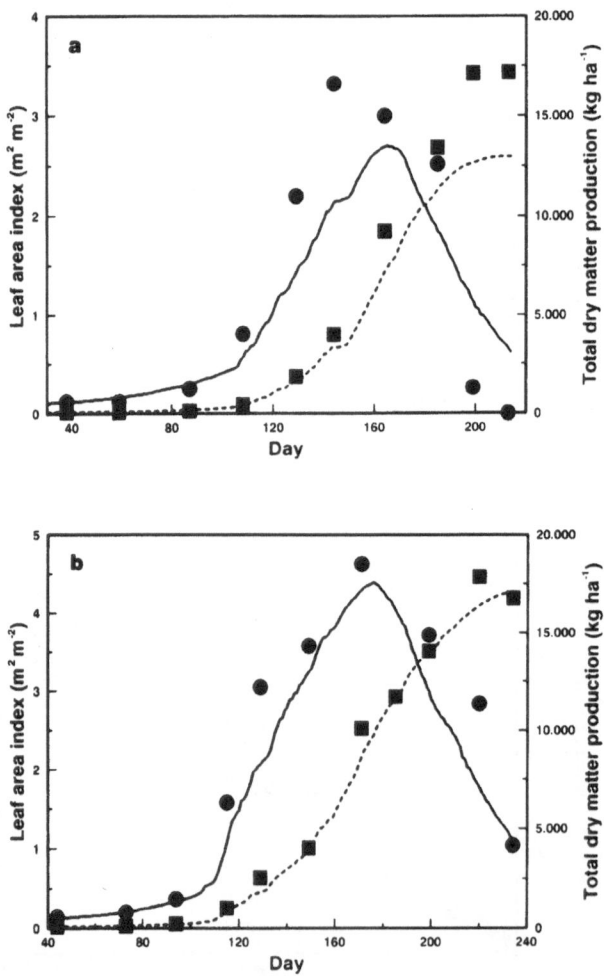

Fig. 3. Time course of leaf area index and total dry matter production for the experiment Bouwing 1983 treatment N3 (a) and for the experiment Eest treatment N2 (b). ●: measurements of leaf area index, ■: measurements of total dry matter production, ——: simulation of leaf area index, ---: simulation of total dry matter production.

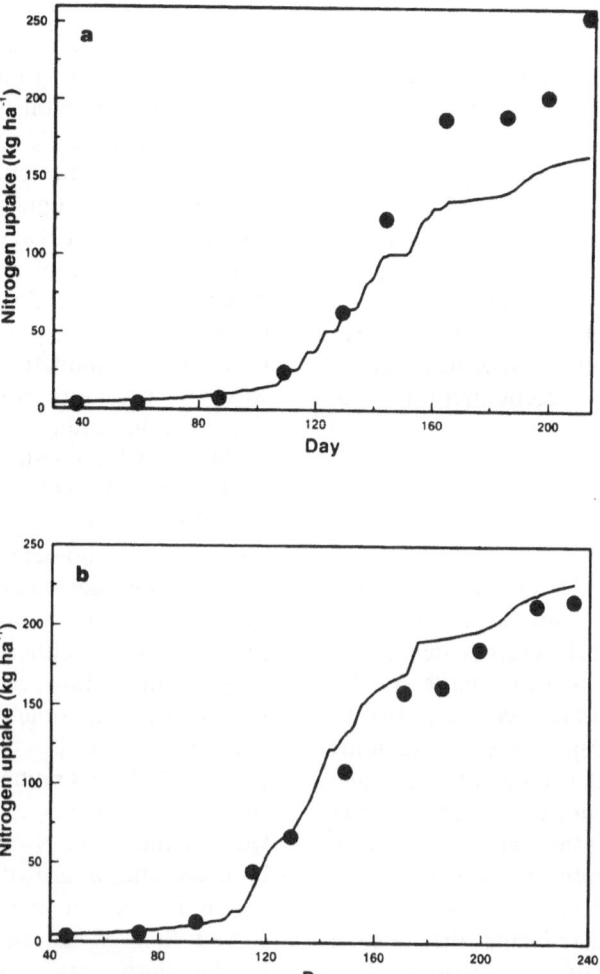

Fig. 4. Time course of N uptake for the experiment The Bouwing 1983, treatment N3 (a) and for the experiment Eest 1984, treatment N2 (b). ●: measurements of N uptake, ——: simulation of N uptake.

To compare measurements and simulations of soil mineral N, the time course of crop N uptake should be simulated accurately. Therefore in the remainder of this paper only 1984 results are considered.

Soil mineral N and crop N uptake

Soil mineral N was not measured at sowing. The initial amounts of N at sowing were chosen so that simulated values at the first sampling date agreed with the measurements. For the experiment PAGV 1984 treatment N1, in which only a single N application was given early in the grow-

ing season, the measured and simulated dynamics of the total amount of soil mineral N in the 0–100 cm layer and the measured and simulated crop N uptake during the growing season are in good agreement (Fig. 5a). Fig. 6a shows also that the distribution of mineral N in the soil profile throughout the season was simulated accurately. In treatment N3, however, in which more fertilizer N was given later in the season, the simulated sharp increase in total soil mineral N was not reflected in the measurements (Fig. 5b). According to Fig. 6b the amount of mineral N is overestimated mainly for the layers 0–20 and 20–40 cm.

Soil moisture content

The simple soil moisture module simulated the relative fluctuations in soil moisture content for the layers 0–20 and 20–40 cm reasonably well (Fig. 7), but throughout the whole growing season the absolute value of simulated moisture content was too high. According to the model, the soil dries out in the deeper soil compartments (40–60, 60–80 and 80–100 cm, Fig. 7), but this drying is not at all reflected in the measurements. Presumably, capillary moisture movement occurred, which is not accounted for in the model.

Discussion

To simulate soil N dynamics in a crop-soil system, it is essential that N uptake is simulated accurately. In our approach, crop N demand is related to the simulated total dry matter of the standing crop, which in itself is very sensitive to the accuracy of leaf area index calculations. Simulation of leaf area development, especially for winter crops, is difficult, due to the interactions between leaf extension and leaf photosynthesis. At low temperatures leaf extension is limited, and due to the absence of sinks carbohydrates may accumulate, reducing the rate of photosynthesis, and thus dry matter production. To include these effects in a crop growth model, a detailed description of leaf emergence and areal growth of individual leaves is required. However, we doubt if such an approach will improve the accuracy of simulation of leaf area index, as it requires extra parameters and thus increases the uncertainty of the model outcome. One might as well adopt a much more simplified description of N uptake. Whitmore and Addiscott [32] fitted the N uptake using a site-specific maximum N uptake capacity, that is modified for the effects of temperature and mineral N availability during the growing season.

Soil mineral N and N uptake during the growing season could be simulated accurately for the 1984 experiments in which nitrogen was applied early in the season, as illustrated for PAGV 1984 treatment N1 (Fig. 5a). In treatments where nitrogen was applied late in spring, the amount of soil mineral N after the application was overestimated by the model (Fig. 5b). It could be argued that the long intervals between measurements of soil mineral N preclude detection of a sudden increase followed by a steep decline due to N uptake by the crop. N was applied during periods in which uptake rates were high (about $3 \, kg \, ha^{-1} \, day^{-1}$ between day 113 and 147, Fig. 5b), but this cannot explain the soil mineral N dynamics observed. In Fig. 5b only one example of this 'loss' of N is presented, but in the data set used for the simulations [11], this phenomenon was observed more often after fertilizer applications late in spring. In a series of fertilizer experiments with potatoes in which nitrogen was applied as NH_3NO_4, Neeteson et al. [15] observed that up to 80% of the N applied in April 'disappeared' immediately, but this fraction progressively decreased, and five weeks after application virtually all the N applied could be accounted for in either the soil or the crop. This reappearance, however, did not occur in the PAGV 1984 experiments (Fig. 5b).

Nitrogen can be immobilized by microbial growth, but to immobilize $100 \, kg(N) \, ha^{-1}$ approximately 3000 kg carbon per ha is required if we assume a C/N-ratio of 10 for microbial biomass and a growth efficiency of 30%. In recent incubation experiments it has been observed that N can be immobilized almost instantaneously after application, without increased metabolic activity of the microbial biomass measured as CO_2-release (J. Hassink, pers. comm.). Nitrogen immobilization through a change in C/N-ratio of microbial biomass might be a cause of the 'disappearance'.

Another possible explanation for the loss of fertilizer N in Fig. 5b was given by Wehrman and Coldewey-Zum Eschendorf [31], who applied ammonium fertilizer to rape plants grown in pots filled with a silty loam. Near the roots a concentration gradient of recently fixed NH_4^+ was measured, indicating that recently fixed NH_4^+ was taken up. Recently fixed NH_4^+ will not be extracted with 1 M KCl, the extraction agent generally used to determine soil mineral N.

Soil moisture content at field capacity was generally overestimated by the model (Fig. 7). Moisture content at field capacity was defined as the moisture content at a moisture tension of

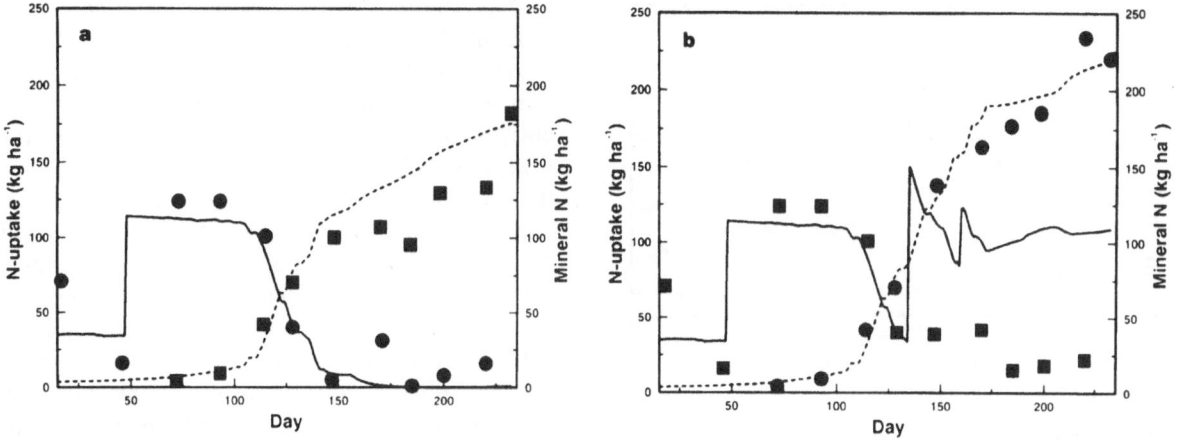

Fig. 5. Time course of soil mineral N in the 0–100 cm layer and crop N uptake for the experiment PAGV 1984, treatment N1 (a) and treatment N3 (b). ●: measurements of soil mineral N, ■: measurements of crop N uptake, ——: simulation of soil mineral N, – – –: simulation of crop N uptake.

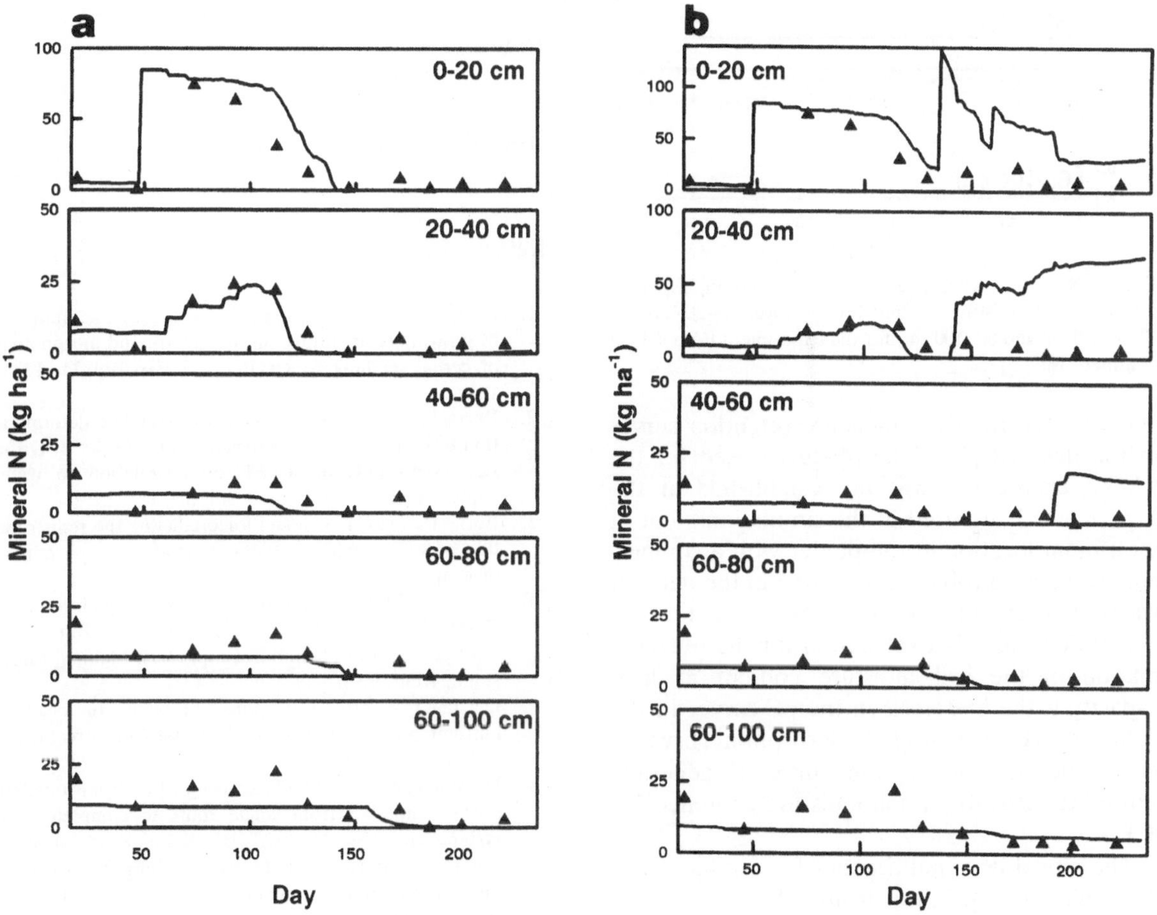

Fig. 6. Simulated (lines) and measured (triangles) time course of soil mineral N for the layers 0–20, 20–40, 40–60, 60–80 and 80–100 cm, for the experiment PAGV 1984 treatment N1 (a) and treatment N3 (b).

270

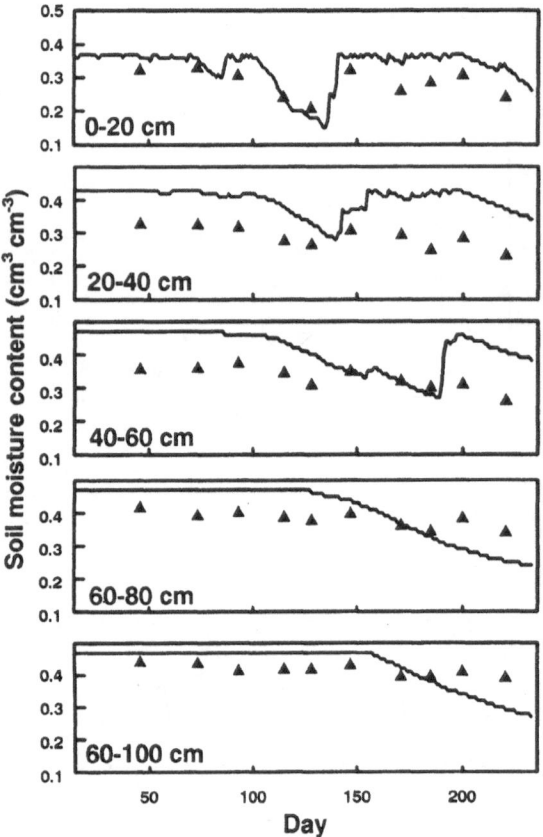

Fig. 7. Simulated (lines) and measured (triangles) time course of soil moisture content for the layers 0–20, 20–40, 40–60, 60–80 and 80–100 cm, for the experiment PAGV 1984 treatment N3.

100 cm, read from the moisture retention curves in the data set [11]. The pF-curve used for the PAGV experiment was not established at the experimental site used in the simulations, but at a different location in the polder with a comparable soil type. Probably, the curve in the data set was not representative of the experimental site.

Another possible explanation for the overestimation of the soil moisture content at field capacity is the hysteresis in the pF-curve. In the data set [11], a so-called 'desorption curve' is given, but the 'adsorption curve' is generally shifted towards lower soil moisture contents [27, 33].

The model does not account for capillary rise. According to Fig. 7, this is unrealistic for polder locations with a fluctuating water table.

Fig. 5 illustrates that, in spite of the inaccuracy of simulated soil moisture contents, the uptake

of N is simulated satisfactorily. Calculations with a detailed model for nitrogen uptake in relation to soil moisture content and nitrogen distribution, showed that as long as the soil was sufficiently moist, and roots were present, there was little physical limitation to N uptake [10]. Thus the accuracy of the simulated distribution of mineral N and the accuracy of simulations of soil moisture content are only of minor importance for accurate simulations of nitrogen uptake under the growing conditions as presented in the data set [11]. However, to calculate leaching losses of nitrate from the rooting zone, not only an accurate calculation of nitrogen uptake but also a proper simulation of mineral nitrogen distribution and soil moisture content are necessary.

Note

A listing of the simulation model is available from the authors upon request.

References

1. Abbas Al-Ani MK and Hay RKM (1983) The influence of growing temperature on the growth and morphology of cereal seedling root systems. J Exp Bot 34: 1720–1730
2. Barraclough PB and Tinker PB (1981) The determination of ionic diffusion coefficients in field soils. I. Diffusion coefficients in sieved soil in relation to water content and bulk density. J Soil Sci 32: 225–236
3. Burns IG (1974) A model for predicting the redistribution of soils applied to fallow soils after excess rainfall or evaporation. J Soil Sci 25: 165–178
4. Cooper AJ (1973) Root temperature and plant growth. Research Review No. 4, Commonwealth Bureau of Horticulture and Plantation Crops, East Malling, Maidstone, Kent, 73 p
5. De Jager A (1985) Response of plants to localized nutrient supply. PhD thesis, University of Utrecht, the Netherlands, 137 p
6. De Willigen P and Van Noordwijk M (1990) Modelling nutrient uptake: from single roots to complete root systems. In: Van de Broek BJ, Kabat P, Marshall B, Vos J and Van Keulen H (eds.) Modelling the growth of the potato crop. Simulation Monographs, PUDOC, Wageningen. In press
7. Frissel MJ and Van Veen JA (1980) Simulation of nitrogen behavior of soil-plant systems. Papers of a workshop. PUDOC, Wageningen. 249 p

8. Gregory PJ, McGowan M, Biscoe PV and Hunter B (1978) Water relations of winter wheat. J Agric Sci (Camb) 91: 91–102

9. Groot JJR (1987) Simulation of nitrogen balance in a system of winter wheat and soil. Simulation Reports CABO-TT no. 13, Centre for Agrobiological Research and Department of Theoretical Production Ecology, Wageningen, 69 p

10. Groot JJR and Van Keulen H (1990) Prospects for improvement of nitrogen fertilizer recommendations for cereals: A simulation study. In: Van Beusichem ML (ed.), Plant nutrition – physiology and applications, pp 685–692

11. Groot JJR and Verberne ELJ (1991) Response of wheat to nitrogen fertilization, a data set to validate simulation models for nitrogen dynamics in crop and soil. Fert Res 27: 349–383

12. Jenkinson DS and Rayner JH (1977) The turnover of soil organic matter in some of the Rothamsted classical experiments. Soil Sci 123: 298–305

13. Johnsson H, Bergström L, Jansson P-E and Paustian K (1987) Simulated nitrogen dynamics and losses in a layered agricultural soil. Agric Ecosys Environ 18: 333–356

14. Kortleven J (1963) Kwantitatieve aspecten van humusopbouw en humusafbraak. Versl Landbouwk Onderz nr. 69.1, PUDOC, Wageningen. 109 p

15. Neeteson JJ, Greenwood DJ and Habets EJMH (1986) Dependence of soil mineral N on N-fertilizer application. Plant Soil 91: 417–420

16. Penman HL (1956) Evaporation: an introductory survey. Neth J Agric Sci 4: 9–29

17. Penning de Vries FWT (1974) Substrate utilization and respiration in relation to growth and maintenance in higher plants. Neth J Agric Sci 2: 40–44

18. Penning de Vries FWT (1975) The cost of maintenance processes in plant cells. Ann Bot 39: 77–92

19. Penning de Vries FWT (1975) Use of assimilates in higher plants. In: Photosynthesis and productivity in different environments. International Biological Programme, No. 3, pp 459–480

20. Reinink K, Jorritsma I and Darwinkel A (1986) Adaptation of the AFRC wheat phenology model for Dutch conditions. Neth J Agric Sci 34: 1–13

21. Sofield I, Evans LT, Cook MG and Wardlaw IF (1977) Factors influencing the rate and duration of grain filling in wheat. Aust J Plant Physiol 4: 785–797

22. Sofield I, Wardlaw IF, Evans LT and Zee SY (1977) Nitrogen, phosphorus and water contents during grain development and maturation in wheat. Aust J Plant Physiol 4: 799–810

23. Spitters CJT, Van Keulen H and Van Kraalingen DWG (1989) A simple and universal crop growth simulator: SUCROS87. In: Rabbinge R, Ward SA and Van Laar HH (ed.) Simulation and Systems Management in Crop Production. pp 147–181. Simulation Monographs, PUDOC, Wageningen

24. Stapper M (1984) SIMTAG: A Simulation Model of Wheat Genotypes. Model Documentation. University of New England, Armidale, Australia and International Center for Agricultural Research in the Dry Areas, Aleppo, Syria. 108 p

25. Van der Linden AMA, Van Veen JA and Frissel MJ (1987) Modelling soil organic matter levels after long-term applications of crop residues, and farmyard and green manures. Plant Soil 101: 21–28

26. Van Keulen H and Seligman NG (1987) Simulation of water use, nitrogen nutrition and growth of a spring wheat crop. Simulation Mongraphs, PUDOC, Wageningen, 310 p

27. Van Vuuren WE (1984) Validation of the agrohydrological model DEMGEN (Demand Generator) on point data from the Hupselse Beek area in the Netherlands. In: Udluft P et al. Proc Int Symp on recent investigations in the zone of aeration, Munich, W Germany. October 1984. pp 829–839. Technical University Munich

28. Verberne ELJ, Hassink J, De Willigen P, Groot JJR and Van Veen JA (1990) Modelling organic matter dynamics in different soils. Neth J Agric Sci 38: 221–238

29. Verbruggen J (1985) Simulatie van het denitrificatieproces in de bodem. Doctoraatproefschrift nr. 140, Faculteit Landbouwwetenschappen, K.U. Leuven. 180 p

30. Vos J (1981) Effect of temperature and nitrogen supply on post anthesis growth of wheat: measurements and simulations. Agric Res Rep 811, PUDOC, Wageningen, 164 p

31. Wehrman J and Coldewey-Zum Eschenhoff (1986) Distribution of nitrate, exchangeable and non-exchangeable ammonium in the soil-root interface. In: Lambers H, Neeteson JJ and Stulen I (eds.), Fundamental, ecological and agricultural aspects of nitrogen metabolism in higher plants. pp 447–450. Martinus Nijhoff Publishers, Dordrecht, The Netherlands

32. Whitmore AP and Addiscott TM (1987) A function for describing nitrogen uptake, dry matter production and rooting by wheat crops. Plant Soil 101: 51–60

33. Wösten JHM, Schuren CHJE, Bouma J and Stein A (1990) Functional sensitivity analysis of four methods to generate soil hydraulic functions. Soil Sci Soc Am J 54: 832–836

Fertilizer Research **27**: 273–281, 1991.
© 1991 *Kluwer Academic Publishers.*

Modelling nitrogen dynamics in a plant-soil system with a simple model for advisory purposes

K.C. Kersebaum & J. Richter
Institute of Geography & Geoecology, Technical University of Braunschweig, Langer Kamp 19c, DW-3300 Braunschweig, Germany

Key words: N model, nitrate transport, mineralization, wheat growth, microbial immobilization

Abstract

A simple functional computer model for advisory purposes is described. Results of simulation indicate some limitations of the model especially in handling the water regime in soils with fluctuating water tables. A major problem seems to be the 'disappearance' of fertilizer N. Measurements by the fumigation-extraction method of microbial N during the growing season show that disappearance of fertilizer N can partly be explained by immobilization by the microbial biomass.

Introduction

Ecological as well as economical reasons are driving farmers to a more efficient use of nitrogen fertilizers in plant production. To optimize fertilization it is necessary to adapt the nitrogen supply to the nitrogen demand of the plants taking into account the N dynamics of the soil. Measurements of mineral N content in the soil are expensive, require much labour and they give only a random indication of the N supply in the soil. Simulation models which are able to describe nitrogen transformations and interactions in the soil-plant system are a useful tool considering the dynamic aspect of the nitrogen behaviour.

Models for practical advisory purposes have to be so simple that their input requirements can be fulfilled by the information available from the farmers. Our aim was to create a model which can be used in combination with commercial field data bases installed at the farmers' Personal Computer. In a first step we tried to simulate the N dynamics from harvest to the following spring [28]. Later the model was extended to simulate the whole growing season of winter wheat [14].

The model was applied, without any recalibration, to simulate an existing data set [9].

Description of the model

A scheme of our model is shown in Fig. 1. The model takes into account the processes of net mineralization, transport of water and nitrate as well as growth and N uptake by winter wheat. Nitrogen immission is assumed to be $20 \, kg(N) \, ha^{-1} \, yr^{-1}$ and is distributed uniformly over the year. Losses by volatilization are set at 10–15% of the ammonia or urea in mineral fertilizer. The model has been described in detail by Kersebaum [14].

Submodel water balance

We use a simple plate theory model to describe the water balance of the soil. Therefore the soil profile has been divided into 10 layers of 10 cm thickness. Compared to recently published versions of the model [13] using the potential concept to simulate Darcy flux, this simple approach has the advantage that parameters can be easily

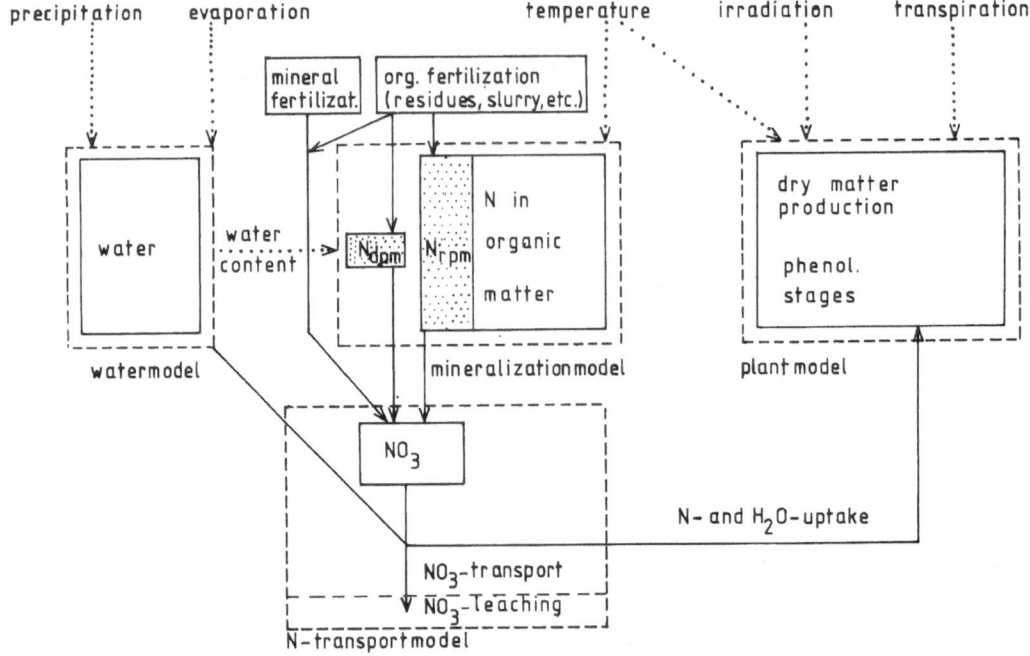

Fig. 1. Scheme of the model.

derived from soil texture classes. Additionally, the capacity parameters used in this type of model show much less variability in the field than intensity parameters such as hydraulic conductivity. Values for the water content at field capacity (pF 1.8) and wilting point (pF 4.2) are taken from literature [1].

Potential evapotranspiration (PET) is calculated by a simple empirical method developed by Haude [10] for the climatic conditions of Germany using plant specific coefficients from Heger [11]. This method requires only the daily vapour pressure deficit as input data which can be derived from temperature and relative humidity. Potential evaporation can be calculated using a relationship of Goudriaan [7] depending on the leaf area index. Actual evaporation depends on the water content of the upper layer [17] and is distributed over depth according to van Keulen [15]. Water uptake by plant roots is driven by the potential transpiration which is distributed over depth taking into account the root length density and a soil-moisture-dependent root effectivity factor [8]. Potential transpiration of each layer is reduced if the water supply is not sufficient. The deficit between potential and actual water uptake

is then distributed to deeper rooted layers according to the above mentioned relationship. So the plants can compensate water stress in upper layers until also the deeper layers become dry.

Transport of nitrate

Fluxes between layers calculated by the water model are used to simulate transport of nitrate with the convection-dispersion equation:

$$\frac{\partial \Theta c}{\partial t} = \frac{\partial}{\partial z}\left(\Theta D \, \frac{\partial c}{\partial z}\right) - \frac{\partial qc}{\partial z} \qquad (1)$$

with the water content Θ (cm^3 cm^{-3}), the nitrate concentration in soil solution c (mg cm^{-3}), the water flux q (cm day^{-1}) and the dispersion coefficient D which is a combination of the diffusion and hydrodynamic dispersion:

$$D = \tau^{-1}D_0 + D_v\left|\frac{q}{\Theta}\right| \qquad (2)$$

where $D_0(= 2.14 \, \text{cm}^2 \, \text{day}^{-1})$ is the diffusion coefficient for nitrate in water, τ is the tortuosity and $D_v(= 25 \, \text{cm})$ is the dispersion factor.

Mineralization of nitrogen

Simulation of mineralization includes nitrification in one step and is based on first order kinetics dividing the potential mineralizable nitrogen into 2 pools with different decay coefficients: a relatively small fraction (N_{dpm}) of fast and easily decomposable organic compounds from plant residues and a larger fraction (N_{rpm}) of slowly decomposable organic matter in the soil [23, 26]. The amount of N_{dpm} can be derived from the yield using empirical yield to residue relationships. Mean N contents are used for different plant residues and also the percentage of the easily decomposable nitrogen compounds depends on the previous crop. The slowly decomposable N_{rpm} fraction is derived from the total nitrogen content in the soil and can be assumed as about 13% of N_{total} [24].

The mineralization coefficients k_{rpm} and k_{dpm} of the two fractions are mainly dependent on temperature and soil moisture. The effect of temperature is taken into account by Arrhenius functions derived from incubation experiments [23]:

$$k_{rpm}(T) = 4.0 \cdot 10^9 \, e^{-8400/(T+273)} \qquad (3)$$

$$k_{dpm}(T) = 5.6 \cdot 10^{12} \, e^{-9800/(T+273)} \qquad (4)$$

The influence of soil moisture on mineralization is described by a reduction function according to Myers et al. [20]. The depth of mineralization is limited to the upper 30 cm.

N uptake by plants

N uptake of the plants is assumed to be driven by the difference between actual and a maximum N content in the plant. Maximum N content depends on the phenological stage of the plants. Transport of nitrogen is calculated by mass flow with the amount of water taken up in each layer. If convective transport does not fulfil the potential demand of the plants the maximum amount of nitrate which can be delivered by diffusion (N_{max}) is calculated according to Baldwin et al. [2]:

$$N_{max} = \sum_{i=1}^{ZR} N_{max_i} = \sum_{i=1}^{ZR} 2\pi r_r RD_i D \, \frac{(c_i - c_{min})}{r_i}$$
$$\times \Delta z \, \Delta t \qquad (5)$$

with ZR the rooting depth, r_r the root radius, RD_i the root length density and c_{min} the minimum concentration of nitrate at the root surface. Assuming a homogenous distribution of the roots $r_i = (\pi RD_i)^{-0.5}$ is half of the mean distance between roots. The dependence of the diffusion coefficient on water content is taken into account by an empirical equation of Olsen & Kemper [25] for the tortuosity. Dispersion is neglected in this case.

Growth of winter wheat

The submodel for plant growth is based on the 'SUCROS' model of van Keulen et al. [16]. The daily dry matter production by photosynthesis is calculated from global radiation and temperature. Temperature response to photosynthesis is taken from Groot [8]. It is assumed that dry matter production is reduced in the same proportion as transpiration of the plants during dry periods. Nitrogen stress is considered if the nitrogen content in the aboveground plant falls below a critical value depending on the stage of plant development. This will cause a reduction of dry matter production. Critical values are derived from literature [30].

The calculation of crop development is based on the model described by Weir et al. [31] considering temperature, photoperiod and vernalization. The phenological effective thermal time resulting from this calculation is used to partition the assimilates to different plant organs [8, 31].

Dry matter production of the roots is distributed empirically over depth according to Gerwitz & Page [6]. Time dependence of rooting depth and distribution is considered by an empirical function of Whitmore & Addiscott [32] using the thermal time from sowing date. To calculate root length we assumed 7% of dry matter in roots and a mean radius of the roots of 0.015 cm [3].

276

Results and discussion

The following figures will show some examples of simulations from the given data set [9]. Simulation always starts at harvest of the previous crop and only the mineral N content in the soil ($NO_3 + NH_4 = N_{min}$) was reinitialized at the date of first measurement.

Fig. 2 shows the comparison of measured and

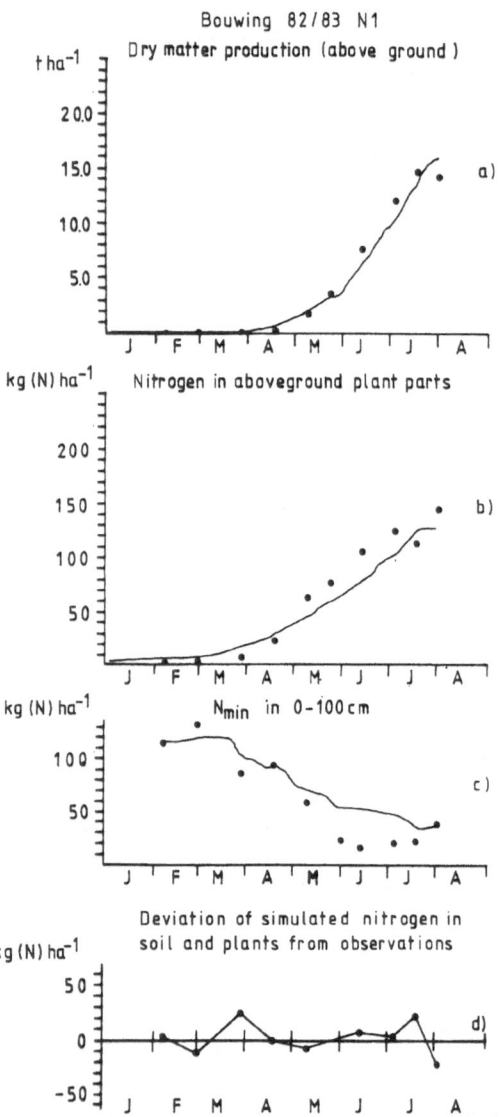

Fig. 2. Measured (dots) and simulated (lines) courses of (a) dry matter production, (b) N uptake, (c) soil mineral nitrogen in 0–100 cm and (d) deviation of the model from observed nitrogen in plants and soil together for experiment Bouwing N1 in 1983.

simulated values for the field Bouwing in 1983 without any nitrogen fertilization (N1). Fig. 2a shows the course of the aboveground dry matter of winter wheat. The simulated curve and the observed values are in good agreement. The simulated N uptake in Fig. 2b underestimates the observed values during the last 3 months. This is reflected by a corresponding overestimation of mineral nitrogen in the soil. Comparing observed and simulated nitrogen available for plant uptake by summing mineral N in soil and nitrogen already taken up by the plants, the model in most of the time during the growing period shows good agreement with the measurements. The total deviation between simulation and observations as shown in Fig. 2d is mostly within a range of $\pm 20\,kg(N)\,ha^{-1}$ which is assumed to be the mean confidence level of the nitrogen estimation in the field [27]. Looking at the N_{min} distribution in the profiles shown as a time sequence in Fig. 3 it becomes clear that the underestimation of plant uptake occurs mostly in the lower parts of the profile. The reason might be a distinct underestimation of root length density by the model especially in the lower parts of the profile. This can be pointed out comparing simulated to measured root length density in other years and sites. The N2-plot of Bouwing (Fig. 4) was fertilized with $60\,kg(N)\,ha^{-1}$ during May. Most of this nitrogen must have been taken up immediately by the plants because there is no difference to be observed in the soil mineral N between the N1 and the N2 plot. The agreement between observed and simulated nitrogen supply is similar to the unfertilized field.

The amount of N_{min} in Bouwing soil in February 1984 was relatively low. Therefore even the N1 experiment was fertilized with $70\,kg(N)\,ha^{-1}$. Although simulated dry matter production at Bouwing falls below the measured one on the N1 plot in 1984 the simulated N uptake as well as the mineral N content is in an acceptable agreement with the observations over a wide period. This is due to a relatively low proportion of N_{min} in the lower part of the profile. The simulation of the total N supply of the plants again is in good agreement with the measurements.

At the experimental site Eest the model shows its limitations on soils with fluctuating

277

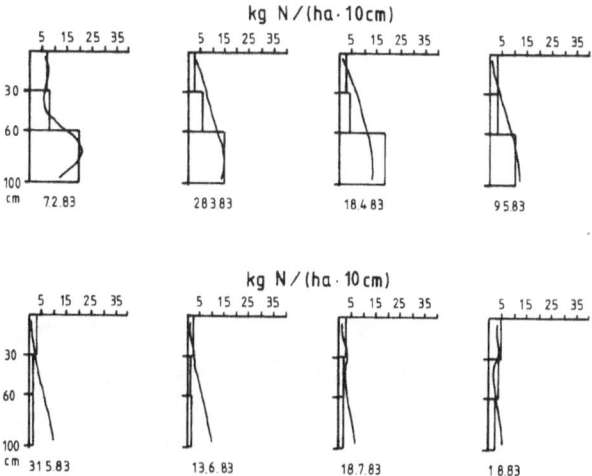

kg N/(ha·10cm)

kg N/(ha·10cm)

Fig. 3. Measured (bars) and simulated (curves) distribution of mineral nitrogen in the soil profile of Bouwing N1 plot at different times in 1983.

groundwater table. Groundwater level varied in 1984 from 59 to 150 cm below groundsurface and volumetric water contents in 1983 indicate fluctuations within the investigated profiles also for 1983 (Fig. 6c). As an example Fig. 5 shows simulated and measured values for the N2 experiment in Eest in 1983. Whereas simulations of dry matter production and N uptake are in good agreement with the observations up to mid June, mineral N content in the soil profile is distinctly overestimated by the model. One reason is that field capacity which is derived from the texture is inaccurate especially when the groundwater level changes within the profile. Another point is that simulated mineral nitrogen is reinitialized at the first measurement when groundwater level seems to be high or the water has not been equilibrated as assumed in the model. So in reality drainage and nitrogen leaching might have taken place after the measurement whereas the transport of nitrate is already considered in the simulation before reinitialisation. The sharp decrease of N_{min} in the lower part of the profile in April coupled to a decrease of water content in the same layer indicates that a high amount of nitrate has been leached out of the profile by draining groundwater (Fig. 6c). Fig. 7 demonstrates the 'disappearance' of fertilizer nitrogen applied in February 1983 at the experimental site PAGV N2. While the model indicates an increase of soil mineral nitrogen, in reality, a

corresponding increase can be observed neither in soil nor in plants. Looking at the deviation in N supply between simulation and observation in Fig. 7d, it can be pointed out that the disappeared fertilizer reappears successively during a period of about 2 or 3 months. Also the second dressing can not be detected by the measurements. The same phenomenon can be observed on the other plots receiving different fertilizer dressings. This temporary disappearance is similar to observations reported by Neeteson et al. [21] and Nielsen & Jensen [22]. The reappearance of fertilizer nitrogen indicates that a reversible process must be responsible for this phenomenon. In our opinion the 'disappearance' of fertilizer nitrogen is due at least partly to microbial immobilization which is not included in our model. This argument is supported by Fig. 8 which shows the deviation of simulation and measurements for the nitrogen in soil and plants for a field on loess soil in Germany and the additionally measured nitrogen in microbial biomass during the growing season of winter wheat [14]. Nitrogen in microbial biomass as determined by the fumigation-extraction method [4] increased especially during May and June and decreases again during July. The fluctuation within the observed period was about 80 kg (N) ha^{-1}. Similar courses in plant covered soils are reported by other authors [5, 18, 29]. Probably C released from plant roots may cause

278

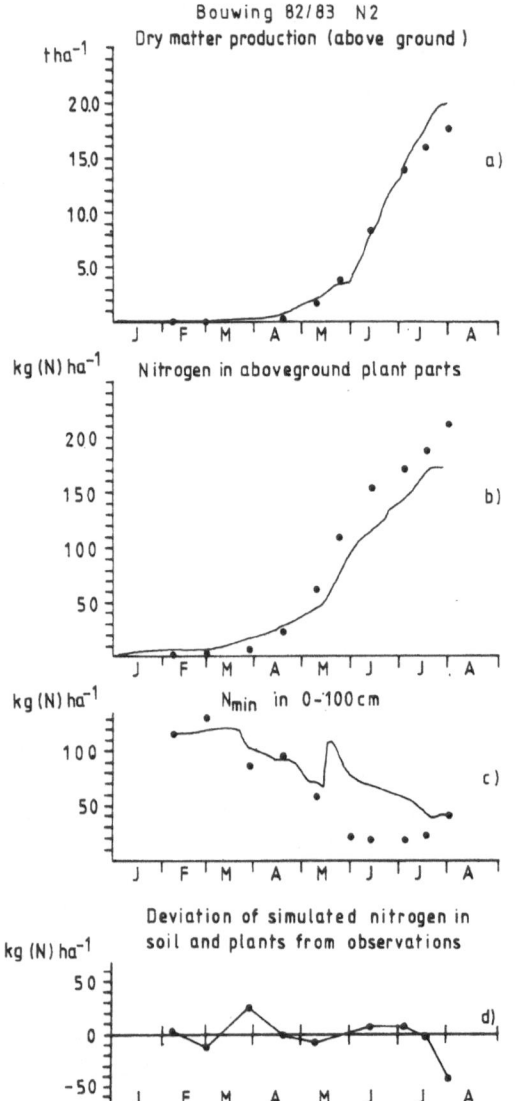

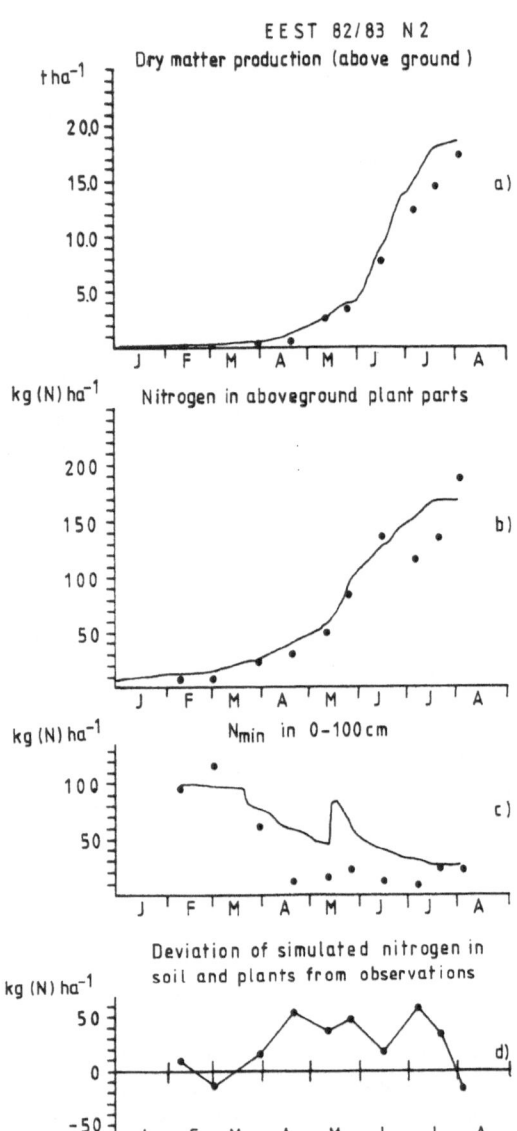

Fig. 4. Measured (dots) and simulated (lines) courses of (a) dry matter production, (b) N uptake, (c) soil mineral nitrogen in 0–100 cm and (d) deviation of the model from observed nitrogen in plants and soil together for experiment Bouwing N2 in 1983.

Fig. 5. Measured (dots) and simulated (lines) courses of (a) dry matter production, (b) N uptake, (c) soil mineral nitrogen in 0–100 cm and (d) deviation of the model from observed nitrogen and plants and soil together for experiment Eest N2 in 1983.

the increase in microbial biomass during the main growing period. Investigations with labelled CO_2 indicate amounts of 2–5% of gross assimilated carbon which remained in the soil after translocation below ground and after respiration by roots and soil organisms [12, 19]. Assuming a C:N ratio of 6 of the microbial biomass

this may be responsible for an immobilization of about 30 to 75 kg (N) ha^{-1} within the growing period. In the case of the early disappearance of nitrogen fertilizer at PAGV in 1983 other processes such as N fixation to clay minerals seems to be more probable, because carbon assimilation is too low.

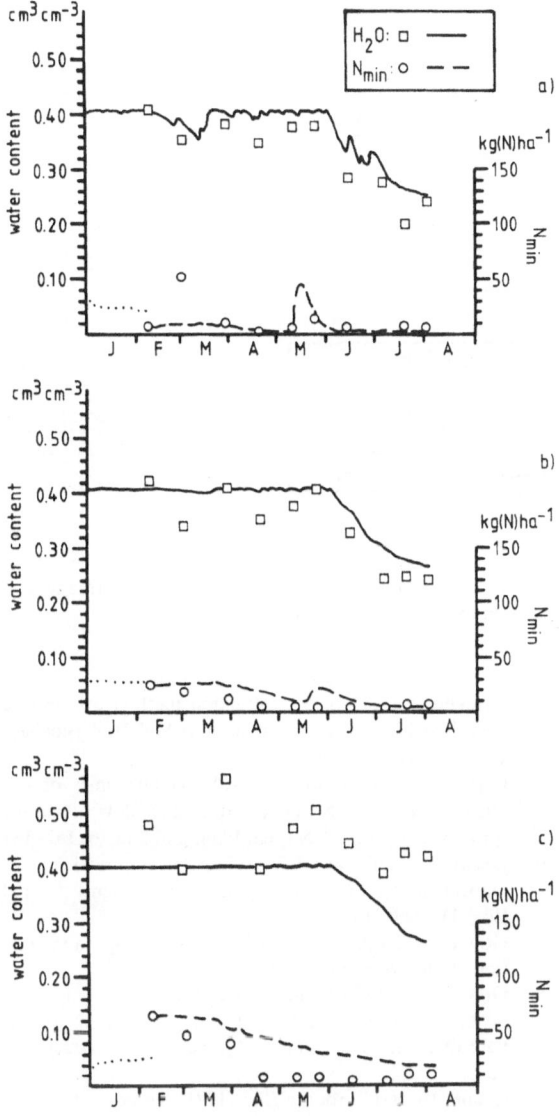

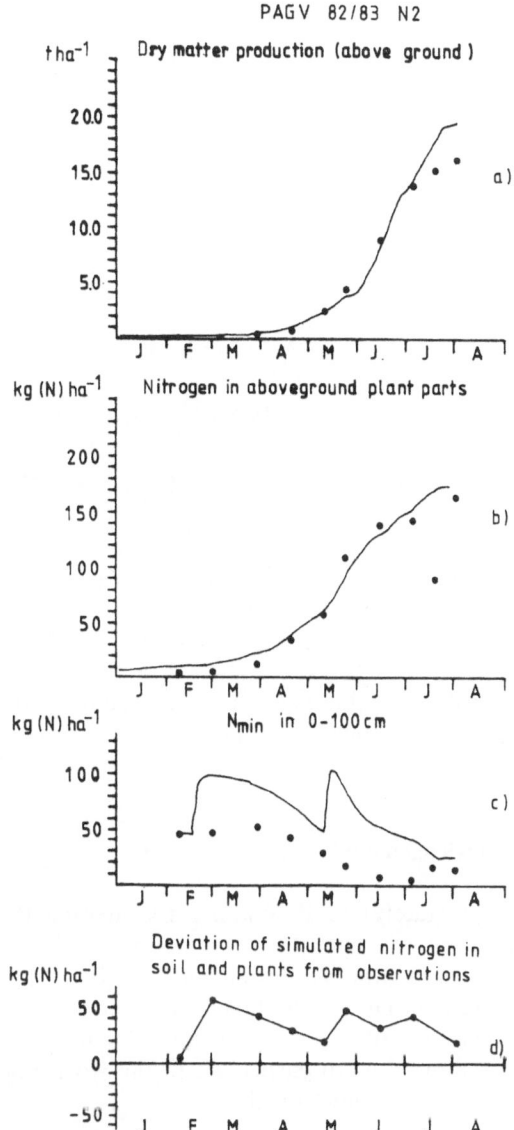

Fig. 6. Measured (symbols) and simulated (lines) courses of volumetric water content and mineral nitrogen in (a) 0–30 cm depth, (b) 30–60 cm depth and (c) 60–100 cm depth for experiment Eest N2 in 1983.

Fig. 7. Measured (dots) and simulated (lines) courses of (a) dry matter production, (b) N uptake, (c) soil mineral nitrogen in 0–100 cm and (d) deviation of the model from observed nitrogen and plants and soil together for experiment PAGV N2 in 1983.

Conclusions

The results of our simulations have shown some clear limitations of the model. One is the handling of fluctuating water tables within the soil profile. Generally there are some other uncertainties as nitrogen losses by volatilization or denitrification and nitrogen deposition from the

atmosphere. The main problem in modelling nitrogen dynamics in a soil-plant system seems to be the behaviour of mineral nitrogen fertilizer with regard to the mineralization-immobilization-turnover. The reasons for temporal 'disappearance' of fertilizer nitrogen are still not clear. Whether short-term fixation of ammonia or microbial immobilization or probably both is

280

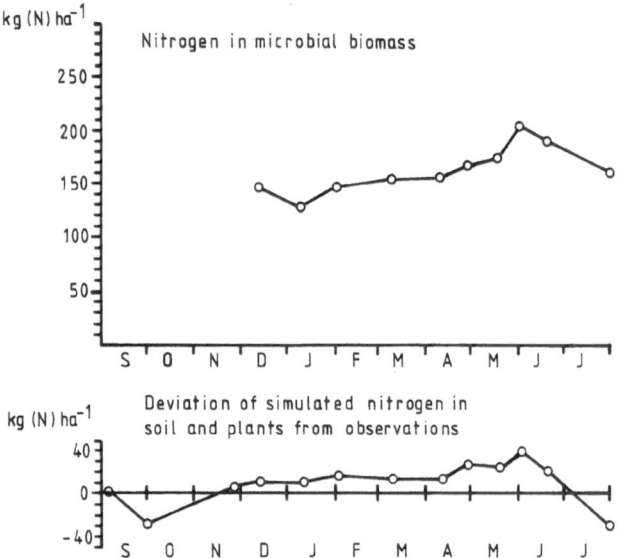

Fig. 8. Deviation of the model from observed nitrogen in plants and soil together for a loess soil in W. Germany and corresponding nitrogen in microbial biomass determined by the fumigation-extraction method [14].

responsible for this phenomenon should be the goal of further investigations.

Acknowledgements

We are grateful to P. Widmer for making the investigations by the fumigation-extraction method and Mrs. A. Freimann-Kersebaum for plotting the figures. The financial support of the Commission of the European Communities which allowed us to participate in the workshop is gratefully acknowledged.

References

1. AG Bodenkunde (1982) Bodenkundliche Kartieranleitung Stuttgart: Schweizerbart
2. Baldwin JP, Nye PH and Tinker PB (1973) Uptake of solutes by multiple root systems from soil. III. A model for calculating the solute uptake by a randomly dispersed root system developing in a finite volume of soil. Plant and Soil 38: 621–635
3. Barraclough PB (1986) The growth and activity of winter wheat roots in the field: nutrient inflows of high yielding crops. J Agric Sci Camb 106: 53–59
4. Brookes PC, Landman A, Pruden C and Jenkinson DS (1985) Chloroform fumigation and the release of soil nitrogen: a rapid direct extraction method to measure microbial biomass nitrogen in soil. Soil Biol Biochem 17: 837–842
5. Carter MR and Rennie DA (1984) Dynamics of soil microbial biomass N under zero and shallow tillage for spring wheat, using ^{15}N urea. Plant and Soil 76: 157–164
6. Gerwitz A and Page ER (1974) An empirical mathematical model to describe plant root systems. J Appl Ecol 11: 773–781
7. Goudriaan J (1977) Crop micrometeorology: a simulation study. Wageningen: Pudoc, 257 pp
8. Groot JJR (1987) Simulation of nitrogen balance in a system of winter wheat and soil. Simulation Report CABO-TT 13. Wageningen: Agricultural university, 195 pp
9. Groot JJR and Verberne ELJ (1991) Response of wheat to nitrogen fertilization, a data set to validate simulation models for nitrogen dynamics in crop and soil. Fert Res 27: 349–383
10. Haude W (1955) Zur Bestimmung der potentiellen Verdunstung auf möglichst einfache Weise. Mitt d Dt Wetterdienst 11
11. Heger K (1978) Bestimmung der potentiellen Evapotranspiration über unterschiedlichen landwirtschaftlichen Kulturen Mitteilgn Dtsch Bodenkundl. Gesellsch 26: 21–40
12. Helal HM and Sauerbeck DR (1984) Influence of plant roots on C and P metabolism in soil. Plant and Soil 76: 175–182
13. Kersebaum KC, Richter J and Utermann J (1987) Die Simulation der Stickstoff-Dynamik von Ackerböden unter Getreidevegetation. Mitteilgn Dtsch Bodenkundl Gesellsch 55/II: 613–618
14. Kersebaum KC (1989) Die Simulation der Stickstoff-

Dynamik von Ackerböden. Ph.D. Thesis University Hannover

15. Keulen H van (1975) Simulation of water use and herbage growth in arid regions. Wageningen: PUDOC, 184 pp

16. Keulen H van, Penning de Vries FWT and Drees EM (1982) A summary model for crop growth. In: Penning de Vries FWT and van Laar HH (eds.) Simulation of plant growth and crop production. pp 87–97. Wageningen: PUDOC

17. Keulen H van and Seligman NG (1987) Simulation of water use, nitrogen nutrition and growth of a spring wheat crop. Wageningen: PUDOC, 310 pp

18. Lynch JM and Panting LM (1980) Cultivation and the soil biomass. Soil Biol Biochem 12: 29–33

19. Martens R (1990) Contribution of rhizodeposits to the maintenance and growth of soil microbial biomass. Soil Biol Biochem 22: 141–147

20. Myers RJK, Campbell CA and Weier KL (1982) Quantitative relationship between net nitrogen mineralization and moisture content of soils. Can J Soil Sci 62: 111–124

21. Neeteson JJ, Greenwood DJ and Habets EJMH (1986) Dependence of soil mineral N on N-fertilizer application. Plant and Soil 91: 417–420

22. Nielsen NE and Jensen HE (1986) The course of nitrogen uptake by spring barley from soil and fertilizer nitrogen. Plant and Soil 91: 391–395

23. Nordmeyer H and Richter J (1985) Incubation experiments on nitrogen mineralization in loess and sandy soils. Plant and Soil 83: 433–445

24. Nuske A (1983) Ein Modell für die Stickstoff-Dynamik von Acker-Lößböden im Winterhalbjahr – Messungen und Simulationen. Ph. D. Thesis University Hannover

25. Olsen SR and Kemper WD (1968) Movement of nutrients to plant roots. Adv Agron 20: 91–151

26. Richter J, Nuske A, Habenicht W and Bauer J (1982) Optimized N-mineralization parameters of loess soils from incubation experiments. Plant and Soil 68: 379–388

27. Richter J, Nordmeyer H and Kersebaum KC (1984) Zur Aussagesicherheit der Nmin-Methode. Z Acker- und Pflanzenbau 153: 285–296

28. Richter J, Nordmeyer H and Kersebaum KC (1985) Modelling of the nitrogen regime in loess soils in the winter half-year: comparison between field measurements and simulations. Plant and Soil 83: 419–431

29. Ritz K and Robinson D (1988) Temporal variations in soil microbial biomass C and N under a spring barley crop. Soil Biol Biochem 20: 625–630

30. Vielemeyer HP, Neubert P, Hundt I, Vanselow G and Weissert P (1983) Ein neues Verfahren zur Ableitung von Pflanzenanalyse-Grenzwerten für die Einschätzung des Ernährungszustandes landwirtschaftlicher Kulturpflanzen. Arch Acker-Pflanzenbau Bodenkd 27: 445–453

31. Weir AH, Bragg PL, Porter JR and Rayner JH (1984) A winter wheat crop simulation model without water or nutrient limitations. J Agric Sci Camb 102: 371–382

32. Whitmore AP and Addiscott TM (1987) A function for describing nitrogen uptake, dry matter and rooting by wheat crops. Plant and Soil 101: 51–60

Fertilizer Research **27**: 283–291, 1991.
© 1991 *Kluwer Academic Publishers.*

Simulation of nitrogen in soil and winter wheat crops: modelling nitrogen turnover through organic matter

A.P. Whitmore, K.W. Coleman, N.J. Bradbury & T.M. Addiscott
AFRC Institute of Arable Crops Research, Rothamsted Experimental Station, Harpenden, Hertfordshire, AL5 2JQ, UK

Key words: Nitrogen, computer model, organic matter turnover, winter wheat

Abstract

A computer model is described that simulates leaching, organic matter turnover and nitrogen uptake by a winter wheat crop. The model is assessed against a data set from the Netherlands where winter wheat was grown in two seasons (1982–3 and 1983–4) on three different soils in two different parts of the country. The model satisfactorily simulated the growth, N uptake and production of grain. It also simulated the dynamics of indigenous soil N well but it did not always account for the fate of applied fertilizer N. Some possible reasons for this and ways of improving the model are discussed.

Introduction

Simple models based on sound mechanistic principles but none the less built from experimentally derived relationships are often quite successful in simulating spatially variable soil and crop data from the field. Simple models are not necessarily more accurate nor more realistic than detailed mechanistic models, but because they need fewer inputs to initialize and run them, they are more reliable on a field scale and more 'portable' in application from field to field. Even so, such models cannot normally predict nitrogen in the soil more accurately than to the nearest $20 \, kg \, N \, ha^{-1}$ [3, 7, 17] nor do they normally estimate grain production more accurately than to the nearest tonne ha^{-1} [22]. This paper briefly describes and evaluates a fairly simple, dynamic model of nitrogen in soil and crop constructed from empirical descriptions of the processes most relevant to arable agriculture. It also discusses the main ways in which such a model might be improved in the light of its performance in simulating a published data set [8].

The model

There are three main parts to our computer simulation model: (i) water movement (that is, percolation, leaching and evaporation), (ii) crop growth, development and nitrogen uptake, and (iii) organic matter turnover in the soil. The model has already been described elsewhere [3, 25] and so only the most important details will be described here.

Leaching

The model divides the soil into a series of uniform horizontal layers each 50 mm thick. Within each is a mobile and immobile compartment containing amounts w_m or w_r of water per unit area. Incoming rain fills the mobile compartment until it displaces w_m (and the nitrate it contains). When complete this displacement is followed by sideways flow of water between the mobile and immobile compartments. Nitrate in soil diffuses between mobile and immobile water, so we do not allow it to equilibrate between compartments in the model as rapidly as the water; the effect of diffusion on retarding nitrate movement in peds of different sizes was estimated by Addiscott et al. [2]. Evaporation takes place from both compartments, taking nitrate with it. As the soil dries this evaporative demand cannot be met entirely from the liquid phase, so that vapour movement becomes more important, reducing the amount of nitrate transported. The model uses daily rainfall and evaporation data.

Addiscott [1] assumed that half the water held more tightly than -15 bar excludes anions; the division between mobile and immobile water was made at -2 bar so that for a 50 mm layer w_m and w_r are calculated in mm as follows:

$$w_m = 50(\theta_{0.05} - \theta_2) \tag{1}$$

$$w_r = 50\left(\theta_2 - \frac{1}{2}\theta_{15}\right) \tag{2}$$

where $\theta_{0.05}$, θ_2 and θ_{15} are the percentage amounts of water (by volume) held by the soil at tensions greater than -0.05, -2 and -15 bar, respectively. The values of w_m and w_r were estimated from the pF curves in the dataset [8].

The turnover of organic matter

The model separates the soil organic matter into four active compartments: easily decomposable plant material (DPM), resistant plant material (RPM), microbial biomass (BIO), and humified organic matter (HUM). There is also an inert compartment which is completely inactive but is required to obtain the correct radiocarbon age of the soil [10]. Each compartment is given a fixed C:N ratio with the exception of the DPM which varies to reflect the N content of the crop residues. The material in all active compartments decomposes by first-order kinetics yielding biomass and humus as products. Some carbon is always respired because heterotrophic organisms dominate; inorganic N may be immobilized if the C:N ratio of the decomposing material is much wider than the organic matter produced, otherwise N is mineralized. Sørenson [18] has shown that during the decomposition of ^{14}C labelled cellulose, the ratio of $^{14}CO_2$ respired to ^{14}C remaining in the soil (as humus or biomass)

alters with the texture of the soil. Ladd et al. [12] showed that the proportions of ^{14}C respired, and of ^{15}N mineralized by soils amended with labelled plant material decreases with increasing clay content; accordingly clay content determines these proportions in the model. Rate constants were kept at the same values during all computer runs, and were established for the soil growing winter wheat on Broadbalk at Rothamsted [10, 25]. Use of these rates is intuitively reasonable if the composition and activity of the microflora and -fauna is comparable in roughly similar agricultural soils in roughly similar climates. These base rate constants are modified to the same extent for all organic matter compartments by soil temperature and soil moisture content. Full details of the fitting procedures for these constants and the setting-up of the model are given elsewhere [9, 10, 25].

Crop growth and N uptake

Crop growth is estimated using a simple function [24]:

$$Y = (A^{-(1/n)} + e^{-kx})^{-n} \tag{3}$$

where Y is the nitrogen uptake, production of dry matter, rooting depth or growth of root mass, n is a shape factor set at 1.5 for processes here, k is a rate constant in $(day.degree)^{-1}$ and x thermal time (the accumulation of measured soil temperature above 0 degrees centigrade each day at 20 cm depth). The parameter A is the maximum value Y is allowed to take: $1900\,g\,m^{-2}$ in these experiments for dry matter, $200\,kg\,N\,ha^{-1}$ for N uptake, $200\,g\,m^{-2}$ for root mass. The rooting depth was set according to the data provided [8], but the amount of root in each layer declines exponentially with depth in the manner proposed by Gerwitz and Page [6]. The parameter k was predicted according to the empirical relationship derived by Whitmore and Addiscott [24] for crop growth and N uptake, for rooting it was set at 0.003.

The model also predicts crop development, grain production and nitrogen translocation into grain. By partitioning N between grain and the rest of the plant we can estimate the contribution of N from straw and roots to the turnover of soil organic matter; grain data provide a valuable check on the working of this part of the model. Crop development follows the description proposed by Weir et al. [22] and we have used the same parameters for the different stages of growth except that we modified the emergence and vernalization parameters for Dutch conditions [16]. The number of grains surviving to maturity was set at 1000 grains for every 33 g dry matter produced by the crop in the last 400 degree days (corrected for day length) immediately before anthesis [22]. The amount of dry matter which becomes grain during grain-fill is calculated from equation (3). Nitrogen uptake also follows the form of equation (3) but is moved into grain from the rest of the crop tissue more rapidly at higher temperatures [20]. Senescence follows an exponential decay function fitted to data of Thorne and Wood [19] and begins shortly after ear initiation. After anthesis, root growth ceases and half of the nitrogen in senescing roots is made available to the growing grain. Much of the remainder appears in the soil as the roots decompose.

When nitrogen is in short supply, the crop takes up only what is available. If the shortfall is made good before anthesis, with fertilizer for example, then the model allows the crop to extract up to 30% more N from the soil than estimated by equation (3). When the crop is short of water, dry matter production is reduced by the ratio of the amount of water available in soil to the amount of transpiration that day. Restraints to crop N uptake and dry matter production are independent except that the crop N may not exceed 10% of dry matter nor may the N content of straw fall below 0.2%. Neither extreme was encountered during these simulations.

Running the model

The simulations presented here were derived largely without first fitting the model to any subset of the data. They can therefore be said to be a genuine test of the applicability of the model developed in the UK to Dutch conditions. A few points must be made before considering the simulations. Simulation starts at harvest of the previous crop. However the dataset contains neither a measurement of soil mineral N at this time, nor one of the mass and N content of the residues. Accordingly we estimated the size of the residues and the amounts of mineral N likely to be present in the soil. Where these estimates did not produce good simulations of the first measurement of mineral N in the soil in autumn or spring we adjusted the estimate to get the simulated quantity of soil mineral N to agree more or less with this first measurement. Initial soil moisture deficits for September were calculated from the meteorological data and from the rooting depths in the data set [8].

Although very detailed measurements were made, little replication is given in the data set and so we have confined ourselves to qualitative assessments only of the success of the model in simulating the data. For this reason we have presented all the simulated results leaving judgement up to the reader.

Results

Figures 1–3 display results from all the computer simulations. There are fairly important differences between soil types at the three sites and in the weather between both sites and years. The rainfall at Bouwing was slightly greater in 1982–3 than at Eest and PAGV, but this was reversed in 1983–4. In 1983 much rain fell in May resulting in some leaching of applied N. Despite this, the model did not simulate well the fate of the nitrogen fertilizer applied to the experimental fields in either year. It is possible that this N was 'missed' in the field sampling for some reason, yet this is unlikely three or four times in succession. It appears that sufficient N was available to the crop because good yields were obtained. Looking at the simulations of soil mineral N in Figs. 1–3 (a), (d), (g), (j), (m), (p) it is quite apparent that where no fertilizer (N_1 treatment) was applied the model simulated the behaviour of soil mineral N well (see for example Fig. 1 (a)). However where fertilizer N was applied the simulations of soil mineral N were generally less good (Fig. (d) and (g)). As Addiscott et al. have pointed out [4], fertilizer N applied early in the growing season did appear in the measurements and so the model sometimes agreed with meas-

286

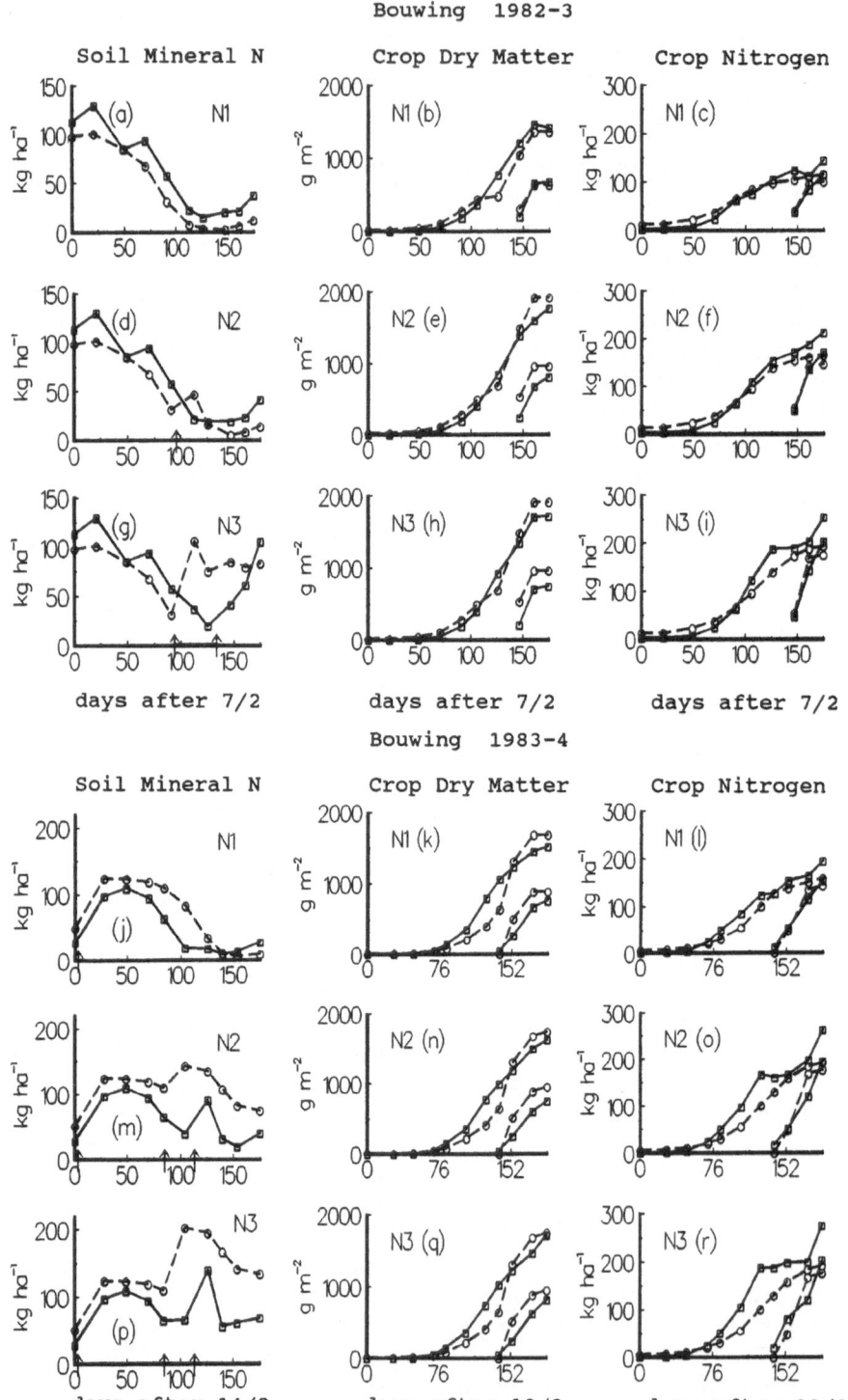

Fig. 1. Results from the Bouwing experimental farm (Wageningen) N_1:0 (70), N_2:60 (170), N_3:160 (230) kg N ha^{-1} fertilizer in 1983 (1984). ↑ indicates time of fertilizer application. Soil Mineral N: ammonium plus nitrate to 100 cm depth. —□— measured values, − −○− − simulated. Lower right insets in Crop Dry Matter and Crop N diagrams show Grain Dry Matter and Grain N respectively. Note different scales for soil mineral N in the two years.

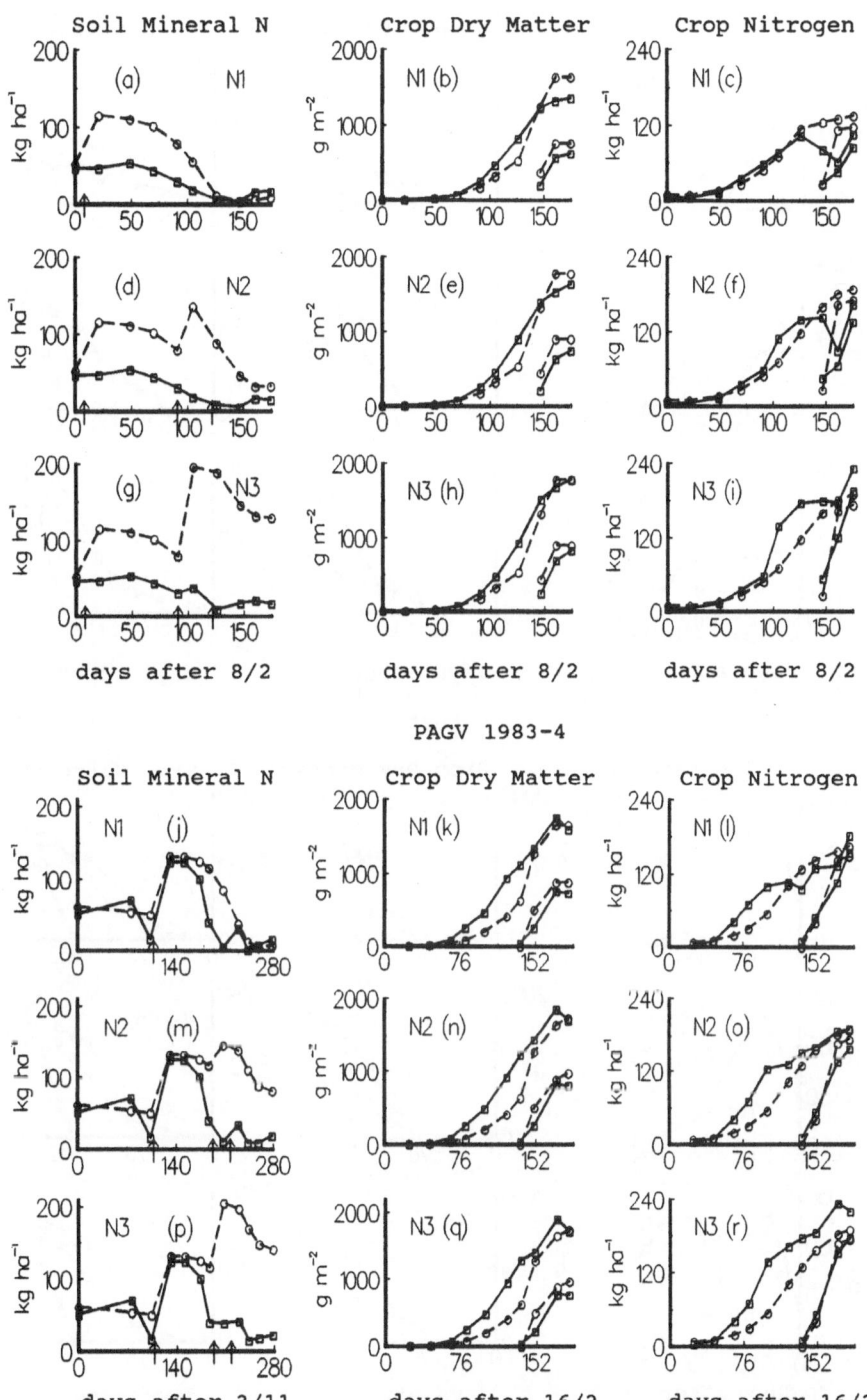

Fig. 2. Results from the PAGV experimental farm (Lelystad) N_1: 60 (80), N_2: 140 (180), N_3: 240 (240) kg N ha^{-1} fertilizer in 1983 (1984). ↑ indicates time of fertilizer application. Soil Mineral N: ammonium plus nitrate to 100 cm depth. —□— measured values, − − O − − simulated. Lower right insets in Crop Dry Matter and Crop N diagrams show Grain Dry Matter and Grain N respectively.

288

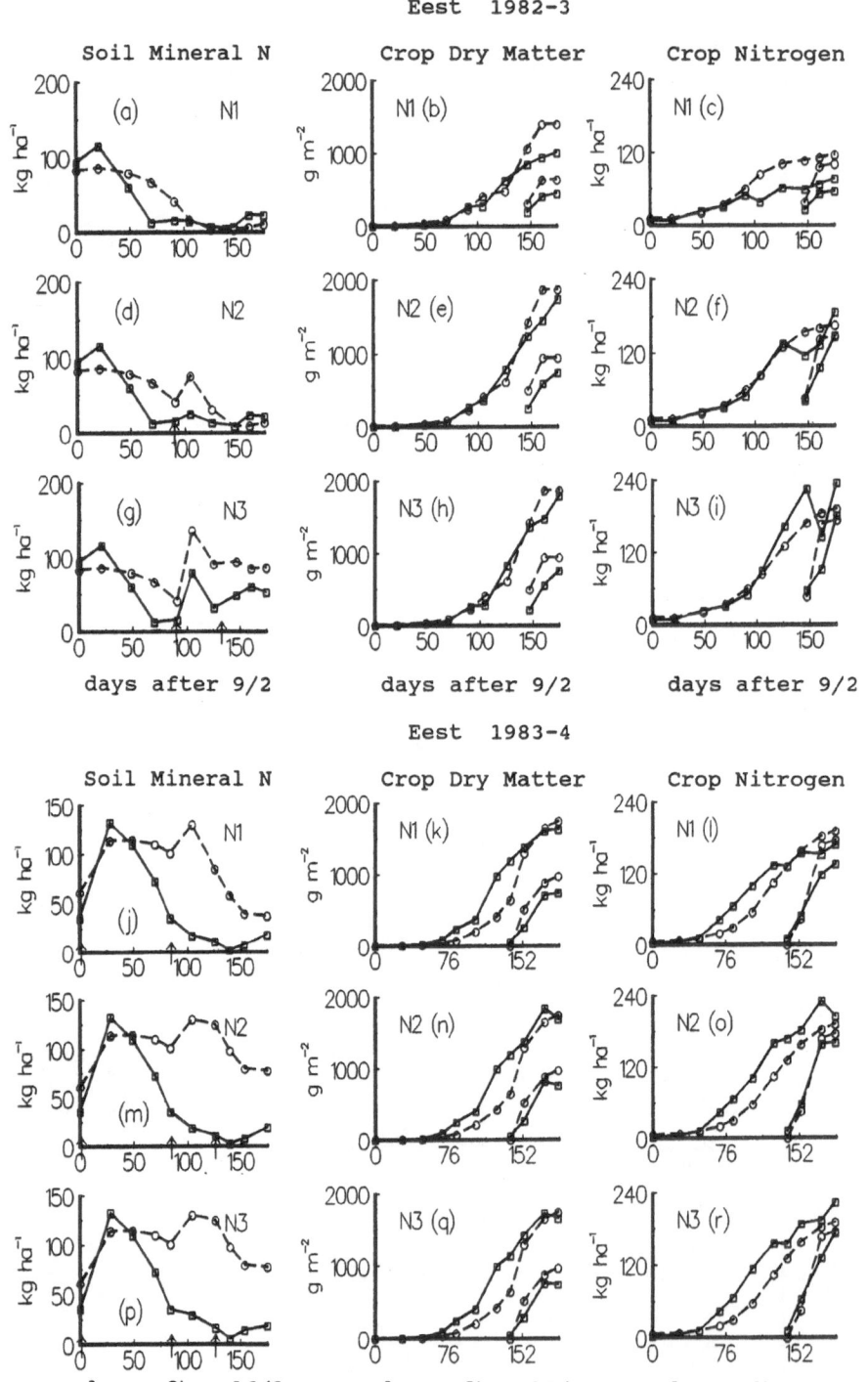

Fig. 3. Results from the Eest experimental farm (Nagele) N_1:0 (110), N_2:60 (150), N_3:160 (150) kg N ha^{-1} fertilizer in 1983 (1984). ↑ indicates time of fertilizer application. Soil Mineral N: ammonium plus nitrate to 100 cm depth. —□— measured values, --○-- simulated. Lower right insets in Crop Dry Matter and Crop N diagrams show Grain Dry Matter and Grain N respectively. Note different scales for soil mineral N in the different years.

ured data up to the time of the second fertilizer dressing; this was more striking in 1984 than in 1983. Note that the fertilizer treatments N_2 and N_3 were the same at Eest in 1984 so that the simulations of this data were identical.

The simulations of crop dry matter and nitrogen uptake are acceptable but two points need mention. The first is that in a few cases the simulations early in the growing season were smaller than the measured values. The model requires that the user estimate target maxima for dry matter and N uptake, but we decided not to alter our normal input values for UK agriculture (see crop growth and N uptake section). For these experimental fields our targets were too low. Second, the growth of the crop in the 6 weeks or so before anthesis is used as an estimator of grain number and so too much crop growth just before anthesis as in these simulations leads to too much grain growth after. The error introduced into the grain production model in this way was fairly small in most cases, but it made the simulations of grain production systematically larger than the measurements. It was more apparent in the amounts of N in grain for the N_1 experiments which were kept short of nitrogen. Otherwise the simulations of grain mass and N content were very good although in one or two instances (Fig. 1 (h), (q) for example) the simulated crop ceased to grow too soon.

Discussion and ways of improving the model

The functions that describe N uptake and dry matter are clearly quite a reasonable fit of the data. They are empirically based and independent of one another so should not be relied upon too heavily. For instance it is possible for grain growth to be estimated well but grain N not, yet at the same time total dry mass may be estimated poorly while total crop N uptake is almost exactly right. Early season growth is important if the model is to avoid producing too much carbohydrate in the weeks before anthesis and so initiating too many florets for filling later. Getting grain growth right will greatly improve the return of C and N to the soil thus improving estimates of the availability of N during the following growing season.

Our model estimates the turnover of organic carbon as well as nitrogen, so we examined the turnover of C in relation to applications of fertilizer N. If much C is present in a decomposable form N should be immobilized by soil microorganisms. Taking the Bouwing N_3 soil receiving $160\,kg\,N\,ha^{-1}$ in 1983 as an example, Fig. 4 shows the gross amounts of C becoming available from the decomposition of all the organic fractions of the model. Ocio and Brookes [14] have shown that as the soil microbial biomass grows its C:N ratio does not change. Kilham et al. [11] have shown that its C:N ratio may widen in response to moisture stress, although there is some doubt that microbial C and thus the C:N ratio can be measured effectively under moisture stress [23]. Thus with a C:N ratio of about 6 or 7 and without taking account of losses through respiration and maintenance, the biomass requires at least $600\,kg$ C to incorporate $100\,kg\,N\,ha^{-1}$. Fig. 4 suggests that even this is far more C than is likely to be available.

We also examined how the incoming material from roots and trash affected C and N availability. In the organic matter turnover model most of the readily decomposable material is acted upon

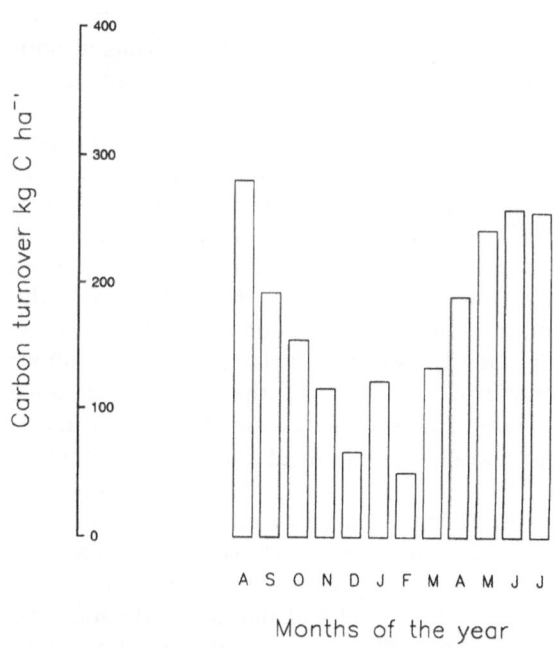

Fig. 4. Modelled amount of carbon becoming available to microorganisms in the soil each month in the Bouwing experiment receiving $160\,kg\,N\,ha^{-1}$ (N_3) in 1982–3.

by the soil microbes within one month. Even enlarging the more recalcitrant fraction (RPM) five times to over 10 tonnes ha^{-1} (much assumed to have been carried over from previous years) there was far too little C available in spring to account for the loss of N. Only a very large input of readily decomposable C (about 2 tonnes ha^{-1}) with a very wide C:N ratio (>100:1) could supply enough C to provide a demand for so much N. Realistically, decomposing humus could supply sufficient N to meet the demands from organism' in this soil.

Powlson et al. [15] correlated the rainfall immediately after ^{15}N application with ^{15}N not recovered in soil (0–23 cm) plus crop at harvest. Their relationship predicts a loss of about 40 kg N ha^{-1} in both years at all three sites, but our model predicts that at least half of this loss could have leached below 23 cm in 1983. Thus in 1984 up to 40 kg N ha^{-1} may have been denitrified. This still leaves about 50 kg N ha^{-1} loss in 1984 and about 70 in 1983 unexplained.

In the N_3 Bouwing experiment in 1983 (but not necessarily in all the others) the model under-simulated the uptake of N by the plant; even at the point of greatest disagreement, however, it could only account for about 30–40 kg ha^{-1} of the 'missing' N, leaving about the same amount to be accounted for. Thus in both years roughly the same amount of applied N (40 kg ha^{-1}) remained unaccounted for.

Wehrmann and Coldewey-Zum Eschenhoff [21] recently demonstrated that plant roots may deplete the soil of recently fixed non-exchangeable ammonium N up to a distance of about 2 mm away, although they are unable to extract native fixed N. Of course, they also deplete the soil of exchangeable ammonium N. With a rooting density for high-yielding crops of about 10 cm cm^{-3} in topsoil [5] the inter-root distance becomes a maximum of about 4 mm. Thus most, but not all, recently fixed ammonium should be available to plants even though it cannot be found during chemical extraction with 1M KCl. Nommik and Vahtras [13] concluded that ammonium fixing capacity is only poorly related to clay content of soil and that silt and even sand fractions may fix some ammonium. The clay in the Bouwing experiment is an open Illite well disposed to fixing ammonium or potassium. The other soils are on a recently reclaimed polder, but as these soils weather, releasing potassium, it is possible that more exchange and fixation sites become able to fix ammonium. Fixation is complex and the interactions with K are important, especially since K is able to collapse clay layers thus trapping recently-fixed ammonium. It appears that ammonium fixation and denitrification may well be able to play a part in improving our computer simulation model.

Acknowledgements

We are grateful to P de Willigen and JJR Groot for drawing our attention to the work of Wehrmann and Coldewey-Zum Eschenhoff [21] and to that of Reinink et al. [16]. NJB thanks the Home-Grown Cereals Authority for financial support.

References

1. Addiscott TM (1977) A simple computer model for leaching in structured soil. J Soil Sci 28: 554–563
2. Addiscott TM, Thomas VH and Janjua MA (1983) Measurement and simulation of anion diffusion in natural soil aggregates and clods. J Soil Sci 34: 709–721
3. Addiscott TM and Whitmore AP (1987) Computer simulation of changes in soil mineral nitrogen and crop nitrogen during autumn, winter and spring. J Agric Sci, Camb 109: 141–157
4. Addiscott TM, Bailey NJ, Bland GJ and Whitmore AP (1990) Simulation of nitrogen in soil and winter wheat crops: a management tool that makes best use of available information. Fert Res 27: 305–312
5. Barraclough PB (1986) The growth and activity of winter wheat roots in the field: nutrient inflows of high-yielding crops. J Agric Sci, Camb 106: 53–59
6. Gerwitz A and Page ER (1974) An empirical mathematical model to describe plant root systems. J Appl Ecol 11: 773–781
7. Greenwood DJ, Verstraeten LMJ, Draycott A and Sutherland RA (1987) Response of winter wheat to N-fertilizer. Dynamic model. Fert Res 12: 139–156
8. Groot JJR, and Verberne ELJ (1991) Response of wheat to nitrogen fertilization, a data set to validate simulation models for nitrogen dynamics in crop and soil. Fert Res 27: 349–383
9. Hart PBS (1984) Effects of soil type and past cropping on the nitrogen supplying ability of arable soils. Ph.D Thesis University of Reading
10. Jenkinson DS, Hart PBS, Rayner JH and Parry LC (1987) Modelling the turnover of organic matter in

long-term experiments at Rothamsted. INTECOL Bull 15: 1–8

11. Kilham K, Schimel JP and Wu D (1990) Ecophysiology of the soil microbial biomass and its relation to the soil microbial N pool. Soil Use Man 6: 86–88

12. Ladd JN, Oades JM and Amato M (1981) Microbial biomass formed from ^{14}C, ^{15}N labelled plant material decomposing in soils in the field. Soil Biol Biochem 13: 119–126

13. Nommik H and Vahtras K (1982) Retention and fixation of ammonium and ammonia in soils. In: Nitrogen in Agricultural Soils, pp 123–171 Agronomy 22. Stevenson FJ (ed) ASA, Madison, Wisconsin

14. Ocio J and Brookes PC (1990) An evaluation of methods for measuring the microbial biomass in soils following recent additions of wheat straw and the characterization of the biomass that develops. Soil Biol Biochem 22: 685–694

15. Powlson DS, Jenkinson DS, Pruden G and Johnston AE (1983) Losses of ^{15}N labelled fertilizer N applied to winter wheat. Rothamsted Report for 1982 Part 1: 263

16. Reinink K, Jorritsma I and Darwinkel A (1986) Adaptation of the AFRC wheat phenology model for Dutch conditions. Neth J Agric Sci 34: 1–13

17. Richter J, Nordmeyer H and Kersebaum K.Chr (1985) Simulation of nitrogen regime in loess soils in the winter half-year: comparison between field measurements and simulations. Plant and Soil 83: 419–431

18. Sørenson LH (1975) The influence of clay on the rate of decay of amino acid metabolites synthesized in soils during decomposition of cellulose. Soil Biol Biochem 7: 171–177

19. Thorne GN and Wood DW (1987) The fate of carbon in dying tillers of winter wheat. J Agric Sci, Camb 108: 515–522

20. Vos J (1984) Aspects of modelling post-floral growth of winter wheat and calculations of the effects of temperature and radiation. In: Day W and Atkin RK (eds) Wheat Growth and Modelling, NATO Series A: Life Sciences 86: 143–148 Plenum Press New York and London

21. Wehrmann J and Coldewey-Zum Eschenhof H (1986) Distribution of nitrate, exchangeable and non-exchangeable ammonium in the soil-root interface. In: Lambers H, Neeteson JJ and Stulen I (eds.) Fundamental, Ecological and Agricultural Aspects of Nitrogen Metabolism in Higher Plants, pp 447–450. Martinus Nijhoff, Dordrecht Boston Lancaster

22. Weir AH, Bragg PL, Porter JR and Rayner JH (1984) A winter wheat crop simulation model without water or nutrient limitations. J Agric Sci, Camb 102: 371–382

23. West AW, Sparling GP, Speir TW and Wood JM (1988) Dynamics of microbial C, N-flush and ATP, and enzyme activities of gradually dried soils from a climosequence. Aust J Soil Res 26: 519–530

24. Whitmore AP and Addiscott TM (1987) A function for describing nitrogen uptake, dry matter production and rooting by wheat crops. Plant and Soil 101: 51–60

25. Whitmore AP and Parry LC (1988) Computer simulation of the behaviour of nitrogen in soil and crop in the Broadbalk continuous wheat experiment. In: Jenkinson DS and Smith KA (eds.) Nitrogen Efficiency in Agricultural Soils, pp 418–432. Elsevier, London

Fertilizer Research **27**: 293–304, 1991.
© 1991 *Kluwer Academic Publishers.*

Simulation of the effects of nitrogen supply on yield formation processes in winter wheat with the model TRITSIM

W. Mirschel[1], H. Kretschmer[1], E. Matthäus[2] & R. Koitzsch[3]
[1] *Research Centre of Soil Fertility Müncheberg, Academy of Agricultural Sciences of the GDR, 1278 Müncheberg, Wilhelm-Pieck-Strasse 72, Germany*
[2] *Central Institute of Cybernetics and Information Processes Berlin, Academy of Science of the GDR, 1080 Berlin, Kurstrasse 33, Germany*
[3] *Research Institute for Agrometeorology Halle of the Meteorological Service of the GDR, Agrometeorological Research Station, 1278 Müncheberg, Wilhelm-Pieck-Straße 72e, Germany*

Key words: Agroecosystem model, nitrogen supply, yield formation, mineralization, simulation, nitrogen uptake

Abstract

An outline of the dynamic winter wheat model TRITSIM is given. The model describes in one-day steps growth, yield formation and development of a crop from post-winter tillering until harvest under various conditions of water and nitrogen supply. TRITSIM is coupled with a simple soil nitrogen model and a soil water model to describe effects of nitrogen and water on yield formation processes. Comparisons between model and experimental results for ontogenesis, grain biomass, nitrogen uptake and soil mineral nitrogen are given for a series of Dutch experiments. Simulations were satisfactory, except for the time course of soil mineral nitrogen.

Introduction

Detailed knowledge of qualitative and quantitative effects of weather and management strategies as irrigation, fertilization and pest control on yield formation processes is an essential prerequisite for sustained yield levels and a clean environment. To reach economic as well as environmental goals, the use of agroecosystem models, which include the major soil and plant processes and take into account the more important environmental and climatic parameters, might be necessary in the near future.

A useful tool for agroecosystem modelling is given by the SONCHES simulation system (Simulation Of Nonlinear Complex Hierarchic EcoSystems) [9]. SONCHES served as the basis for development of the agroecosystem model for winter wheat, AGROSIM-WHEAT (AGROecosystem SIMulation WHEAT) [2, 4], which contains TRITSIM (TRITicum SIMulation) [14, 18] as the central submodel. The soil water status is described by the soil moisture model BOWA, and the soil mineral nitrogen status by the simple soil mineral nitrogen balance model BOST, which are described in this paper.

The agroecosystem model makes it possible to perform simulation experiments under various conditions and to study the reactions of the subsystems in different situations.

Short description of the TRITSIM model

Crop growth

The model TRITSIM describes in one-day steps ontogenesis, yield formation and growth of a winter wheat crop from post-winter tillering until harvest under field conditions. The structure of the winter wheat model TRITSIM is given in Fig. 1.

The ontogenesis subprocess in TRITSIM acts as a time-related control variable on other processes. Other processes are initiated, stopped, accelerated or slowed down by ontogenesis. This controlling function is represented in Fig. 1 by dotted lines. The ontogenesis is influenced by temperature, water and nitrogen stress. In the model, the stage of ontogenesis is described with the decimal code for crop development according to Zadoks [31].

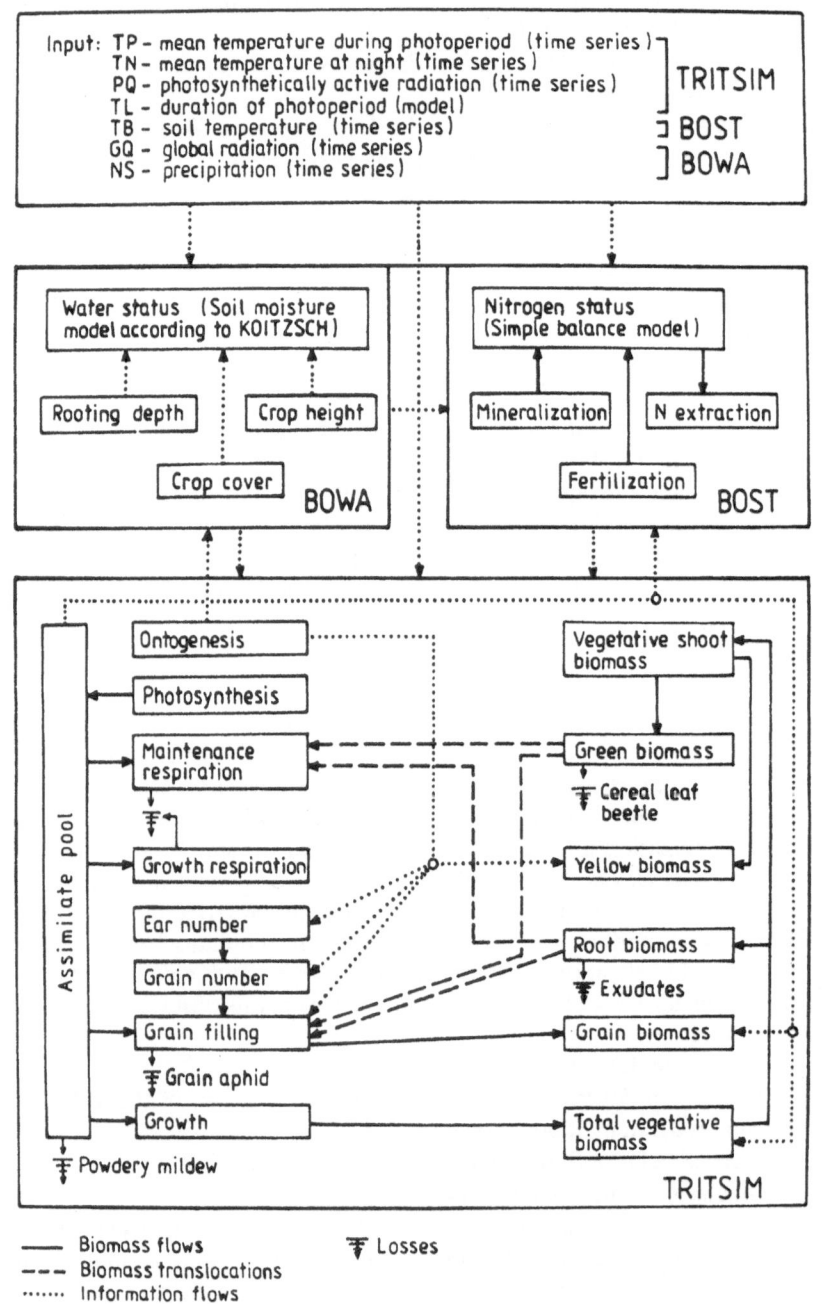

Fig. 1. Structure of the TRITSIM winter wheat model with its auxiliary models for soil water (BOWA) and soil nitrogen (BOST).

Within TRITSIM, a photosynthesis submodel acts as the source of daily biomass production and is based on a maximum photosynthetic rate per unit green biomass. The approach differs from other photosynthesis models [7, 29], where leaf area index (LAI) is used as the main parameter to calculate photosynthesis. Reference to green biomass is essential because not only the leaves but also stems and, especially, ears contribute to assimilate production [20]. The maximum photosynthetic rate is lowered by radiation and temperature characteristics normalized to the interval (0, 1), water stress and nitrogen stress, and is corrected by the length of daily photoperiod and a feedback function for accumulated green biomass.

The assimilates produced daily are initially stored in an assimilate pool from which they are available in the first place for respiration, secondly for grain filling (after flowering), and only lastly for growth. If there are insufficient assimilates for respiration and grain filling, they are supplied by translocation from green biomass or roots. The vegetative biomass is partitioned into shoot and root biomass by using a fixed shoot: root ratio. Losses of the assimilates allocated to roots occur by exudation and by root mortality. Within TRITSIM the total vegetative shoot biomass is divided into green and yellow (dead) biomass. At the end of ontogenesis, senescence occurs, resulting in a loss of green biomass and a reduction in photosynthesis. Senescence is accelerated by water and nitrogen stress.

Nitrogen uptake

The daily nitrogen demand of the above-ground biomass is calculated on the basis of a maximum nitrogen uptake curve for winter wheat (Fig. 2) derived from our own experimental data (un-

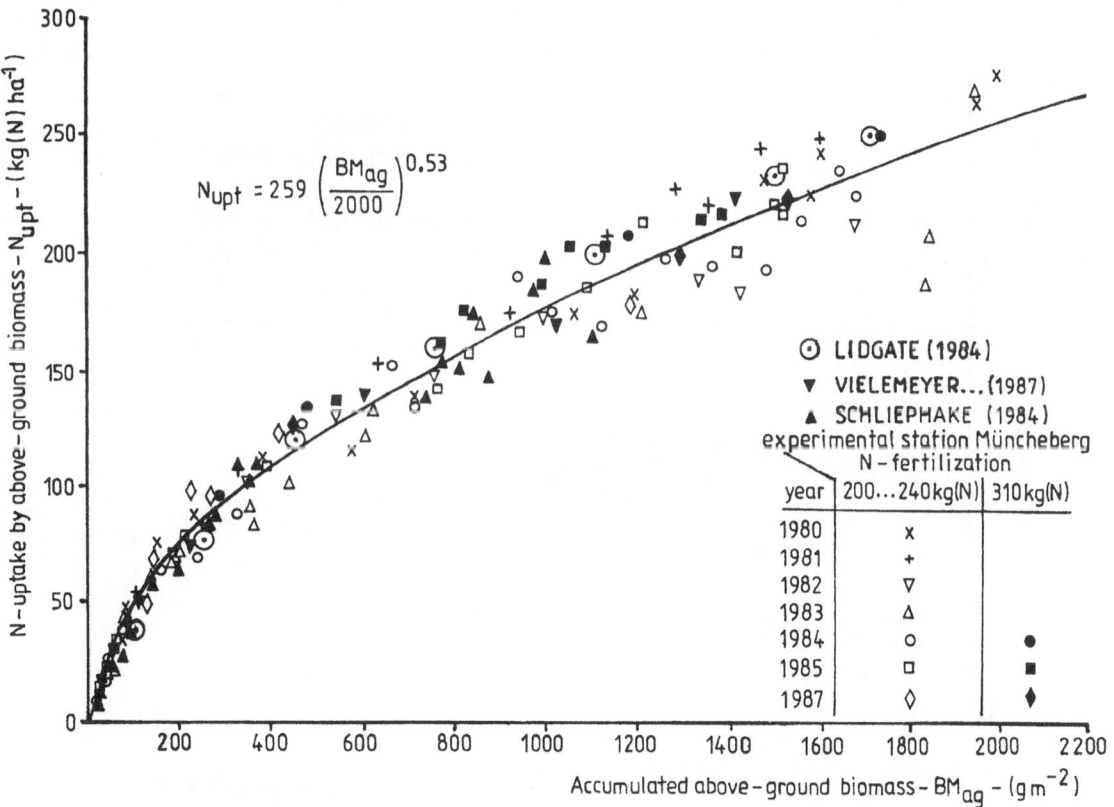

Fig. 2. Above-ground nitrogen uptake as a function of above-ground biomass for winter wheat grown under sufficient water and nitrogen supply (data from the experimental station Müncheberg (cv. Alcedo; unpublished) and from literature [13, 23, 28]). The curve gives the relationship used in the model.

published) and from literature [13, 23, 28]. Potential daily nitrogen uptake is calculated on the basis of a comparison between nitrogen demand and the amount of available soil mineral nitrogen. The actual rate equals the lower of the two. From our own experimental data (unpublished) we derived that all soil mineral nitrogen in excess of $10 \, \text{kg ha}^{-1}$ is available for uptake.

Effects of moisture and nitrogen availability on growth

Water stress is described by reduction factors for short-term water stress and long-term water stress. The calculation of these factors depends on the ratio of actual to potential evapotranspiration (AET/PET), calculated using the soil moisture model BOWA described in this paper. The short-term stress factor depends on the extent to which the ratio AET/PET drops below an ontogenesis-dependent threshold curve for

AET/PET. Green biomass grown under conditions of water stress is more effective in photosynthesis because of a higher stomata density. When a cumulative water stress threshold level and a given duration of stress are exceeded, growth compensation occurs until the end of leaf growth. This positive long-term effect decreases over a longer period and the photosynthesis effectiveness returns to its initial level.

Effects of nitrogen availability on growth are based on the comparison between the actual nitrogen content of the above-ground biomass and an ontogenesis-dependent lower threshold level for nitrogen content, derived from our own experimental data (unpublished) and from literature [27] (Fig. 3). As illustrated in Fig. 3, the nitrogen content of crops with an insufficient nitrogen supply drops below the threshold level indicated by the curve.

The reduction factor for photosynthesis depends on the nitrogen deficit of the above-ground biomass (Fig. 3), calculated as the actual

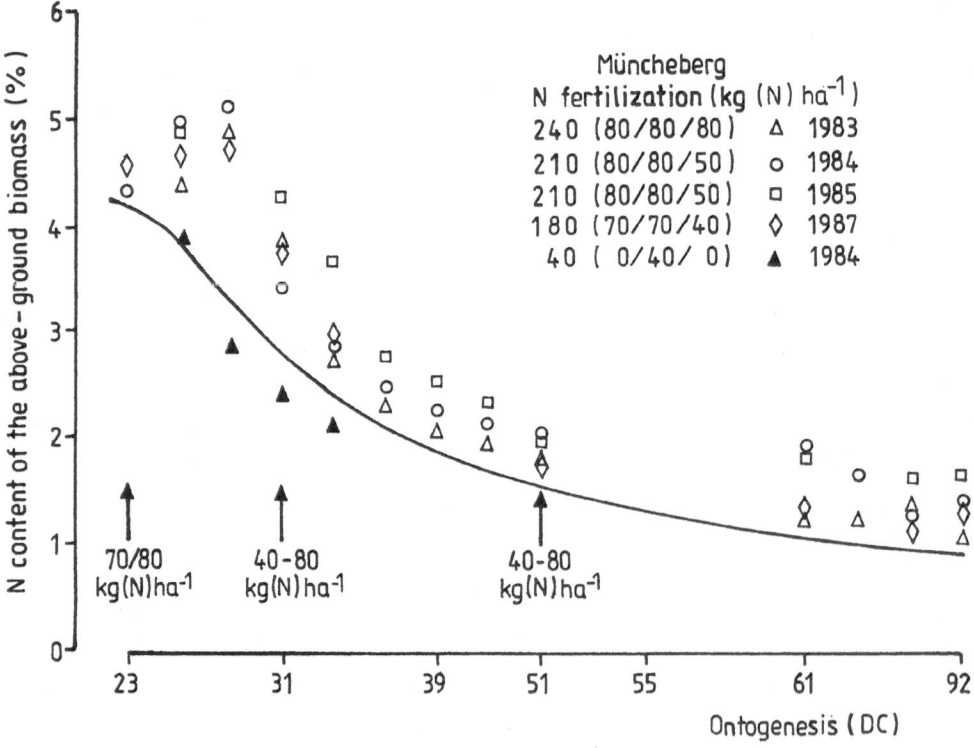

Fig. 3. Nitrogen content of above-ground biomass of winter wheat (cv. Alcedo) in experiments with different nitrogen applications in Müncheberg (GDR) as a function of ontogenesis. The curve indicates the threshold level for nitrogen content used in the model. Codes on the X-axis refer to ontogenesis description by Zadoks et al. [31].

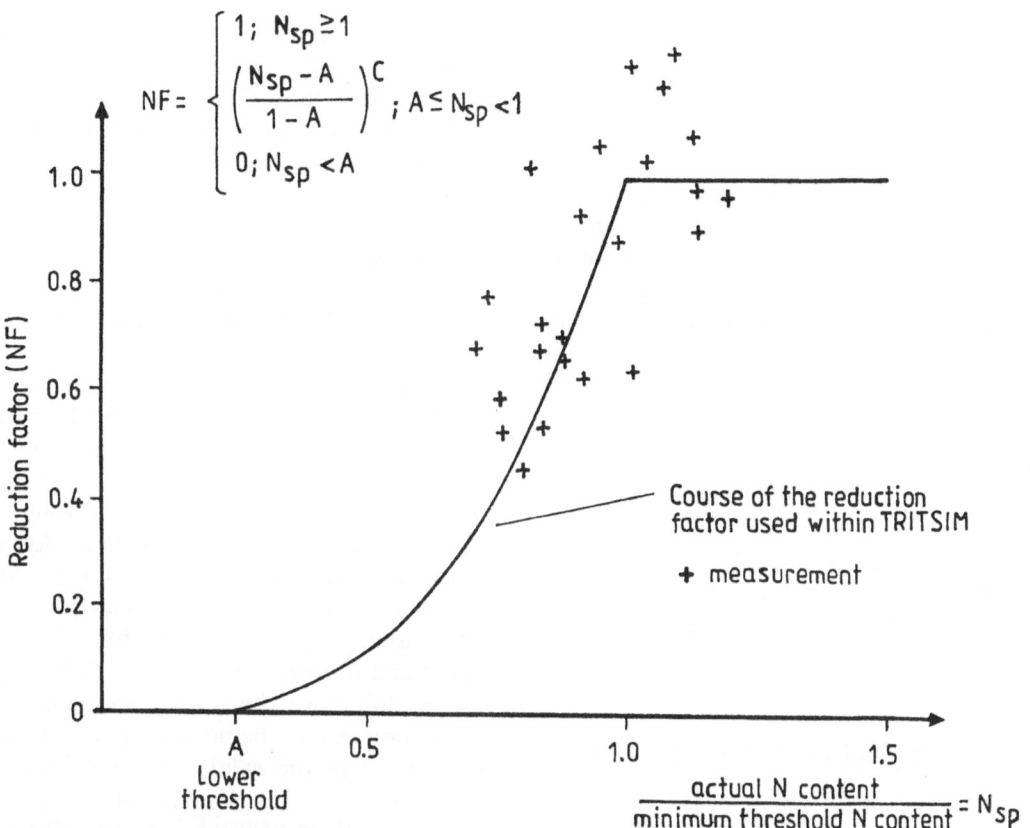

$$NF = \begin{cases} 1; & N_{sp} \geq 1 \\ \left(\dfrac{N_{sp} - A}{1 - A}\right)^{C}; & A \leq N_{sp} < 1 \\ 0; & N_{sp} < A \end{cases}$$

Course of the reduction factor used within TRITSIM

+ measurement

A Lower threshold

$$\frac{\text{actual N content}}{\text{minimum threshold N content}} = N_{sp}$$

Fig. 4. Time course of the nitrogen-stress-dependent growth reduction factor (*NF*) as a function of the nitrogen content of the above-ground biomass divided by the threshold value at the given stage of ontogenesis and a comparison with measurements (Müncheberg, 1984, cv. Alcedo, unpublished data).

nitrogen content divided by the threshold value at a given stage of ontogenesis. The course of the photosynthesis reduction is shown in Fig. 4. A value for nitrogen content in the above-ground biomass is defined below which daily photosynthesis is not sufficient for growth. The values for the reduction of daily photosynthesis rates originate from gas-exchange measurements under field conditions [22] conducted at different levels of nitrogen supply.

A detailed description of the algorithms and mathematical relationships discussed is given by Matthäus et al. [14] and Mirschel et al. [18].

Short description of the soil water model BOWA

BOWA was developed for a horizontal location, not affected by a groundwater table [3, 10, 11]. The soil is subdivided into 10 layers of 15 cm

thickness each, and each layer is characterized by its water content at field capacity (pF 2) and at the wilting point (pF 4.2). The extraction of water depends on potential evapotranspiration, rooting depth, and on vertical distribution of water in the soil profile. When the water content of each of the rooted layers is greater than or equal to a critical value, water extraction equals potential evapotranspiration. From the first layer, extraction is largest and decreases linearly to zero for the deepest rooted layer. If in any of these layers the water content is less than the critical value, then potential extraction from this layer is reduced linearly and reaches zero at the wilting point. The critical value depends on *PET'*, i.e. the potential evapotranspiration from a soil with a completely closed canopy. For *PET'* > 20 mm the reduction begins at field capacity, for *PET'* = 2.5 mm at 75% of field capacity, and for *PET'* = 0 mm at the wilting point. For other values of PET' the critical value is found by linear interpolation. Infiltration of water by precipitation is simulated with the percolation model of Glugla [5], and upward water transport is omitted.

Meteorological input data required are daily precipitations (in mm), daily mean air temperatures (T in °C) and daily totals of global radiation (G in MJ m^{-2}), respectively. Using a modified form of Turc's formula [26], potential daily evapotranspiration from the fraction S_f of soil surface covered by green plants is estimated by

$$PET = FS_f V_{Turc}$$
$$= FS_f 0.31(G + 2.09)\frac{T}{(T + 15)}, \qquad (1)$$

in which F is a crop-dependent conversion factor. For winter wheat after tillering $F = 1.2$ and from tillering onwards evaporation from the soil surface is omitted.

Description of the soil mineral nitrogen balance model BOST

The amount of soil mineral nitrogen (NS_t, in g m^{-2}) is calculated by:

$$NS_t = NS_{t-1} + NM_{t,0-30} + NM_{t,30-60} + FE_t$$
$$- NU_t \qquad (2)$$

in which FE_t is fertilization (in g m^{-2}), $NM_{t,0-30}$ and $NM_{t,0-30}$ are mineralization rates for the layers 0–30 cm and 30–60 cm (in g m^{-2}), respectively, and NU_t is nitrogen uptake by the crop (in g m^{-2}).

Mineralization depends on soil water and soil temperature. The daily soil water values are calculated in the soil moisture model BOWA. The daily mineralization rates for the two layers are calculated according to Rausch et al. [21] and Stanford [24]:

$$NM_{t,i} = \frac{W_i P_i 10^Q}{W_{FC,i}} \qquad (3)$$

$$Q = 6.865 - \left(\frac{2758}{T_i + 273}\right) \qquad (4)$$

in which i is soil layer (0–30 cm or 30–60 cm), W_i

is soil water (mm), P_i is mineralization potential (g m^{-2} d^{-1}), $W_{FC,i}$ is field capacity (mm), and T_i is soil temperature (°C).

The mineralization potential P_i in Eq. (2) was established according to Stanford [25].

Simulations

Müncheberg (GDR) and Keszthely (Hungary)

TRITSIM was originally validated with data from the intensive measuring field at the Research Centre of Soil Fertility at Müncheberg (loamy sand, annual precipitation 533 mm, annual mean temperature 8.2°C) for the winter wheat variety Alcedo during 1980–1985. Agreement between model and experimental data for grain biomass is shown in Fig. 6. The coefficient of correlation is 0.976 ($n = 53$) for different irrigated and fertilized treatments (1983–1985), the mean difference between measurements and simulation results being 42.3 g m^{-2}. A detailed description of the validation for Müncheberg is given elsewhere [14, 19, 17, 15].

A comparison between model predictions and experimental data from a fertilizer experiment with different treatments in Keszthely, Hungary, showed that the difference in total biomass between model and experimental data was below 5% in 40% of all cases, between 5 and 10% in 30% of all cases, and above 10% in 30% of all cases (F. Zemancovics, personal communication).

the Bouwing, the Eest and PAGV (The Netherlands)

The TRITSIM model was tested in its original structure and with original model parameters for three Dutch locations in 1982/83 and 1983/84. A detailed description of experiments and validation data is given by [6].

Because the TRITSIM model needs special input time series (see Fig. 1) it was necessary to make some assumptions. Mean temperature during the photoperiod (TP) and mean night temperature (TN) were estimated using daily maximum and minimum temperatures (T_{max}, T_{min}) according to [8]:

$$TP = T_{max} - \frac{T_{max} - T_{min}}{4} \tag{5}$$

$$TN = T_{min} + \frac{T_{max} - T_{min}}{4} \tag{6}$$

According to [8] and [16], the photosynthetically active radiation was assumed to be 40% of the global radiation. For all simulation runs a sufficient water supply and a mean daily mineralization rate of 1.1 kg (N) ha^{-1} were assumed. Variety specific crop parameters were taken from the variety Alcedo [30].

In the model ontogenesis is the time-related control variable for other processes, and its accurate estimation is important for the whole crop model. A comparison between simulated and experimental ontogenesis data for the Bouwing (1983) is shown in Fig. 5. For both years and all three locations, the mean difference between

simulated and measured data for several stages of ontogenesis ranged from 1.8 to 5.8 days.

TRITSIM underestimated the grain biomass in all cases in 1983 (Fig. 6). For 1984, the agreement between simulations and measurements was better and the accuracy of simulations was in the same range as those from Müncheberg. For all Dutch experimental stations and two years the mean deviation for grain biomass was 97.3 g m^{-2}. Simulations for the experiments on the Bouwing in 1983 and 1984 show that nitrogen uptake was overestimated in June and July (Fig. 7). In general, the absolute values for nitrogen uptake were overestimated by the model, but the time courses of nitrogen uptake agreed with the measurements. For the Bouwing (1984), in Fig. 8 nitrogen uptake rates with and without nitrogen stress are compared (70 and 170 kg (N) ha^{-1}, respectively). Up to the beginning of June, the nitrogen uptake pattern of the stressed variant

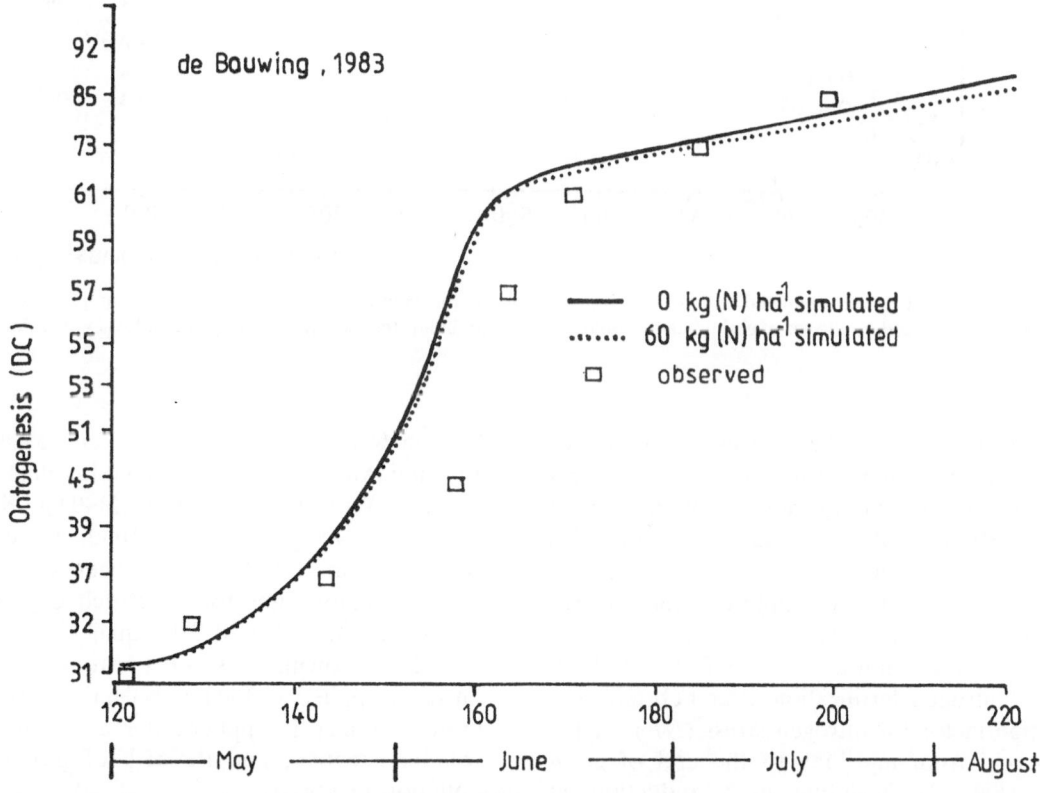

Fig. 5. Measured and simulated ontogenesis for winter wheat (the Bouwing, 1983, cv. Arminda). Simulations show the effect of ontogenesis acceleration (solid line) due to insufficient nitrogen supply (non-fertilized variant). Codes on the Y-axis refer to ontogenesis description by Zadoks et al. [31].

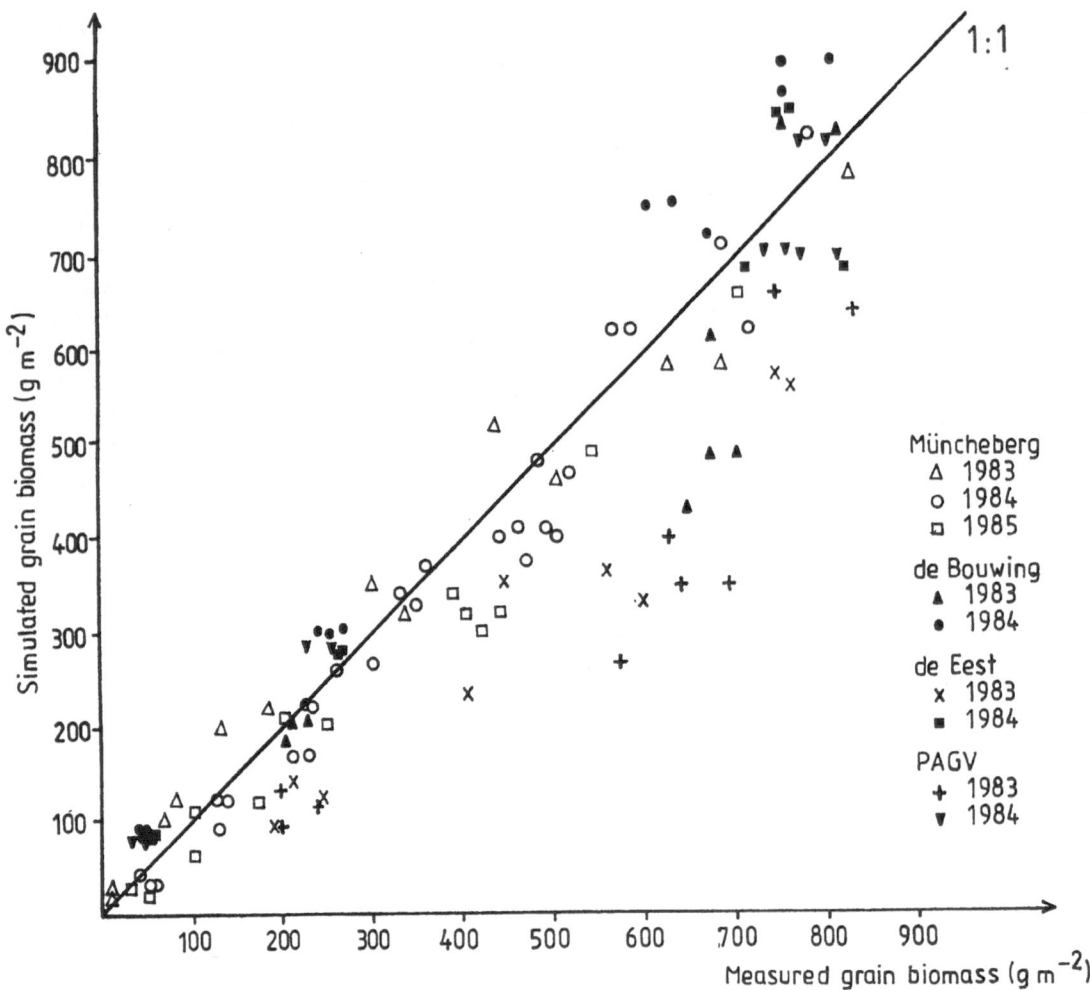

Fig. 6. Comparison between simulated and measured grain biomass for experiments that differ in water and nitrogen supply (Müncheberg (cv. Alcedo), and for experiments the Bouwing (cv. Arminda), the Eest (cv. Arminda) and PAGV (cv. Arminda)).

was similar to that of the unstressed treatment. Later, the soil mineral nitrogen was completely taken up by the crop and uptake became equal to the daily mineralization rate, i.e., $1.1 \, \text{kg (N)}$ $\text{ha}^{-1} \, \text{d}^{-1}$ on average.

The effect of nitrogen supply on biomass production and ontogenesis is introduced into the model by a reduction factor. For PAGV (1983), without nitrogen fertilization after February the reduction factor for nitrogen stress $(NF) < 1$ for the period from mid-May to the end of grain filling (Fig. 9), resulting in a reduction in biomass production and a shorter grain filling period. With an application rate of $80 \, \text{kg (N)}$ ha^{-1} in May, $NF < 1$ for only about 35 days in July and August, and only the grain filling rate is reduced. With two applications ($140 \, \text{kg (N)} \, \text{ha}^{-1}$ in May and $40 \, \text{kg (N)} \, \text{ha}^{-1}$ in June), the value of NF equals one for the whole vegetation period. The simulation runs for all Dutch experiments show that the value of NF equals unity during the whole growing season if the total amount of nitrogen applied is $150 \, \text{kg (N)} \, \text{ha}^{-1}$ or more. If more fertilizer is applied, the economic effect will be negative, and the risk of groundwater pollution increases.

An important prerequisite for an exact description of the effects of nitrogen supply on

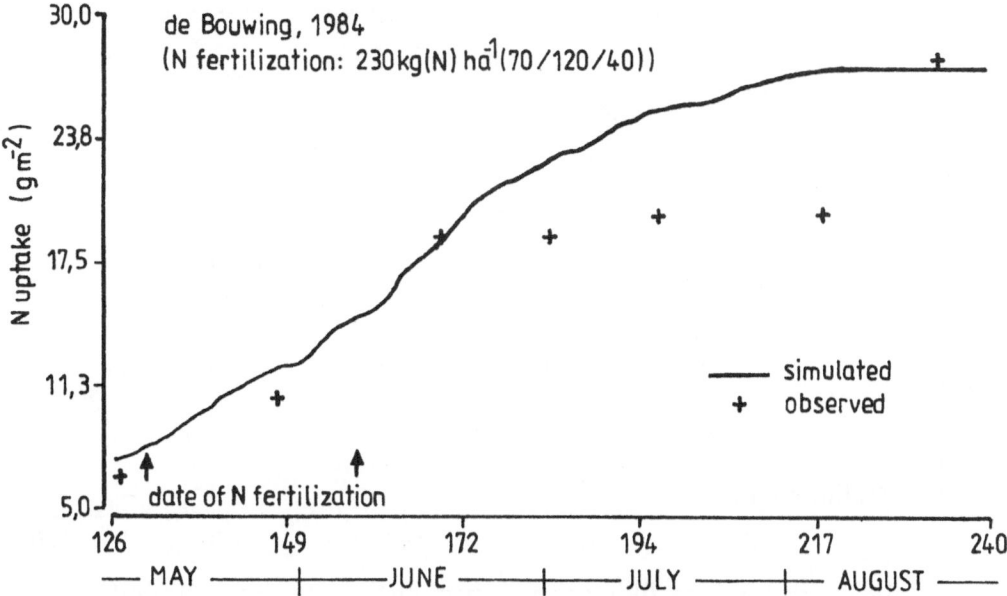

Fig. 7. Simulated and observed cumulative nitrogen uptake, the Bouwing, 1984 (N fertilization: 230 kg (N) ha^{-1}, cv. Arminda).

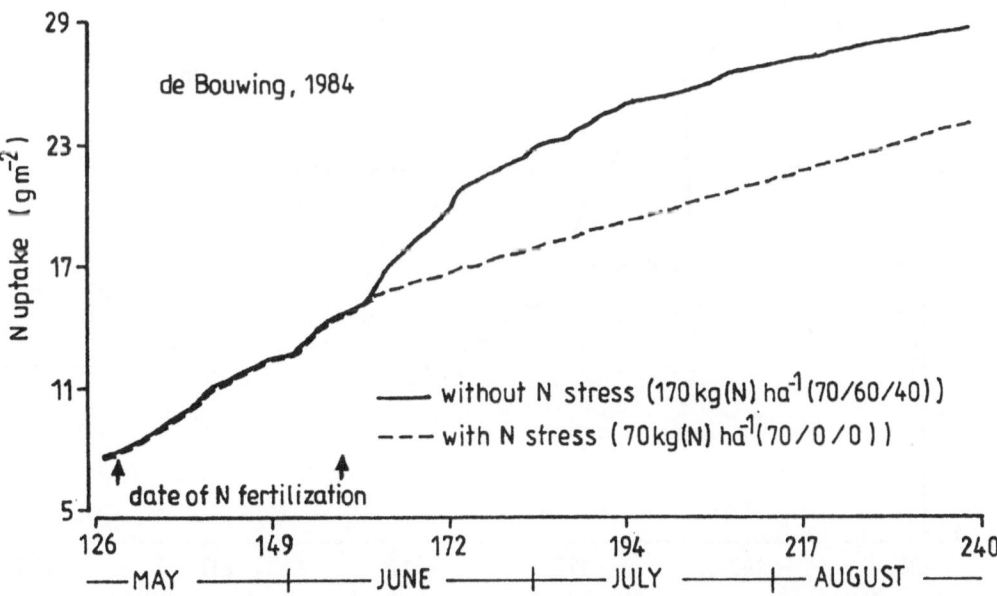

Fig. 8. Cumulative nitrogen uptake curves for a non-nitrogen-stressed and a nitrogen-stressed winter wheat crop (the Bouwing, 1984, cv. Arminda).

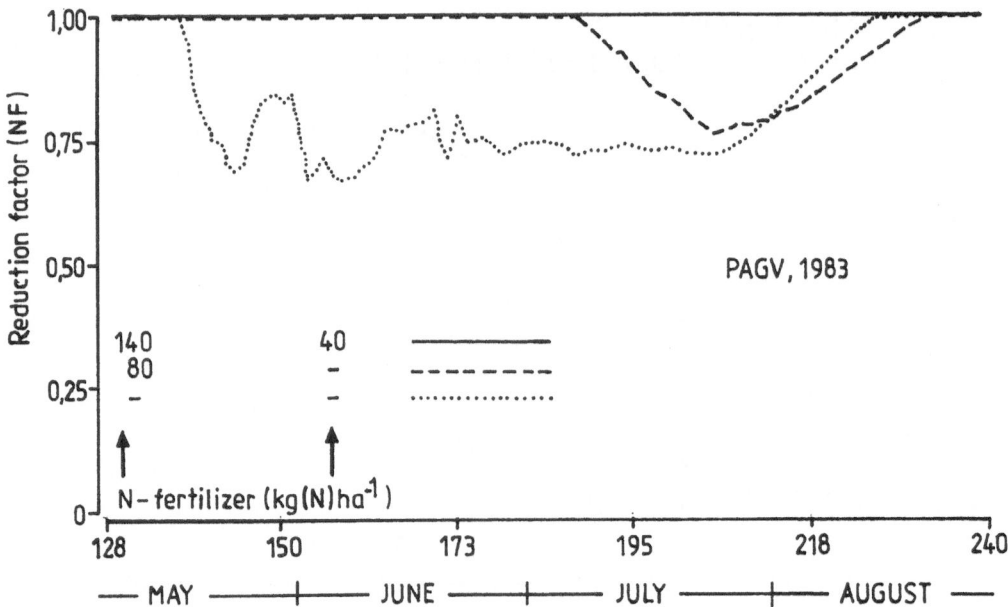

Fig. 9. Nitrogen stress reduction factor (*NF*) for three levels of nitrogen supply (PAGV, 1983).

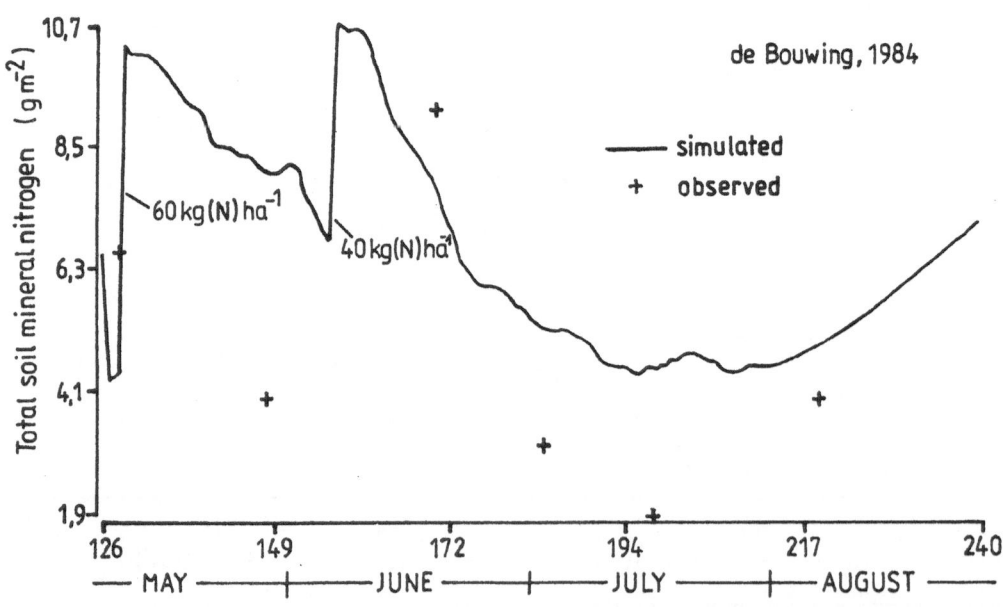

Fig. 10. Simulated and observed soil mineral nitrogen, the Bouwing, 1984 (N fertilization: 170 kg (N) ha^{-1}).

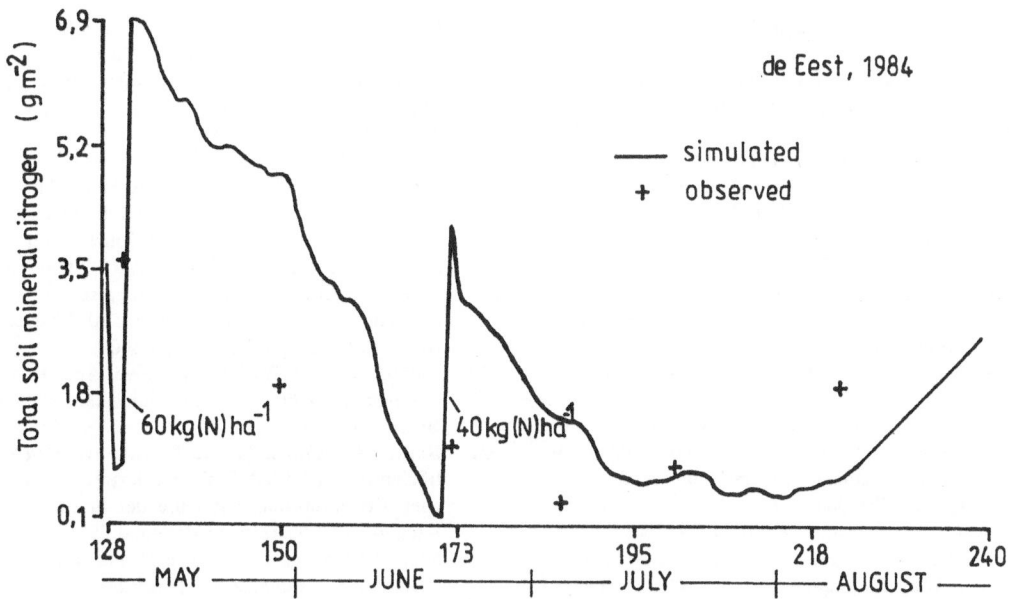

Fig. 11. Simulated and observed soil mineral nitrogen, the Eest, 1984 (N fertilization: 150 kg (N) ha^{-1}).

yield formation processes is an accurate description of the soil nitrogen dynamics throughout the growing season. In TRITSIM, the soil nitrogen balance model is very simple and only accounts for fertilization, mineralization and nitrogen uptake. As in the examples for the Bouwing (1984) (Fig. 10) and the Eest (1984) (Fig. 11), in the other treatments the agreement between simulations and observations is insufficient, although in general the time course of simulations is similar to the measurements. Only after a fertilizer application are the measured soil nitrogen values far below the simulations. The discrepancy is not only typical of the Dutch experiments [6] but also of experiments at Müncheberg (GDR) [12] and elsewhere [1]. One of the reasons for this phenomenon may be immobilization of nitrogen, which is not taken into account in the simple soil nitrogen balance model.

Acknowledgements

We thank Dr. Ir. J.J. Neeteson, Ir. J.J.R. Groot, Ir. E.L.J. Verberne and Dr. Ir. P. de Willigen from the Institute for Soil Fertility Research in Haren, the Netherlands, for arranging our participation in the workshop 'Nitrogen turnover in the soil-crop ecosystem: modelling of biological transformations, transport of nitrogen and nitrogen use efficiency'.

References

1. Aichberger K (1982) Veränderung des pflanzenverfügbaren Bodenstickstoffgehaltes (N_{min}) im Jahresverlauf. Die Bodenkultur 33: 277–288
2. Bellmann K, Matthäus E, Wenzel V and Ebert W (1985) AGROSIM-W, a SONCHES-implemented interactive simulation model for an agroecosystem winter wheat. In: 30. Intern Wiss Koll TH Ilmenau 1985, Vortragsreihe 'Technische Kybernetik/Automatisierungstechnik', GDR, pp 43–47
3. Bohne K and Koitzsch R (1986) Ein digitales Modell zur Simulation des Bodenfeuchteverlaufes für mehrjährige Zeitspannen. Arch Acker Pflanzenbau Bodenkd 30: 395–405
4. Ebert W, Bellmann K, Matthäus E and Wenzel V (1985) Modelling and simulation of agroecosystem – the winter wheat agroecosystem, AGROSIM-W. In: Sydow A, Thoma M and Vichnevetsky R (eds) Systems Analysis and Simulation 1985-II. Applications. Proceedings of the International Symposium Berlin (26–31 Aug. 1985), Akademie-Verlag Berlin, pp 64–77
5. Glugla G (1970) Zur Berechnung des aktuellen Wassergehaltes und des Gravitationswasserabflusses im Boden. Diss. A, Karl-Marx-Univ. Leipzig
6. Groot JJR and Verberne ELJ (1991) Response of wheat to nitrogen fertilization, a data set to validate simulation

304

models for nitrogen dynamics in crop and soil. Fert Res 27: 349–383

7. Hodges T and Kanemasu ET (1977) Modelling daily dry matter production of winter wheat. Agron J 69: 974–978

8. Hunkar M and Zemancovics F (1988) Studies on the adaptation of a winter wheat model. In: Sydow A, Thoma M and Vichnevetsky R (eds) Systems Analysis and Simulation 1985-II. Applications. Proceedings of the International Symposium Berlin (26–31 Aug. 1985), Akademie-Verlag Berlin, pp 233–240

9. Knijnenburg A, Matthäus E and Wenzel V (1984) Concept and usage of the interactive simulation systems for ecosystems SONCHES. Ecological Modelling 26: 51–76

10. Koitzsch R (1977) Schaetzung der Bodenfeuchte aus meteorologischen Daten, Boden- und Pflanzenparametern mit einem Mehrschichtenmodell. Z Meteorologie 27: 302–306

11. Koitzsch R, Helling R and Vetterlein E (1980) Simulation des Bodenfeuchteverlaufes unter Beruecksichtigung der Wasserbewegung und des Wasserentzuges durch Pflanzenbestaende. Arch Acker- Pflanzenbau Bodenkd 24: 717–725

12. Kretschmer H and Groth R (1986) Dynamik des N_{an}-Gehaltes unter Winterweizenbeständen auf einem D2/D3-Standort bei unterschiedlicher Wasser- und N-Versorgung. In: Felddiagnose ertragsbestimmender Eigenschaften von Böden für die Prognose von Boden- und Ertragsentwicklungen und die Entscheidungsfindung für Steuerungsmaßnahmen. Bodenkundl Gesellsch DDR, Vorträge 16. Wiss Tagung. 18–20 March 1986, Erfurt (GDR), pp 53–54

13. Lidgate HJ (1984) Nitrogen uptake of winter wheat. In: Conference 'The nitrogen requirement of cereals.' September 1983, Ministry of Agriculture, Fisheries and Food, UK, pp 177–120

14. Matthäus E, Mirschel W, Kretschmer H, Künkel K and Klank I (1986) The winter wheat crop model TRITSIM of the agroecosystem AGROSIM-W. In: Tag.-Ber. der AdL der DDR (Computer Aided Modelling and Simulation of the Winter Wheat Agroecosystem (AGROSIM-W) for Integrated Pest Management) (242), pp 43–74

15. Mirschel W, Matthäus E and Kretschmer H (1989) Stand und Nutzungsmöglichkeiten des Winterweizenmodells TRITSIM. In: Reiner L, Geidl H and Mangstl A (eds) Agrarinformatik Informationsverarbeitung Agrarwissenschaft (16), pp 231–246, Stuttgart, Verlag Eugen Ulmer

16. Mirschel W and Klank I (1987) Abschaetzung von meteorologisch relevanten Inputgroessen fuer Wachstumsmodelle aus meteorologischen Standarddaten. Zeitschrift f Meteorologie 37: 195–210

17. Mirschel W, Klank I, Kretschmer H and Künkel K (1987) Dynamisches Ertragsbildungs- und Entwicklungsmodell TRITSIM für Winterweizen. 2. Mitteilung: Modellvalidierung und – verifizierung. Arch Acker- Pflanzenbau Bodenkd 31: 259–267

18. Mirschel W, Kretschmer H, Klank I, Matthäus E and Künkel K (1987) Dynamisches Ertragsbildungs- und Entwicklungsmodell TRITSIM fuer Winterweizen 1. Mitteilung: Modellbeschreibung. Arch Acker- Pflanzenbau Bodenkd 31: 249–257

19. Mirschel W, Wenkel K-O and Schultz A (1986) Dynamisches Ertragsbildungs- und Entwicklungsmodell (TRITSIM) für Winterweizen bei limitierter Wasser- und Stickstoffversorgung und Möglichkeiten zur Nutzung des Modells für agrotechnische Steuerentscheidungen. In: Kurzreferate zur 15. Jahrestagung der Modellierung und Simulationstechnik. 10–12 Dec. 1986, Rostock, GDR, pp 39/103 – 39/106

20. Nalborczyk E (1986) Quantitative Analyse der Produktivitaet der Pflanzen. Colloquia Pflanzenphysiologie 10: 19–24

21. Rausch H, Lüttich M and Freytag HE (1985) Quantifizierung der Stickstoffmineralisierung aus der organischen Bodensubstanz mit Hilfe der Stanford-Methode. Arch Acker- Pflanzenbau Bodenkd 29: 77–83

22. Schäfer W, Klank I, Weirauch M and Reiher W (1983) System zur automatischen Erfassung von Messwerten der CO_2-Assimilation und Transpiration unter Feldbedingungen. Arch Acker- Pflanzenbau Bodenkd 27: 361–366

23. Schliephake W (1984) Anforderungen der Wintergetreidearten an die N-Versorgung in ihrer Entwicklung bis zum Schossen. Diss. A, Martin-Luther-Univ. Halle, 127 pp

24. Stanford G (1977) Evaluating the nitrogen supply capacities of soils. In: SEFMIA proceedings of int. Sem. Soil environment and fertility management in intensive agriculture. Tokio, pp 412–418

25. Stanford G and Smith SJ (1972) Nitrogen mineralisation potential of soils. Soil Sci Soc Am J: 465–472

26. Turc L (1961) Evaluation des besoins en eau d'irrigation, evapotranspiration potentielle. Ann Agron Paris 12: 13–49

27. Vielemeyer H-P, Neubert P, Hundt I, Vanselou S and Weissert P (1983) Ein neues Verfahren zur Ableitung von Pflanzenanalyse-Grenzwerten fuer die Einschaetzung des Ernaehrungszustandes landwirtschaftlicher Kulturpflanzen. Arch Acker- Pflanzenbau Bodenkd 27: 445–453

28. Vielemeyer H-P, Weissert P and Podlesak W (1987) Untersuchungen des N-Aufnahmeverlaufes von Wintergetreide in Abhängigkeit vom Ertrag. Arch Acker- Pflanzenbau Bodenkd 31: 647–655

29. Weir AU, Bragg PL, Porter JR and Rayner JH (1984) A winter wheat crop simulation model without water or nutrient limitations. J Agric Sci 102: 371–382

30. Witt H (1974) Sortenpaß für den Winterweizen 'Alcedo' (Stand März 1974). Feldwirtschaft 15: 328

31. Zadoks JC, Chang TT and Konzak CF (1974) Decimal code for the growth stages of cereals. Weed Research 14: 415–431

Fertilizer Research **27**: 305–312, 1991.
© 1991 *Kluwer Academic Publishers.*

Simulation of nitrogen in soil and winter wheat crops: a management model that makes the best use of limited information

T.M. Addiscott[1] N.J. Bailey[1], G.J. Bland[2] & A.P. Whitmore[1]
[1]*Department of Soil Science, Rothamsted Experimental Station, Harpenden, Herts., AL5 2JQ, UK*
[2]*Present address*: *Earth Observation Sciences, Innovation Development Centre, Surrey Research Park, Guildford, Surrey, GU2 5YH, UK*

Key words: Nitrogen, modelling, management, winter wheat, soil, crop

Abstract

A model that simulates changes in mineral N in the soil and N uptake by crops has been adapted to require as little detailed information as possible so that it is useful as an aid to management. The adapted model, which was developed in the UK, was tested against data from six experiments on winter wheat in the Netherlands. It proved reasonably successful in simulating the amounts of mineral N found in the soil in early spring and the changes that resulted from applying small amounts of fertilizer N in February. It was much less successful in simulating the effects of later, larger applications of N, mainly because the mineral N measured in the soil did not seem to respond to these applications. The uptake of N by the crops and their production of dry matter were simulated very well in some cases and rather less so in others.

Introduction

Those who model the behaviour of nitrogen in soils and crops have to balance mechanistic rigour against practical usefulness. The more exactly a model tries to describe the processes involved, the more complex it becomes and the more difficult it is to use. Greater complexity means more parameters and often more difficulty in obtaining values for them. With models for solute transport, and maybe other processes too, there is another facet to the problem: the more mechanistic the model becomes the more sensitive it is likely to be to variation in its parameters. This can mean that ignoring the variances of the parameters when using the model can lead it to give an incorrect result. Models designed to help with the management of nitrogen in farming and the environment must avoid these problems. They are no use if they need detailed information that no farmer is likely to have, or if using them requires a high level of computer expertise. This paper describes a model that simulates the quantities of mineral N in the soil and of N in the crop from simple information. It was developed for management or advisory purposes and we show here the simulations that it gave for several field experiments made with winter wheat in the Netherlands [6].

The model

The model includes leaching, the mineralization of soil organic N and the growth and N uptake of the wheat crop. Its general structure is similar to that of a model published earlier [4] and it has mineralization [2] and crop [8] components that are essentially the same as in the earlier model. The main change is in the leaching section, where an early leaching model [1] has been replaced by the more recent SLIM model [3, 5].

The SLIM model divides the soil into layers, and the water and solute in each layer are separated into mobile and immobile categories. The model has two main parameters. One, the amount of immobile water, is a measure of the soil's capacity to hold back water and solute; the other, a permeability parameter, reflects the ease with which water can pass through the soil and carry solute with it. Both parameters can be derived from the particle size distribution of the soil [5]. This information is not always available, but the description of the soil is, and for the management model we evolved a 'menu' of soil types which were assigned likely particle size distributions on the basis of information published by the Soil Survey and Land Research Centre. This means that values of the parameters can be assigned for each soil type. If necessary the topsoil and subsoil can be treated as different in type from each other. The menu categorizes the soils as: 1 Clay, 2 Silty Clay, 3 Silty Clay Loam, 4 Clay Loam, 5 Sandy Clay Loam, 6 Silty Loam, 7 Sandy Silt Loam, 8 Sandy Loam, 9 Loamy Sand, 10 Sand (subsoil).

The mineralization model assumes simply that at any given temperature there is a steady release of mineral N from the soil organic matter such that the quantity mineralized increases linearly with time [2]. The rate of mineralization, i.e. the slope of the line, is related to temperature by the Arrhenius Equation, and is also slowed down by a shortage of water. Default values of the mineralization rate at a fixed temperature, 20°C, and optimum water content, between 0.05 and 0.33 bars tension, are assigned to the various topsoil types on the basis of previous experiments, but these can be replaced where a specific value is available.

The production of dry matter by the crop and the uptake of N are described by a simple sigmoid relationship that has two parameters: a rate constant and an asymptotic value towards which the upper part of the curve tends. The rate constant for dry matter production depends inversely on the date the crop is sown, expressed as the number of days after September 1, while that for N uptake depends directly on the number of days between sowing and the return of the soil to field capacity [8]. The latter is computed in the leaching model. The asymptotic values for dry matter production and N uptake have to be specified. This might normally be done on the basis of the farmers' previous experience but here we assumed general values of 17 t ha^{-1} of

dry matter (8.5 t ha^{-1} grain) and 190 kg ha^{-1} of N uptake. Dry matter production and N uptake are limited by shortages of water or nitrogen as shown elsewhere [4].

The model needs weather data: rain, evaporation and soil temperature, all on a daily basis. These and other input data needed by the model were supplied by Groot et al. [6]. At the start of each run, usually September 1, we need estimates of the soil moisture deficit and the quantity of mineral N in the soil profile. The deficit is usually available, but unless the mineral N has been measured after harvest we assume standard values that depend on the previous crop: after cereals or rape 50 kg ha^{-1}; after potatoes 90 kg ha^{-1}; after sugar beet 25 kg ha^{-1}; all to 0.9 m. The values for cereals, rape and potatoes are based on measurements in experiments at Rothamsted and sites nearby, while that for sugar beet was provided by J.J. Neeteson (personal communication).

The simulations

The data to be simulated were usually for the period from mid-February to harvest. The model, however, was run from September 1 through to harvest, so that it was tested during autumn and winter as well as during the main period of crop growth. Table 1 summarizes the data used as inputs to the model, and further details of the experiments are given elsewhere [6]. Because we were using the model as far as possible as it would be used for management purposes, we did not use the detailed soil information; we simply assigned the soil to what we judged to be the most appropriate category in the 'soil menu'. The information shown in the table is the number of the category as selected from the menu.

The following points in Table 1 possibly need explanation:
1. The experiment at Bouwing 1983/4. The initial mineral N and the mineralization rate are decreased by one-third and one-half, respectively. This was because the preceding crop of potatoes had received its fertilizer nearly four months before it was planted. There was

Table 1. Main inputs to model

Site	Soil type*		Weather area	Season	Previous crop	Min N Sept 1 (kg ha^{-1})	Mineralization rate* (mg l^{-1} d^{-1})	Soil moisture deficit (mm) Sept 1
	Top	Sub						
Bouwing	2	2	Wageningen	1982/3	Potatoes	90	0.3	182
				1983/4	Potatoes	30*	0.15*	157
Eest	3	2	Polder (Lelystad)	1982/3	Potatoes	90	0.05*	171
				1983/4	Potatoes	90	0.05*	161
PAGV	8	10	Polder (Lelystad)	1982/3	Sugar beet	25	0.15	171
				1983/4	Sugar beet	25	0.15	161

* Please see comments in text.

much rain in those four months, so little nitrate would have remained in the soil at harvest. The crop must have taken up much less nitrate than usual and left residues that were relatively poor in N.

2. Mineralization rates. The rate assigned to the Bouwing site in 1982/3 is the same as that measured in a somewhat similar soil at Rothamsted, while that assigned to the sandy

PAGV site is slightly larger than the rate for the very sandy soil at Woburn, 0.12 mg l^{-1} d^{-1}. At the Eest site the model initially simulated far too much mineral N in the soil, so the mineralization rate was brought down to a small value. There was no real rationale for this, but it was the only case in which an input was given a value purely to improve the fit.

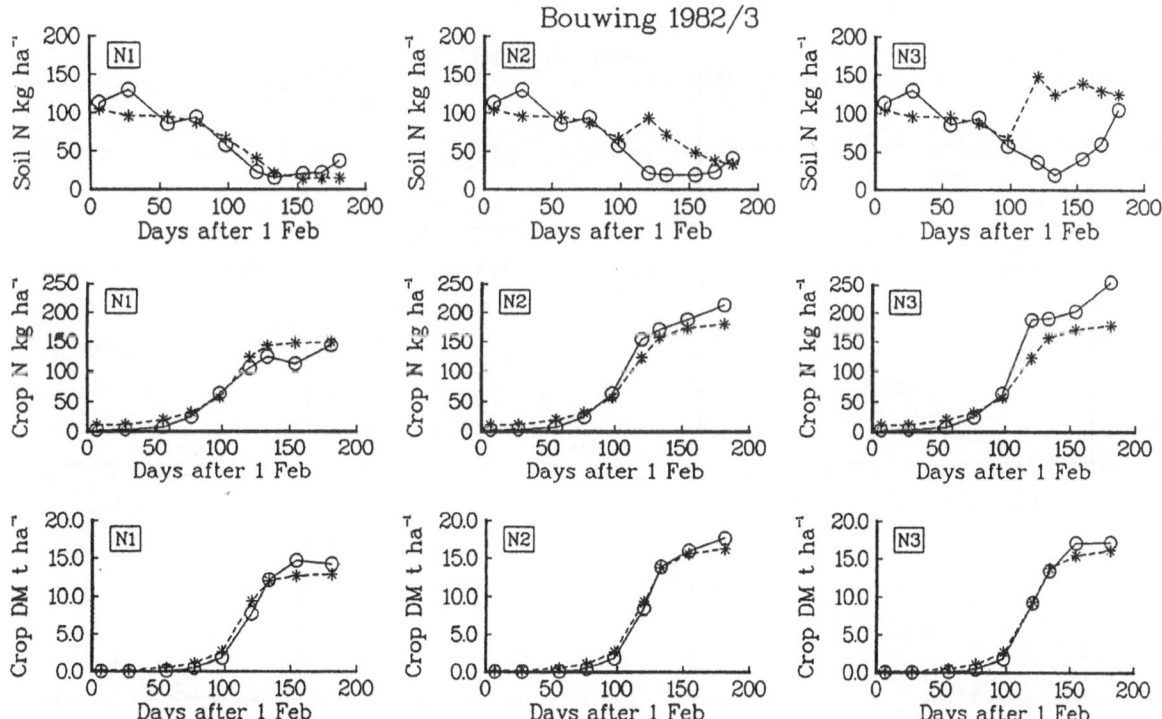

Fig. 1. The Bouwing experiment 1982/3. Soil mineral N ($NH_4^+ + NO_3^-$) to 0.9 m, N in crop, dry matter of crop. −O− measured; -- * -- simulated.

Results

The Bouwing site 1982/3

The model generally performed well in this experiment. Soil mineral N was measured for the first time on February 7 and the simulation that was started on September 1 came within a few kg ha^{-1} of the measured value (Fig. 1). The simulation continued well throughout the growing season for the N_1 treatment (no N fertilizer). For the N_2 (60 kg (N) ha^{-1}) and N_3 (120 + 40 kg (N) ha^{-1}) treatments the simulation was not so good because the simulation showed an increase in mineral N when the fertilizer was applied but the measurements did not. Measured and simulated values came close to each other, however, at the final sampling. The model simulated N uptake well for the N_1 treatment and rather less well for the N_2 and N_3 treatments, but all three simulations of dry matter production were good.

The Bouwing site 1983/4

The model performed rather less well than in the previous season at this site, perhaps because the allowance made for the problems with the previous potato crop was inadequate or inappropriate. The first measurement of mineral N was on February 7 and the simulation started on September 1 came very close to it (Fig. 2). An application of 70 kg (N) ha^{-1} made shortly after this measurement was reflected in almost exactly the same way by both simulation and measurement up to the fourth sampling. After this the model overestimated the mineral N for the N_1 treatment (no N beyond the 70 kg ha^{-1}) partly because it underestimated the amount of N in the crop. It also overestimated the mineral N for the N_2 (60 + 40 kg (N) ha^{-1}) and N_3 (120 + 40 kg (N) ha^{-1}), but this was mainly because the measurements seemed to respond only when the final 40 kg (N) ha^{-1} was applied, whereas the simulation responded to the 60 kg ha^{-1} and the

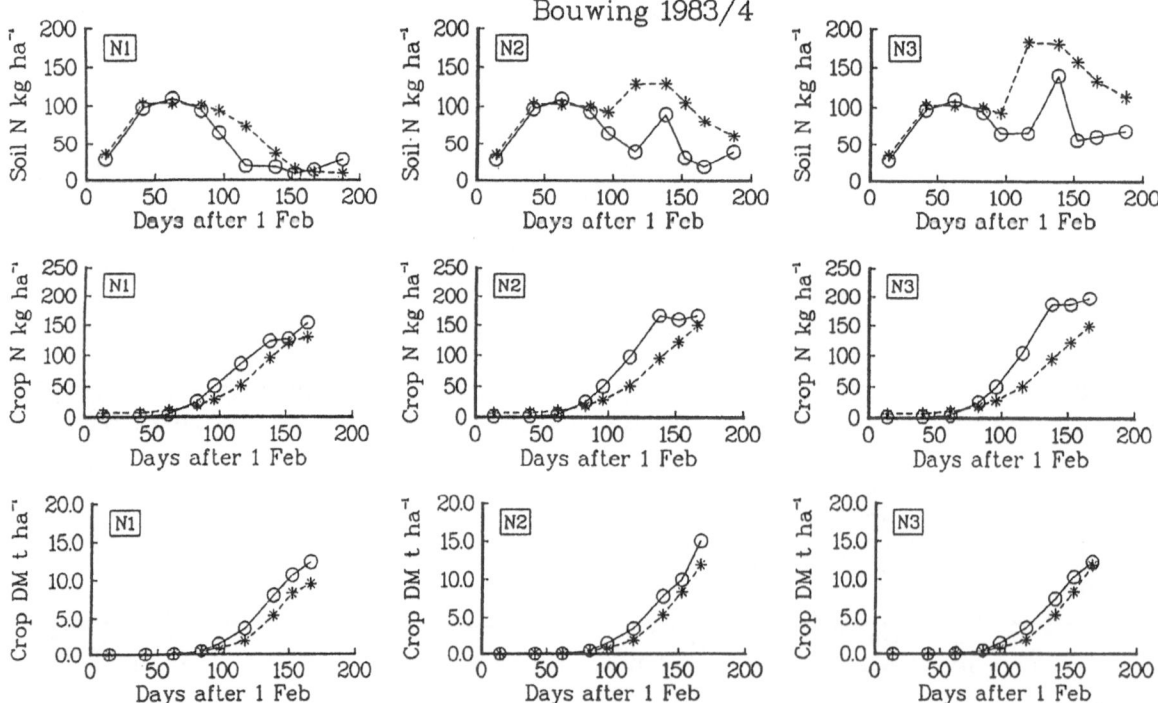

Fig. 2. The Bouwing experiment 1983/4. As for Fig. 1.

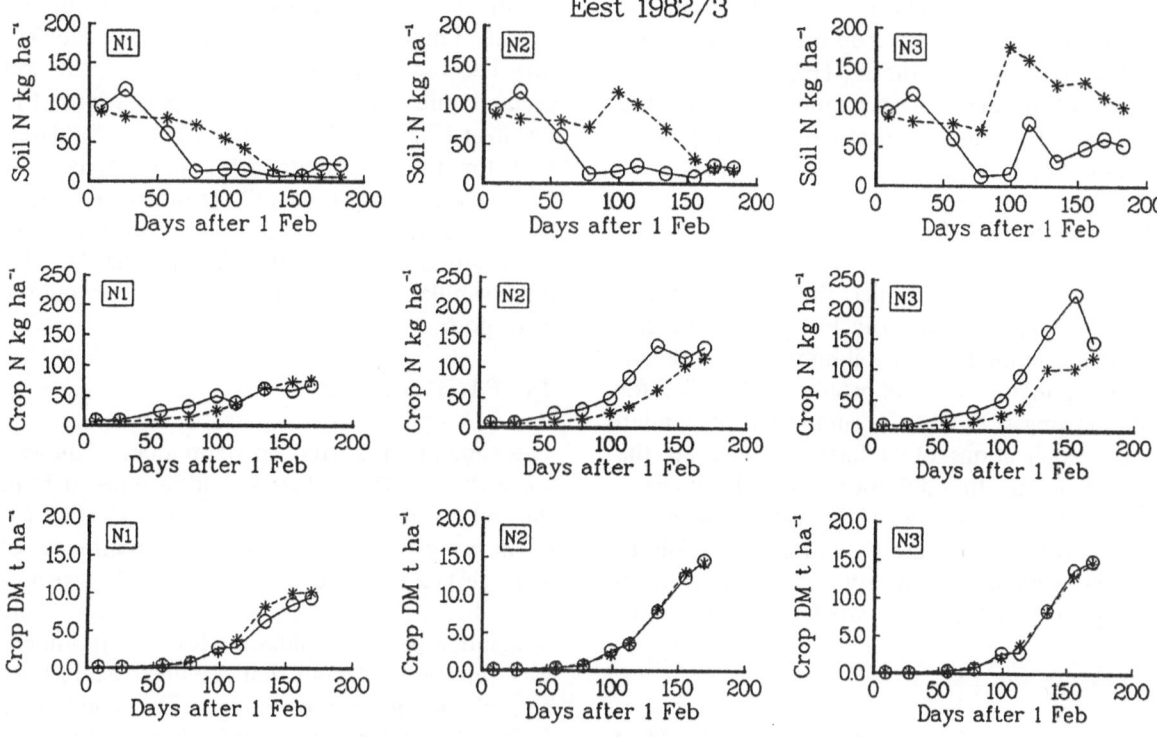

Fig. 3. The Eest experiment 1982/3. As for Fig. 1.

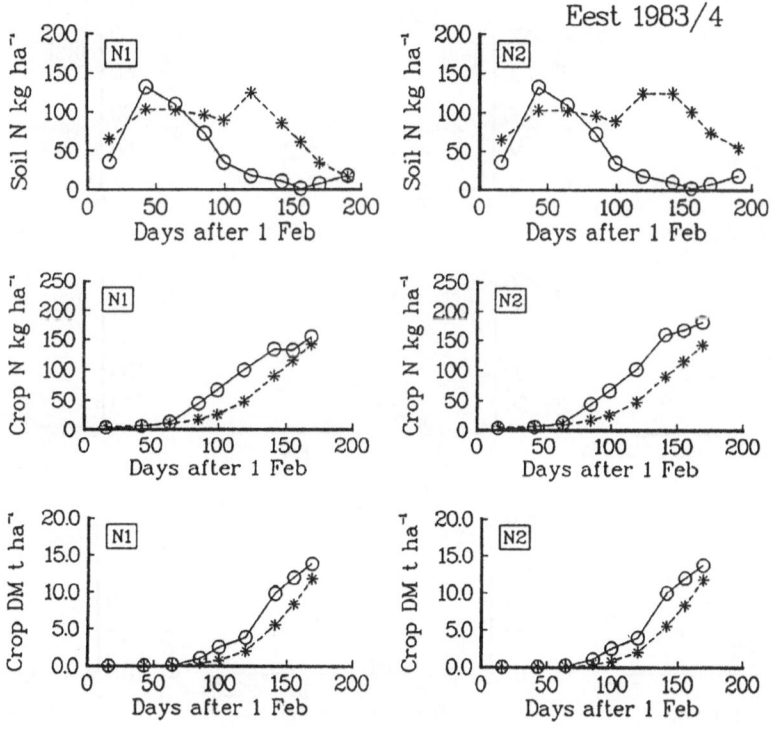

Fig. 4. The Eest experiment 1982/3. As for Fig. 1.

120 kg ha^{-1} as well. The model underestimated the quantity of N in the crop for the N_2 and N_3 treatments as well as the N_1 (Fig. 2). The production of dry matter was simulated best for the N_3 curve. The model seemed to overemphasise the effect of N shortage on the N_1 crop.

The Eest site 1982/3 and 1983/4

This was the site for which the rate of mineralization was set arbitrarily low, so we do not report the simulations of soil mineral N in detail. The main feature was the problem, identified for the Bouwing site, that the simulation responded to the applications of N fertilizer whereas the measurements often did not seem to do so (Figs. 3, 4). The simulations of N uptake generally tended to underestimate the amounts measured, but the simulations of dry matter production were good in 1983/4 and fair the following year.

The PAGV site 1982/3

The simulation of the amount of mineral N at the first sampling (February 8) was not as good as for the other experiments. Once again the overall simulation of mineral N was not satisfactory because the amounts measured in the soil seemed quite unaffected by the applications of N fertilizer (Fig. 5). The uptake of N was simulated well for the N_1 treatment (80 kg (N) ha^{-1} in February), not quite so well for the N_2 treatment (60 + 80 kg (N) ha^{-1}) and rather poorly for the N_3 treatment (60 + 140 + 40 kg (N) ha^{-1}). All three simulations of the production of dry matter were poor.

The PAGV site 1983/4

The simulation started on September 1 showed correctly that there was very little mineral N in the soil at the sampling on February 15 (Fig. 6). It also responded to the N_1 treatment of 80 kg (N) ha^{-1} in the same way as the measurement but not quite to the same extent. The remainder of the simulation for this treatment was quite good except that the decrease in mineral N was not sharp enough. The simulations for the N_2 and N_3 treatments showed the usual problem that the simulations responded to the

PAGV 1982/3

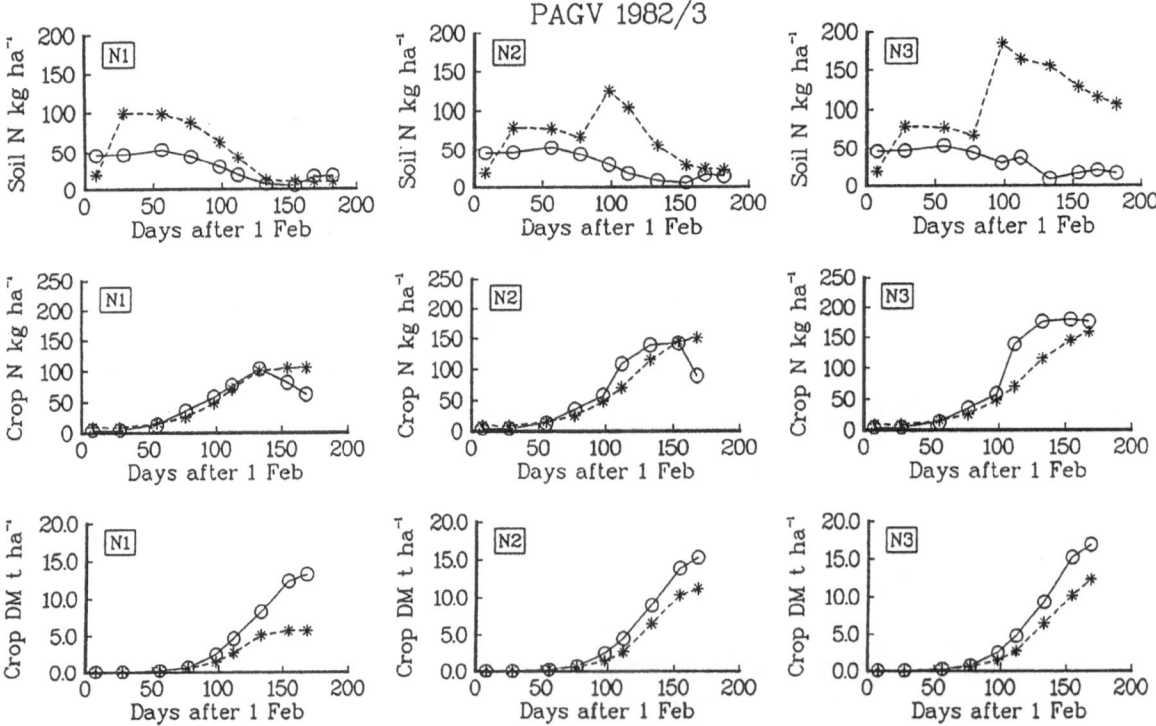

Fig. 5. The PAGV experiment 1982/3. As for Fig. 1.

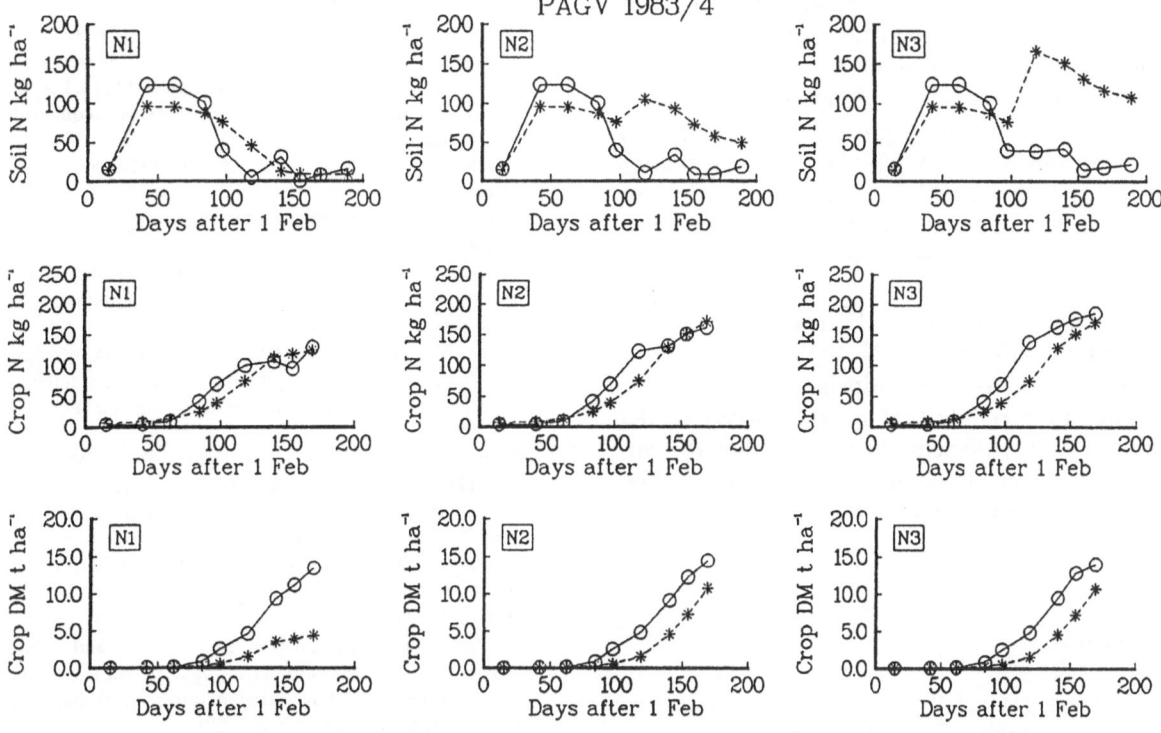

Fig. 6. The PAGV experiment 1983/4. As for Fig. 1.

applications of fertilizer but the measurements did not. The simulations of N uptake from the N_1 and N_2 treatments (80 kg (N) ha^{-1} and 80 + 60 + 40 kg ha^{-1}, respectively) were reasonably good. That for the N_3 treatment, in which the second application was increased to 120 kg ha^{-1}, underestimated the measured uptakes. All three simulations of the production of dry matter were again poor.

Discussion

The model was developed and validated mainly for simulating changes in the mineral N in the soil and the N in the crop during autumn, winter and spring – before the main applications of fertilizer. It has not been used very much for tracing the fate of N applied as fertilizer. The quantity of mineral N in the soil in February was simulated well from minimal information for the two Bouwing experiments and one of the PAGV experiments. It was also simulated reasonably well for the experiments at Eest, but this was partly because of the arbitrarily small rate of mineralization. The main simulation problems came when the measurements of mineral N showed little or no response to N applied as fertilizer. The measurements did respond to three of the four early (February) applications of fertilizer N, and where this happened the simulations followed the measurements, mainly reasonably closely. Later applications of N brought a response in the measurements that was delayed, small or non-existent. The response in the simulations was always larger. Simulation and measurement sometimes, but by no means always, came closer together at the final sampling, as might have been expected if there had been some kind of short-term fixation or immobilization followed by release. Other experiments have shown N applied as fertilizer seemingly to disappear and then reappear in the soil within weeks [7], and possible reasons are discussed in a companion paper [9].

Some of the simulations of N uptake and dry matter production were much better than others. One possible reason for the poorer simulations

312

may lie in the routine in the model that holds back N uptake and the production of dry matter when water or N is simulated to be in short supply. This routine may function too fiercely. Also the general values given to the asymptotes towards which the N uptake and dry matter curves tended may have been inappropriate to some of the combinations of site and treatment.

The idea that the measurements must be right is axiomatic to the concept of validating a model. Measurements of mineral N in the soil that do not respond to mineral N applied as fertilizer place a strain on this principle, but we have perhaps to conclude that we do not yet model mineral N in sufficient detail. Whether the appropriate level of detail, if achieved, will be suitable for a management model seems open to question.

References

1. Addiscott TM (1977) A simple computer model for leaching in structured soils. J Soil Sci 28: 554–563

2. Addiscott TM (1983) Kinetics and temperature relationships of mineralization and nitrification in Rothamsted soils with differing histories. J Soil Sci 34: 343–353

3. Addiscott TM, Heys PJ and Whitmore AP (1986) Application of simple leaching models in heterogeneous soils. Geoderma 38: 185–194

4. Addiscott TM and Whitmore AP (1987) Computer simulation of changes in soil mineral nitrogen and crop nitrogen during autumn, winter and spring. J Agric Sci, Camb 109: 141–157

5. Addiscott TM and Whitmore AP (1991) Simulation of solute leaching in soils of differing permeabilities. Soil Use Management (in press)

6. Groot JRR and Verberne ELJ (1991) Response of wheat to nitrogen fertilization, a data set to validate simulation models for nitrogen dynamics in crop and soil. Fert Res 27: 349–383

7. Weir AH, Addiscott TM, Barraclough PB and Whitmore AP (1987) Computer simulation of N-leaching – the rapid immobilization of fertilizer N. Rep Rothamsted Exp Stn for 1986, pp 149–150

8. Whitmore AP and Addiscott TM (1987) A function for describing nitrogen uptake, dry matter production and rooting by wheat crops. Plant Soil 101: 51–60

9. Whitmore AP, Coleman KW, Bradbury NJ and Addiscott TM (1991) Simulation of nitrogen in soil and winter wheat crops: modelling nitrogen turn-over through organic matter. Fert Res 27: 283–291

Fertilizer Research **27**: 313–329, 1991.
© 1991 *Kluwer Academic Publishers.*

Modelling water flow, nitrogen uptake and production for wheat

H. Eckersten & P.-E. Jansson
Department of Soil Sciences, Swedish University of Agricultural Sciences, P.O. Box 7014, S-75007 Uppsala, Sweden

Key words: Allocation, bypass flow, water table depth, macropores, soil processes, validation

Abstract

Soil water and temperature conditions were simulated for three years at three sites in the Netherlands, using a model named SOIL. Observed water table depths from one site with a sandy loam soil indicated bypass flow in macropores. Nitrogen turnover was simulated using the output of SOIL as input to a nitrogen model. To improve the nitrogen model, a crop-growth submodel was introduced, and simulations were compared with measured data for two seasons and three fertilizer treatments at the three sites. Mineral-N in the soil after application of fertilizer was substantially higher in the simulation than indicated by measurements in 4 out of 18 simulations. Regression analyses showed that simulated mineral-N content in the uppermost metre explained 64% of the observed variation. The corresponding values for nitrogen content (N_{ta}) and biomass (W_{ta}) of aboveground tissues were 86 and 93%, respectively. With a few exceptions annual values of W_{ta} and N_{ta} were simulated with an accuracy of approximately 20%. A sensitivity test showed that growth parameters and especially the light use efficiency parameter strongly influenced biomass production for fertilized treatments whereas the control of nitrogen uptake from soil was most important for non-fertilized treatments.

Introduction

The nitrogen cycle in arable land has been formulated in many different mathematical models. The SOILN model was developed [20] emphasizing the need for a model of soil nitrogen transformations and losses. The model was formulated to agree with the current level of understanding concerning soil processes, but also such that it could be widely applied using information generally available from field investigations of nitrogen turnover.

From independent applications of the model we have found that an accurate estimate of nitrogen uptake by crops is crucial when predicting the nitrogen losses from soil. In these studies [3, 20], nitrogen uptake was simulated using a time dependent empirical function requiring parameter values unique for the site concerned. In this paper, we introduce a new crop growth model which regulates the uptake rate and which can be parameterized from independent data sets. This model is named the CROP-GROWTH submodel.

Driving soil variables in SOILN were supplied by the SOIL model which simulated energy and water processes in the soil. Since both SOIL and SOILN have been described elsewhere [15, 20], we focus the model description on the CROP-GROWTH submodel. Hence, SOIL is only briefly described here, except for the part which deals with water flow in the soil. A description of the original version of SOILN is given by Bergström et al. [4].

Description of the SOIL model

All symbols in the model description are explained and all parameter values used in the simulations are given in the appendix.

A water and heat model [15] provides driving

variables for the nitrogen model (e.g. surface runoff and infiltration, soil water flow between layers and flow to drainage tiles, soil water content, soil temperature and evapotranspiration). The water and heat model predicts these variables using standard meteorological data as driving variables. The water and heat model is based on two coupled differential equations describing heat and water transport (derived from Fourier's and Darcy's laws respectively) in a one-dimensional layered soil profile. Phenomena like freezing/thawing of the soil and soil boundary processes such as snow-melt and interception of precipitation are considered in the model. A detailed description of the model for arable land conditions is given elsewhere [16, 18].

An option to account for bypass flow was included in the model to consider rapid flows in macropores during conditions when smaller pores are only partially filled with water (Fig. 1). The amount of water in the macropores is not accounted for explicitly. Instead, the infiltration flow rate at the soil surface or the vertical flow in the macropores at any depth in the soil profile (q_{in}) determines the partitioning into the ordinary Darcy flow (q_{mat}) and the bypass flow (q_{bypass}). At the soil surface, q_{in} is the infiltration rate. At other depths, q_{in} is the vertical flow rate in the macropores (q_{bypass}) from the layer above. We have:

$$q_{mat} = \max(k(\theta)(d\psi/dz + 1), q_{in})$$
$$0 < q_{in} < S_{mat} \quad (1)$$

$$q_{bypass} = 0 \qquad\qquad 0 < q_{in} < S_{ma} \quad (2)$$

$$q_{mat} = S_{mat} \qquad\qquad q_{in} > S_{mat} \quad (3)$$

$$q_{bypass} = q_{in} - q_{mat} \qquad q_{in} > S_{mat} \quad (4)$$

where $k(\theta)$ is the unsaturated conductivity at a given water content, ψ the water tension and z the depth coordinate. The sorption capacity is defined as:

$$S_{mat} = A_{scale} A_{rel} k_{mat} pF \qquad (5)$$

where k_{mat} is the maximum conductivity of smaller pores (i.e. matrix pores), pF is ^{10}log of ψ and A_{rel} is the ratio between the vertical area of the

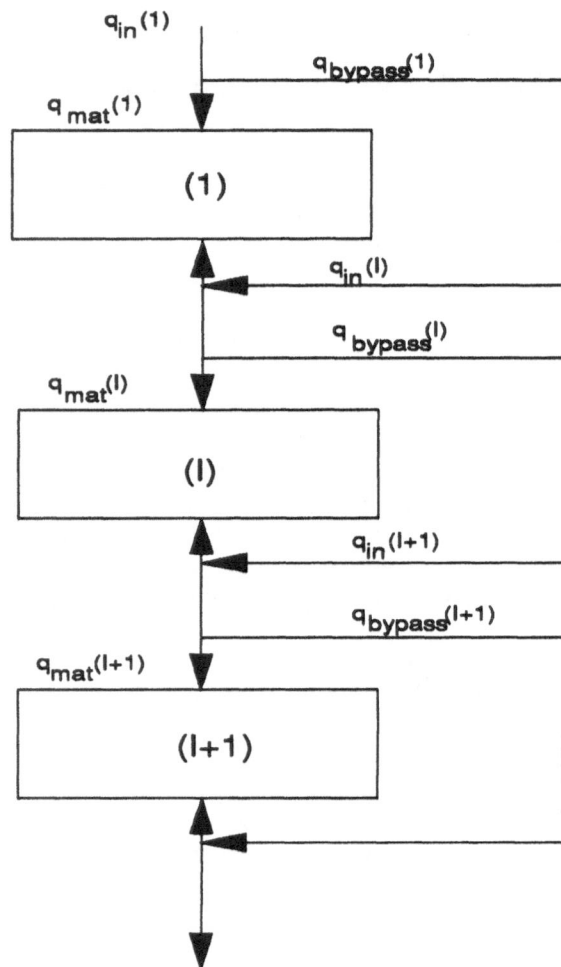

Fig. 1. Flow chart for bypass flow in the SOIL model. For explanation of the symbols is referred to the appendix.

compartment and the unit horizontal area. The vertical area has a depth equal to the compartment thickness and a horizontal length equal to that of the unit horizontal area. A_{scale} is an empirical scaling coefficient accounting for the geometry of aggregates.

The calculated water flow in the matrix (q_{mat}) is used to update the water contents and the water tensions in the numerical solution, whereas q_{bypass} is directed without delay to the next soil compartment. However, q_{bypass} can never reach layers below the water table depth, which is the lower boundary condition for the use of Richards equation.

Only the simulated flow rate is considered in the nitrogen model. Hence, the partitioning between bypass and matrix water flow does not influence the solute transport.

In the saturated zone of the soil profile, a sink flow to drainage pipes (q_{pipe}) is calculated for all compartments above the drain depth, z_{pipe}. The flow is calculated using Darcy's law as:

$$q_{pipe} = k_s A_{rel}((z_{pipe} - z_{gw})/L) \qquad (6)$$

where k_s is the saturated conductivity, z_{gw} the depth to the water table and L a characteristic distance between drainage pipes. The vertical water flow in the saturated zone of the profile is calculated assuming water contents equal to the soil porosity in all layers except for the compartment where the water table depth is located. Conservation of mass is satisfied by redistributing between compartments.

Description of the CROP-GROWTH submodel

CROP-GROWTH is a submodel which is included in the nitrogen turnover model SOILN although it is treated separately from the SOILN model in the description below. The amount of nitrogen in the soil simulated by SOILN is input to CROP-GROWTH. In turn the nitrogen up-

take predicted by CROP-GROWTH is input to SOILN. This interaction is made on a daily basis.

CROP-GROWTH can conceptually be divided into two submodels (Fig. 2). A biomass submodel (i) simulates production, based on the conversion of absorbed light into biomass [21] and empirical allometric functions. The growth equations and the functions determining the grain development were originally developed for field crops [1, 2] while the allometric functions (except for grain development) originate from measurements in short rotated willow stands [8]. A nitrogen submodel (ii) simulates plant uptake based on the idea that growth is the driving force for uptake [12], although it may be limited by the availability of nitrogen in the soil. The equations for nitrogen allocation in the plant (except for grain development) are similar to those used for willow [8] while the relations between uptake rate and root distribution were originally derived for field crops [20].

The two submodels are linked in the sense that: (i) the biomass submodel uses the leaf nitrogen content as an input variable taken from the nitrogen submodel and (ii) the nitrogen submodel, in turn, uses the daily growth of leaf, straw, root and grain as input taken from the

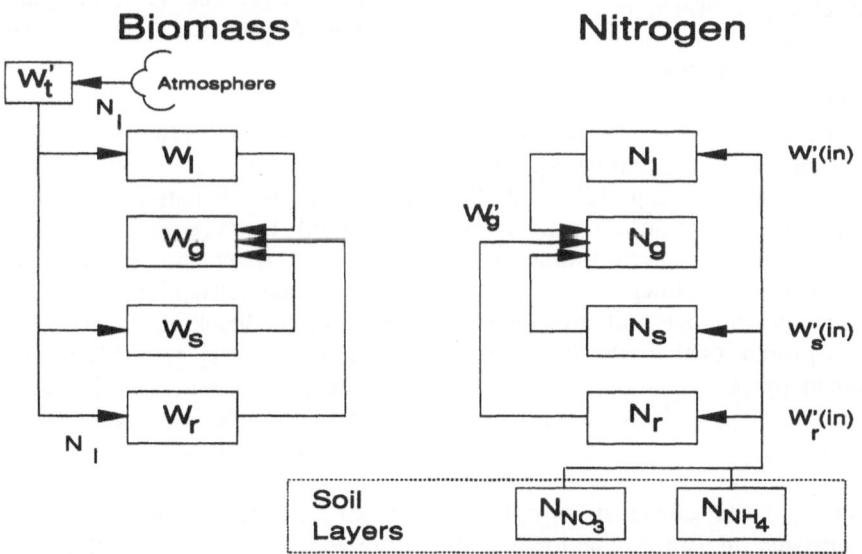

Fig. 2. Schematic description of the biomass submodel (left) and the nitrogen submodel (right). W and N are biomass and nitrogen respectively. Symbols outside boxes indicate how the submodels are connected. Indices are as follows: l = leaf; r = root; s = straw; g = grain and t = total. Inverted commas denote a daily change and (in) denotes a positive change.

biomass submodel (see Fig. 2). The combined model is relevant for horizontally uniform stands where growth is not altered by flowering, frost damage, insect attacks or similar disturbances. Growth is assumed not to be limited by nutrients other than nitrogen. Furthermore, aboveground litter fall is not explicitly accounted for but instead lumped together in the 'grain compartment'. This approach, instead of separating grain and litter in two compartments, has only a minor effect on the simulated development of the other tissues, since the grain compartment does not affect growth or nitrogen uptake. For simulations longer than one season or predictions of grain yield the litter must be distinguished from the grain.

The submodel runs with a daily time step. For each day, the calculations of the biomass submodel are made before those of the nitrogen submodel. Except for the vertical distribution of roots and the ratio between actual and potential evapotranspiration, environmental variables are the only required inputs. These variables are day length, daily sum of global radiation, daily mean air temperature, and soil nitrogen content. The latter is predicted by the SOILN model.

Biomass submodel

Growth starts at day t_0 which is the first day when the daily mean air temperature exceeds the minimum temperature for photosynthesis (T_{Min}). An initial amount of biomass is developed from assimilates stored in the seeds. The assimilates are partitioned between leaf, straw and root according to allocation equations (Eqs. 12–20, below). The newly formed tissues are assumed to receive maximum nitrogen concentrations from the seed (cf. Eqs. 27–29, below).

The daily total growth per unit area of soil surface (W'_t) is proportional to the light intercepted by the canopy (I_i):

$$W'_t = \varepsilon I_i f_T f_N f_W \tag{7}$$

where ε is the light use efficiency at optimal temperature, nitrogen and water conditions and f_T, f_N and f_W are response functions ranging between zero and unity depending on temperature, plant nitrogen and water status respectively. Inverted commas denote a daily change in the state variable.

I_i is calculated from the global radiation (I) according to Beers' law using the radiation extinction coefficient (k) and the leaf area index (A_{li}):

$$I_i = I(1 - \exp(-kA_{li})) \tag{8}$$

The temperature response function (f_T) increases linearly from zero to unity as the daily mean air temperature (T) increases from T_{Min} to T_{Max}:

$$f_T = (T - T_{Min})/(T_{Max} - T_{Min}) \quad 0 \leqslant f_T \leqslant 1 \tag{9}$$

The nitrogen response function (f_N) is unity when the leaf nitrogen concentration (n_l) equals its maximum value (n_{lMax}) and decreases linearly to zero as n_l decreases to the minimum nitrogen concentration (n_{lMin}).

$$f_N = (n_l - n_{lMin})/(n_{lMax} - n_{lMin}) \quad 0 \leqslant f_N \leqslant 1 \tag{10}$$

Finally, the water response function (f_W) equals the ratio between the actual evapotranspiration (E) and the potential evapotranspiration (E_p).

$$f_W = E/E_p \quad 0 \leqslant f_W \leqslant 1 \tag{11}$$

where E_p is calculated according to the Penman-Monteith formula and E is a function of E_p. It should be noted that E and E_p are calculated in SOIL using a 'standard' leaf area development. Thus no feedback mechanism exists between leaf area predicted by CROP-GROWTH and f_W.

The plant biomass is divided into four compartments: root (W_r), straw (W_s), leaf (W_l) and grain (W_g). Allocation of assimilates among the compartments depends on the grain development. The grain starts to receive assimilates from the other tissues when the accumulated daily sum (i_v) of an index becomes unity [2] (i.e. i_v acts as a switch that is used for starting the grain development).

$$i_v = \sum_{t=t_0}^{t} c_0(1 - \exp(c_1(T - C_2)))$$
$$\times (1 - \exp(c_3(D - c_4))) \qquad (12)$$

where D is the day length, T the daily mean air temperature and c_0, c_1, c_2, c_3 and c_4 are coefficients. In the case of grain development (i.e. $i_v \geqslant 1$) the daily increase of grain biomass (W_g) is a fraction (b_g) of the biomass in the other tissues:

$$W'_g = b_g(W_l + W_s + W_r) \quad b_g = 0 \text{ if } i_v < 1 \quad (13)$$

The amount of assimilates allocated to roots (W_r') is a fraction (b_r) of the total daily growth, minus translocation of assimilates to grain during the grain period.

$$W'_r = b_r W'_t - b_g W_r \quad b_g = 0 \text{ if } i_v < 1 \qquad (14)$$

b_r is dependent on the leaf nitrogen concentration (n_l), having a minimum value (b_{ro}) at the maximum leaf nitrogen concentration (n_{lMax}) and increasing as n_l decreases [7].

$$b_r = b_{ro} + 1 - (1 - ((n_{lMax} - n_l)/n_{lMax})^2)^{0.5} \qquad (15)$$

The remaining part of the daily total growth is allocated to the aboveground biomass, W_{ta} ($= W_l + W_s + W_g$).

$$W'_{ta} = (1 - b_r)W'_t \qquad (16)$$

W'_{ta} is partitioned between leaf and straw in accordance with the leaf area index development (A'_{li}) and the specific leaf area (a_{ls}). We assume that a balance exists between A_{li} and W_{ta}, expressed as the ratio b_i ($= A_{li}/W_{ta}$) which decreases with plant size (i.e. $b_i = b_{io} - b_{i1} \ln(W_{ta})$). Then A'_{li} (in) is given according to [7]:

$$A'_{li}(\text{in}) = W'_{ta}(b_{io} - b_{i1}(1 + \ln(W_{ta})))$$
$$A'_{li}(\text{in}) \geqslant 0 \quad (17)$$

where b_{io} and b_{i1} are coefficients. It is necessary to restrict A'_{li} (in) such that it cannot attain larger values than that possible from the daily shoot growth and the specific leaf area (i.e. the condition A'_{li} (in) $\leqslant a_{ls} W'_{ta}$ must be met). A decrease of A_{li} is considered by the loss of leaf biomass to the grain compartment. If we denote the leaf biomass increase with W'_l(in) then the net leaf growth is given by:

$$W'_l = W'_l(\text{in}) - b_g W_l \quad b_g = 0 \text{ if } i_v < 1 \qquad (18)$$

where

$$W'_l(\text{in}) = A'_{li}(\text{in})/a_{ls} \qquad (19)$$

The straw growth (W'_s) is the rest of the total daily increase minus the amount translocated to grain.

$$W'_s = W'_{ta} - W'_l(\text{in}) - b_g W_s \quad b_g = 0 \text{ if } i_v < 1 \qquad (20)$$

Nitrogen submodel

Nitrogen available for root uptake increases by the supply of nitrogen from mineralization, fertilization and atmospheric deposition and decreases by immobilization, leaching, denitrification and plant uptake. The combined effect of these factors gives the amount of nitrate (N_{NO_3}) and ammonium (N_{NH_4}) in different soil layers readily available for plant uptake. All factors affecting these pools, except plant uptake, are calculated by routines in the SOILN model other than the CROP-GROWTH submodel and are described by Bergström et al. [4].

Daily plant uptake of nitrogen from a soil layer (X_{Nu}) is the sum of uptake from the pools of N_{NO_3} and N_{NH_4}. The uptake from each pool is the available fraction (c_u) of the amount of nitrogen in the pool concerned. However, in both cases the uptake is proportional to the relative amounts of N_{NO_3} and N_{NH_4}, x_{fra} ($= N_{NO_3}/(N_{NO_3} + N_{NH_4})$) respectively $1 - x_{fra}$. If the daily uptake from each pool is N'_{NO_3} (out) and N'_{NH_4} (out) respectively, we have for a given layer (i):

$$X_{Nu}(i) = N'_{NO_3}(\text{out})(i) + N'_{NH_4}(\text{out})(i) \qquad (21)$$

where

$$N'_{NO_3}(out)(i) = min(c_u N_{NO_3}(i), x_{fra} X_{Np}(i))$$

$$(22)$$

$$N'_{NH_4}(out)(i) = min(c_u N_{NH_4}(i),$$

$$(1 - x_{fra}) X_{Np}(i)) \qquad (23)$$

The potential root uptake (X_{Np}) in each layer is dependent on the relative fraction (a_r) of the total root surface in that layer, and the total daily nitrogen demand of the plant (X_{Nd}). However, if X_{Nu} is lower than X_{Np} for the layer above ($i - 1$), a fraction (c_{um}) of the deficit can be added to the potential root uptake of layer i. If this amount is denoted by X_{Nm}, we have:

$$X_{Np}(i) = a_r(i) X_{Nd} + X_{Nm}(i) \qquad (24)$$

where

$$X_{Nm}(i) = c_{um}(X_{Np}(i-1) - X_{Nu}(i-1)) \qquad (25)$$

The plant demand for nitrogen depends on the maximum nitrogen concentrations in root (n_{rMax}), straw (n_{sMax}) and leaf (n_{lMax}) in the sense that the plant tries to supply the gross daily growth with nitrogen corresponding to these maximum concentrations.

$$X_{Nd} = n_{rMax} W'_r(in) + n_{sMax} W'_s(in)$$

$$+ n_{lMax} w'_l(in) \qquad (26)$$

where $W'_l(in)$ is given by Eq. 19 whilst $W'_r(in)$ and $W'_s(in)$ are the right hand expressions of Eqs. 14 and 20, respectively, excluding the last term which represents allocation to grains.

Allocation of the daily total nitrogen uptake to root, straw and leaf is based on the idea that the roots receive nitrogen first, until they reach their maximum concentrations, followed by the straw and finally the leaf [8]. If N_r, N_s and N_l are the amounts of nitrogen in root, straw and leaf, respectively, the daily uptake rates for the different plant parts are given by:

$$N'_r(in) = min(X_{Nu}, n_{rMax} W'_r(in))$$

$$N'_r(in) \geq 0 \quad (27)$$

$$N'_s(in) = min(X_{Nu} - N'_r(in), n_{sMax} W'_s(in))$$

$$N'_s(in) \geq 0 \quad (28)$$

$$N'_l(in) = min(X_{Nu} - N'_r(in) - N'_s(in),$$

$$n_{lMax} W'_l(in)) \qquad N'_l(in) \geq 0 \quad (29)$$

In the case of grain development, the allocation of nitrogen to grain (N'_g) is the nitrogen concentration of the tissue concerned (n_r for root, n_s for straw and n_l for leaf) multiplied by the corresponding translocation of assimilates. Hence, the daily net changes in nitrogen contents of the tissues are as follows:

$$N'_r = N'_r(in) - b_g n_r W_r \qquad b_g = 0 \text{ if } i_v < 1 \quad (30)$$

$$N'_s = N'_s(in) - b_g n_s W_s \qquad b_g = 0 \text{ if } i_v < 1 \quad (31)$$

$$N'_l = N'_l(in) - b_g n_l W_l \qquad b_g = 0 \text{ if } i_v < 1 \quad (32)$$

$$N'_g = b_g(n_r W_r + n_s W_s + n_l W_l)$$

$$b_g = 0 \text{ if } i_v < 1 \quad (33)$$

Application of the models

Validation data and input variables

The models were tested on experiments with winter wheat crops at three nearby locations in the Netherlands (the Bouwing, the Eest and PAGV). The validation data and input data and other information concerning the experiments can be found in Groot and Verberne [10]. The different fertilizer treatments are referred to as N1 (none or low N supply), N2 (intermediate) and N3 (highest) respectively.

The SOIL model was tested against measurements of water table depth and volumetric water content in different horizons of the uppermost metre. Climatic data sets from two locations, the Bouwing and the Eest, were used as driving variables for all three sites (the Eest was also used for PAGV). The data sets contained daily values of air temperature, air humidity, wind speed, precipitation and global radiation.

The SOILN model (and hence the CROP-GROWTH submodel) was tested against measurements of biomass production (W_s, W_l and W_g), plant nitrogen contents (N_s, N_l and N_g) and soil mineral-N content for different layers down to one metre depth. The test was performed by first setting some parameter values, not available elsewhere, according to measured data at the Bouwing. Then the simulations were made for all sites using the same parameter set. Concerning the comparison between simulated and measured values, the Eest and PAGV was treated separately from the Bouwing in order to test the model on an independent data set. The test at the Bouwing then gave a test of the model with a parameter setting probably more correct than for the other sites. The fraction of total root surface found in each layer (a_r) was set to the corresponding values for rooting density measured three times a year [10] and assumed constant during the periods between measurements.

Parameterization of the SOIL model

Only some of the necessary input to the SOIL model was determined from the available site specific information. Instead, most parameter values were taken from previous applications with the model to similar arable land conditions [14, 15, 16, 17, 19]. Only a few output variables were measured that could be used either for testing or calibration.

Soil properties (i.e. the water retention curves and hydraulic conductivities) were not available for the experimental sites concerned but for the same sites as for the climatic data [10]. These data were fitted to slightly modified forms of the analytical functions of Brooks & Corey [5] and Mualem [22] according to Jansson & Thoms-Hjärpe [18]. The results for the sandy loam soil at the Eest are shown in Fig. 3. Similar results were obtained for the clay soil at Wageningen used for the Bouwing site, except that the unsaturated conductivity was very low in the subsoil in this case. The lower boundary condition for all sites was simulated assuming a network of drainage pipes at a depth of 1 m and a spacing of 10 m. The only parameter value that was adjusted by calibration was the factor for sorption capacity (A_{scale}; Eq. 5) which was varied for the Eest site until a reasonable agreement was obtained between simulated and measured water table depths.

Parameterization of the CROP-GROWTH submodel

The parameter values for the CROP-GROWTH submodel are given in the appendix. Parameter values in SOILN were taken from Bergström et al. [4].

The specific leaf area (a_{ls}) was determined for the Bouwing site as a mean of all treatments; however, some extreme values in the autumn were neglected. The fraction of biomass translocated to grains (b_g) was arbitrarily assumed.

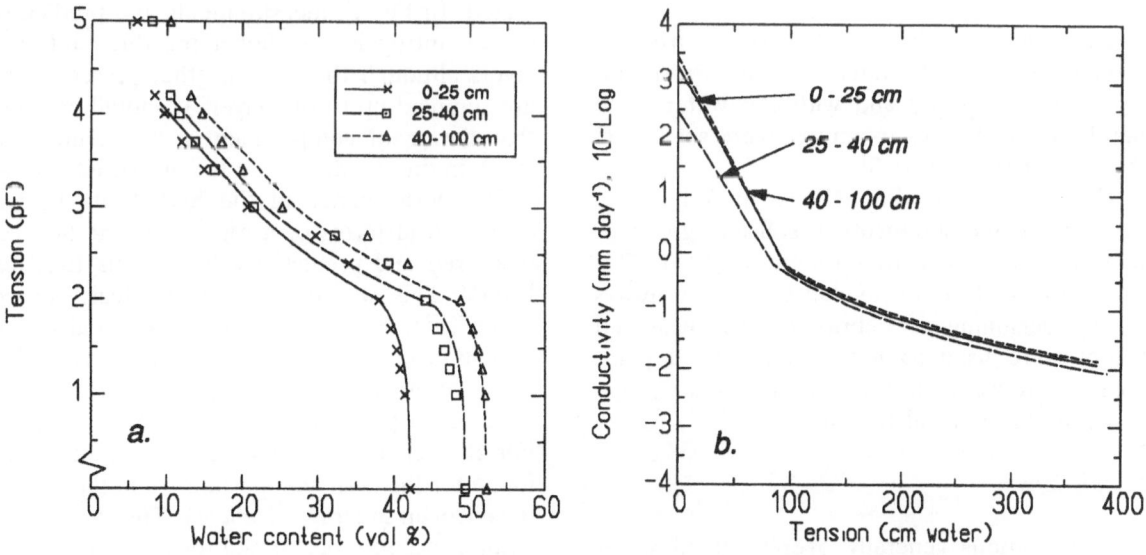

Fig. 3. Water retention curves (a) and unsaturated conductivity curves (b) used for the Eest soil.

Good estimates of the allocation coefficients b_{io} and b_{il} (used for calculating leaf area as a function of aboveground biomass) were achieved from the measured data for the three treatments at the Bouwing. The minimum fraction of biomass allocated to roots (b_{ro}) was taken from data for barley and willow [11, 24, 9]. The coefficients (c_0, c_1, c_2, c_3 and c_4) for determining the start of grain development were taken from Angus et al. [2]. The light extinction coefficient (k) was set equal to 0.6. The temperature limits for no growth (T_{Min}) and optimal growth (T_{Max}) were roughly estimated from the literature [13, 23]. The light use efficiency (ε) was estimated from a literature review [6].

The fraction of N in the soil-nitrogen pool which can be taken up each day (c_u) was set according to Bergström et al. [4]. The fraction (c_{um}) of a deficit in nitrogen availability in one soil layer that can be replaced by nitrogen from the layer above was put equal to unity, i.e. the limitation by the parameter was cancelled. The maximum and minimum nitrogen concentrations (n_{lMax}, n_{sMax} and n_{lMin}) of aboveground plant tissues were estimated as mean values of the annual maximum and minimum values of the N3 and N1 treatments at the Bouwing. An estimate for the maximum root nitrogen concentration (n_{rMax}) was taken from the literature [11, 8, 9].

Initial values

The simulations with the SOIL model started on January 1st 1982 with water tensions corresponding to field capacity and with the water table depth at 1 m. Soil temperatures were set at 5°C throughout the soil profile.

The simulations with SOILN started at the time of first measurements of soil nitrogen content and total aboveground biomass (W_{ta}). This time differed for each simulation but was always in the beginning of February. The measured values were used as initial values, and plant tissues were assumed to contain maximum nitrogen concentrations at this time.

Simulations

The simulations generally overestimated water content at the Bouwing site, especially in the subsoil between 40 and 100 cm depth. This was caused by the very low unsaturated conductivity of the subsoil, set in the model according to the soil properties at Wageningen, which were assumed to be representative for the Bouwing soil [10]. Instead, results from the Eest site during 1984 are presented (Fig. 4) since they represent a period with relatively frequent measurements of both soil water contents and water table depth. Observed drying and wetting patterns were reasonably simulated independent on the setting of the bypass option (Fig. 4a). However, the range of variation in the water content agreed better with the observations, especially in the topsoil, when assuming bypass to have occurred (Fig. 4b). The water table depth was a more sensitive indicator of bypass flow than the water content. Only the simulation including bypass showed reasonable similarities with the observations. Recharge at infiltration events was not possible to simulate without the bypass model approach.

One example of the simulations with SOILN is given for the N3-simulation at the Eest in 1983 (Fig. 5). The agreement between measured and simulated values was fairly good although leaf and straw nitrogen were overestimated from mid July to end of August. The decrease in total aboveground nitrogen in July was not reflected by the simulation and for about the same period soil nitrogen in the deepest layer was underestimated. In Fig. 5, measurements and predictions of soil nitrogen are shown for three different layers although they, in all other presentations, are summed up to one layer. It should be noted that the grain compartment includes litter fall (both in the simulated and the measured values).

The performance of the SOILN model was evaluated in terms of a regression line between measured and predicted values at the Eest and PAGV. This was done by linear least squares fitting (Fig. 6). The predictions of total aboveground production (W_{ta}) were a little better than those of nitrogen uptake (N_{ta}) which in turn were better predicted than soil mineral nitrogen. For the separate plant tissues, the best fit was found for the grain compartment and the poorest for nitrogen in straw (Table 1). The same correlation procedure performed for the Bouwing site gave similar results, indicating that the separa-

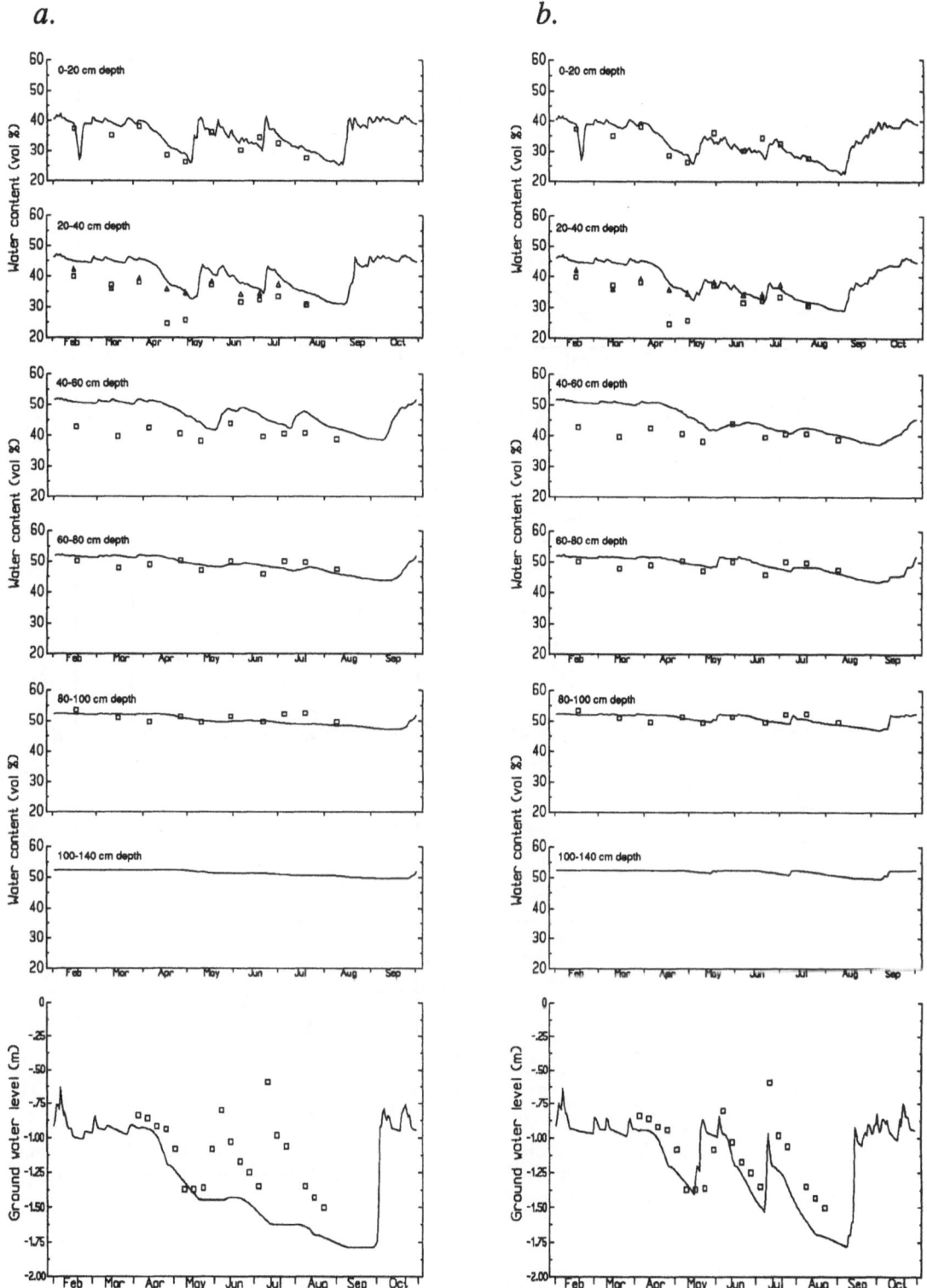

Fig. 4. Simulated and measured soil water contents and groundwater depths for the Eest site 1984. Simulations were done without (a) and with the bypass option (b). Lines are simulations and open symbols are measured values. In layer 20–40 cm squares represent measurements in the 20–30 cm layer and triangles represent measurements in the 30–40 cm layer.

322

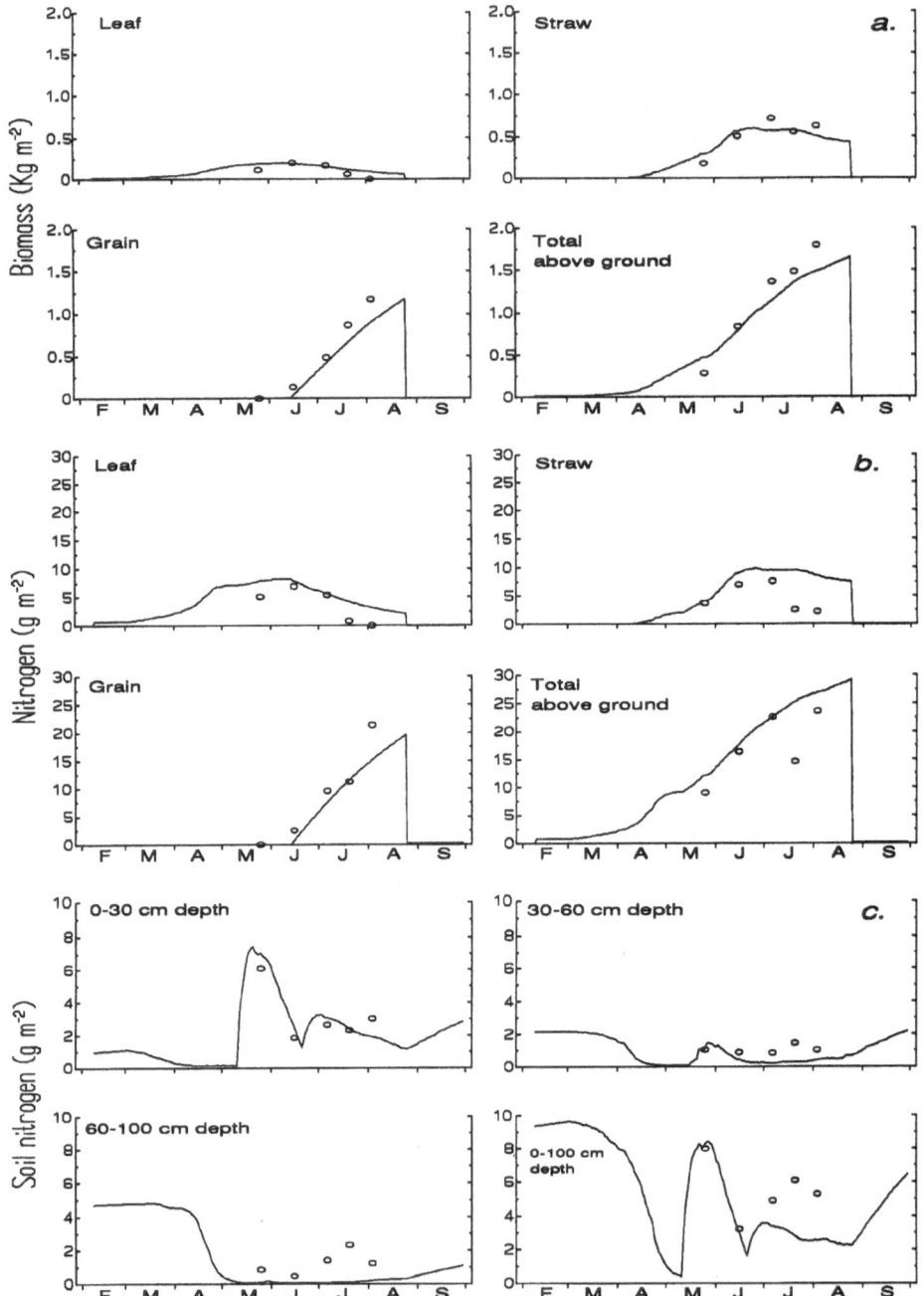

Fig. 5. Simulations (lines) and measurements (open circles) for the Eest in 1983 (N3-fertilizer treatment). The subpictures show leaf, straw, grain (+ litter) and total aboveground biomass (a), leaf, straw, grain (+ litter) and total aboveground nitrogen (b) and soil nitrogen in layers 0–30 cm, 30–60 cm, 60–100 cm and 0–100 cm (c).

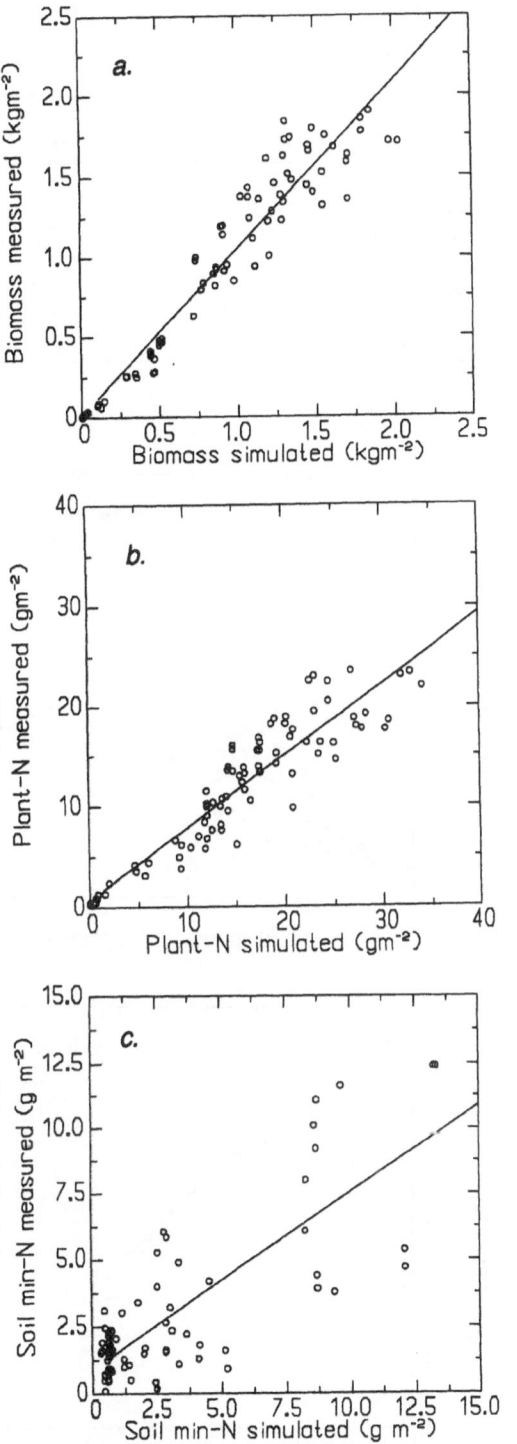

Table 1. Regression lines between the simulated and measured values of biomass (W) and nitrogen (N) for the Eest and PAGV. The regression follows: Measured value = a + b * Simulated value

Variable	R^2 (−)	a ($g\,m^{-2}$)	b (−)	Sample
W_{ta}	0.93	10	1.04	86
W_l	0.61	−30	1.04	80
W_s	0.74	50	0.97	69
W_g	0.93	20	1.10	55
N_{ta}	0.86	0.7	0.72	85
N_l	0.62	−2.5	0.96	74
N_s	0.36	1.0	0.42	68
N_g	0.89	0.4	1.04	62
N_{Soil} (0–1 m)	0.64	0.9	0.67	78

Fig. 6. Regression between simulated and measured values at the Eest and PAGV for aboveground biomass (a), aboveground plant nitrogen (b) and soil mineral nitrogen (c). Regression parameters for each of the lines are given in Table 1.

tion of the Bouwing site from the others is of minor importance.

The seasonal maximum value of W_{ta} was normally simulated with an accuracy of ±20% (Table 2). For N_{ta} the accuracy was of the same order, except for some PAGV simulations and, the non-fertilized polder at the Eest, where the error was up to 50%. This was mainly due to the poor predictions of soil nitrogen content immediately after fertilization.

The amount of biomass produced per unit of nitrogen taken up by the plant (nitrogen productivity) was estimated for the total of all aboveground plant parts, i.e. the nitrogen productivity here equals W_{ta}/N_{ta}. For measurements throughout the growing season the nitrogen productivity increased with the amount of nitrogen taken up but for seasonal maximum values it decreased with N_{ta} (Fig. 7). This was true both for simulated and measured values although this was much more pronounced for the latter. The correlations between nitrogen productivity and W_{ta} were much poorer in all cases.

Sensitivity of the SOILN model

A sensitivity test was made, using the data set from the Bouwing site, to examine the effect of specified changes or errors in the values of parameters and input variables, on the simulated production. The parameters were split into four groups on the basis of which process they mainly

Table 2. Measured seasonal maximum values of the total aboveground biomass (W_{ta}) and the total aboveground nitrogen (N_{ta}). Δ is the relative difference between simulated and measured value

Year	Treatm.	the Bouwing W_{ta} Kgm^{-2}	Δ (%)	N_{ta} gm^{-2}	Δ (%)	the Eest W_{ta} Kgm^{-2}	Δ (%)	N_{ta} gm^{-2}	Δ (%)	PAGV W_{ta} Kgm^{-2}	Δ (%)	N_{ta} gm^{-2}	Δ (%)
1983	N1	1.47	−10	14.4	18	1.01	34	7.6	102	1.36	41	10.5	76
	N2	1.77	−21	21.3	5	1.74	−14	18.7	12	1.63	18	16.3	49
	N3	1.72	−9	25.4	16	1.80	−8	23.5	−11	1.78	14	23.1	47
1984	N1	1.52	−21	19.5	−20	1.63	−20	16.9	−23	1.75	−1	18.2	4
	N2	1.62	−11	26.3	−5	1.84	−20	23.0	7	1.86	8	19.2	49
	N3	1.71	−9	27.5	9	1.73	−15	22.5	−10	1.90	9	23.4	47

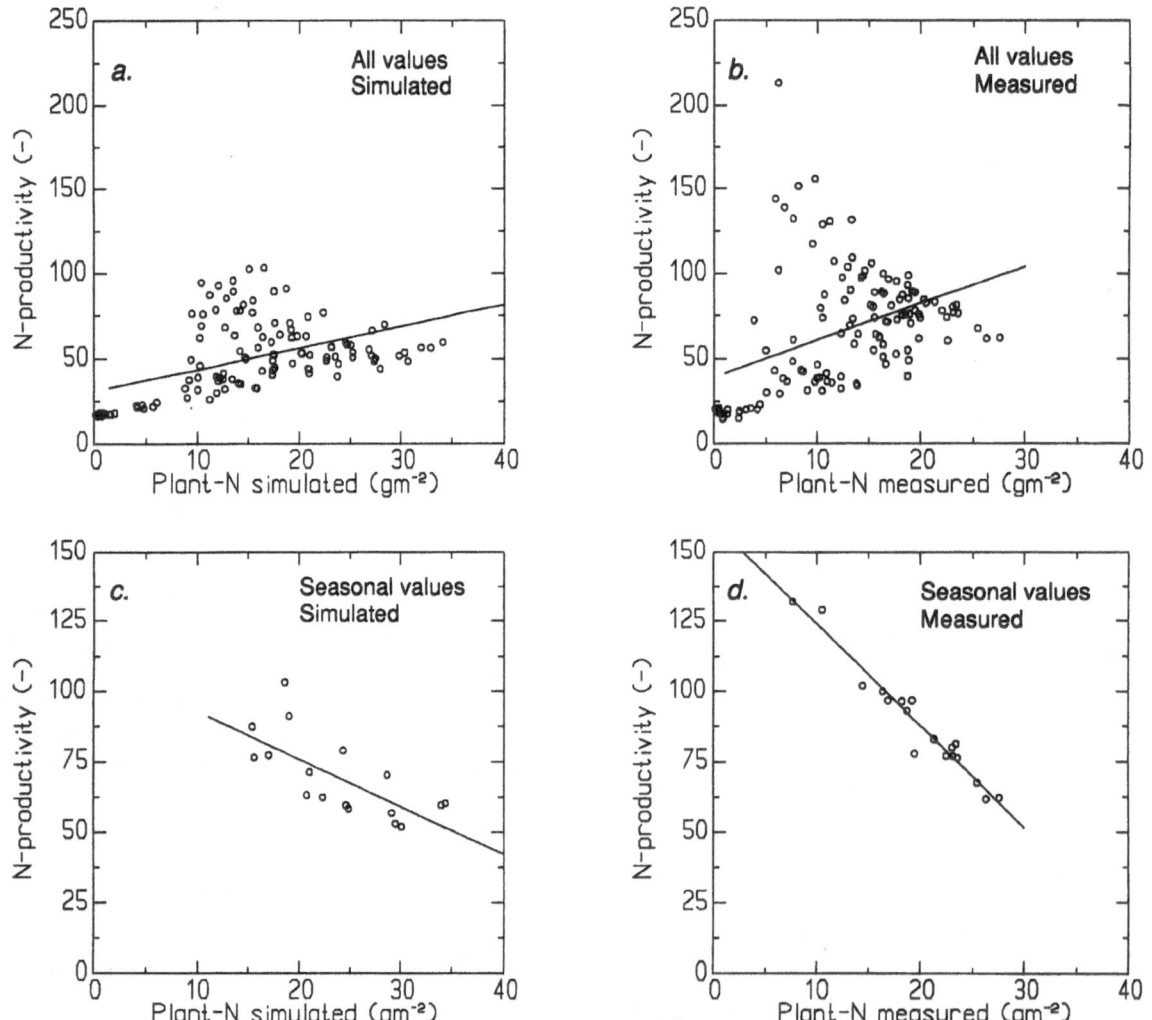

Fig. 7. Simulated and measured nitrogen productivity, defined as the total aboveground biomass divided by the total aboveground nitrogen (N_{ta}) as a function of N_{ta}. The subpictures present all measurements throughout the growing season (a and b) and the maximum values at harvest (c and d).

affect: growth, partitioning of biomass, partitioning of nitrogen, and plant nitrogen uptake. The input variables and initial values of state variables were placed in a fifth group. The values of each parameter were modified separately to cause an increase in the total production (W_t). The parameters were changed by 10%, except for temperature values which were changed by 1°C. The relative differences (compared to the original simulation) in the maximum values of W_t and total plant nitrogen (N_t), respectively, were chosen as indicators of sensitivity. The test was made both for the non-fertilized treatment (N1 in 1983) and treatment N3 in 1983, with a high application rate.

In the fertilized case, the effects of changes on total production (W_t) were largest for the growth parameters (especially the light use efficiency, ε) but also for i_v (i.e. c_0 in Eq. 12 was changed) which determines the start of grain development (Table 3). The lowest sensitivity was found for nitrogen partitioning parameters. N_t was affected about the same by all parameters, although it was much less sensitive than W_t. For the non-fertilized treatment the sensitivity was in general much smaller. W_t was most sensitive to the parameters controlling the partitioning of biomass but those dealing with partitioning of nitrogen were also important. N_t was very insensitive to the growth parameters. Instead, the low soil nitrogen status meant that the uptake of nitrogen was much more important as a regulator of growth.

Making a similar parameter modification, but of the opposite sign (the values in brackets in Table 3), showed whether the response to the change was proportional to the change in the parameter value (or input variable) or not. A clear non-proportional response was found for both ε and air temperature (T).

Some special situations were also analyzed (see Table 3) such as (i) a uniform distribution over depth of root surface area (a_r = constant), (ii) a total availability for root uptake (in each layer) of the soil nitrogen ($c_u = 1$) and (iii) no possibility of compensating a deficit in uptake from one layer by uptake from the layer just below ($c_{um} = 0$). As the fertilization supply decreases the effect of (ii) is to dramatically in-

Table 3. The relative change during one growing season in the maximum values of simulated total biomass (W_t) and total plant nitrogen content (N_t) when values of parameters and input variables are changed by 10% (or those specified). The original values are those given in the appendix. ( ) is the effect when making the same relative change but of opposite sign. The original simulations are those of the Bouwing in 1983. N3 is high fertilizer treatment and N1 is no fertilization

Para-		Relative change (%) in seasonal−max of:			
		W_t		N_t	
meter	Change	N3	N1	N3	N1
Growth					
k	+10%	6 (−8)	1 (−1)	2	+0
T_{Min}	−1°C	3 (−4)	1 (+0)	1	+0
T_{Max}	−1°C	4 (−4)	1 (1)	2	+0
ε	+10%	13 (−17)	4 (−4)	4	+0
Partitioning of biomass					
a_{ls}	+10%	2 (−4)	4 (−3)	−1	1
b_g	−10%	2 (−2)	2 (−2)	3	1
b_i	+10%	4 (−5)	−3 (3)	3	−1
b_{ro}	−10%	1 (−1)	1 (−1)	+0	0
i_v	−10%	6 (−8)	7 (−5)	2	2
Partitioning of nitrogen					
n_{1Max}	−10%	+0 (−0)	3 (−3)	−2	−1
n_{rMax}	−10%	+0 (−0)	2 (−2)	−1	2
n_{sMax}	−10%	+0 (−1)	2 (−2)	−3	+0
n_{1Min}	−10%	+0 (−0)	4 (−4)	+0	1
Uptake of nitrogen					
c_u	a) +10%	+0 (−0)	2 (−3)	1	1
	b) = 1	1	89	3	109
c_{um}	a) −10%	−2 (−4)[a]	−1 (−2)[a]	−1	−0
	b) = 0	−13	−37	−11	−28
Input variables and initial values					
I	+10%	3 (−3)	3 (−4)	+0	+0
T	−10%	4 (+0)	4 (−2)	+0	1
$w_t(t_0)$	+10%	1 (−1)	+0 (−0)	+0	−0
a_r	+10% in layer 1 & −10% in layer 2	2	−0	+0	−0
	const. in all layers i.e. = 0.2	−28	−4	−36	−0

[a] The modification is −20%

crease the production whilst (iii) causes a pronounced decrease in production. The vertical distribution of roots (i) was important to consider only in the fertilized treatments.

Discussion and conclusions

In total, 18 simulations were made with the CROP-GROWTH submodel. Only one is shown here (Fig. 5) but, by comparing all of them with measured data, three different patterns of discrepancies were found. Firstly, the measurements of plant nitrogen above ground (N_{ta}) showed (in about 70% of all cases) that after the grain development started, N_{ta} decreased during a period (cf. Fig. 5). However, it is not possible to simulate a decrease in N_{ta} with this version of the CROP-GROWTH submodel, since processes of nitrogen leaching from the plant and nitrogen translocation to roots are neglected. About simultaneously with these overestimates of N_{ta}, the soil nitrogen was underestimated. Secondly, in three non-fertilized cases there was a tendency for a too rapid decrease of N_{Soil} and a too high increase of N_{ta} during the early part of the growing period. This indicated that plant uptake was overestimated which might be explained by the vertical distribution of roots (a_r), the efficiency of plants to take up nitrogen (c_u and c_{um}) or the plant demand (Eqs. 27–29), not being correctly simulated. Thirdly, soil nitrogen in the upper layer (N_{Soil}; 0–30 cm) was considerably overestimated in four PAGV simulations, which in turn caused an overestimation of N_{ta} from June to the end of the season (thus decreasing the slope of the line in Fig. 6). The model simulated an increase of N_{Soil} after fertilization whilst the measured increases in the soil and plant could only account for 0–50% of the supplied nitrogen.

Nitrogen productivity (W_{ta}/N_{ta}) was found to depend on the time period considered. The measured data showed that W_{ta}/N_{ta} increased during the season while for measurements at final harvest an opposite relationship existed (Fig. 7) which corresponded to a parabolic function between N_{ta} and W_{ta} (i.e. $W_{ta} = 160.12 N_{ta} - 3.615 N_{ta}^2$), however, is not valid for high values of N_{ta}. The simulated data gave similar results, indicating that the CROP-GROWTH simulations reflected the observed behaviour concerning nitrogen productivity. However, this effect was underestimated especially for low values of N_{ta}, perhaps due to errors in the parameters controlling partitioning of biomass and nitrogen or ε. Nitrogen productivity was much better correlated with nitrogen uptake than biomass.

The introduction of the light use efficiency concept (εI_i in Eq. 7) results in a remarkable change of the effect of the light extinction coefficient (k; Eq. 8), compared with commonly used growth models based on photosynthesis. Increasing the light absorption by increasing k, increases the production (Table 3). However, in the photosynthetic growth model the effect is the opposite [8], since a reduction of light on the leaves in the lower part of the canopy decreases the photosynthesis of the whole canopy. Production is, at high fertilizer treatments, remarkably sensitive to variations in ε itself (Table 3).

The reliability of the simulations of the physical conditions in the soil is not known. However, this was not thought to be an important reason for the discrepancies appearing in the crop growth simulations. The coupling between crop growth and the physical conditions in the soil is made via the mineral-N storage in the root zone. The mineral-N is of course strongly influenced by temperature, water content and water flows but during the intensive period of crop growth, large amounts of N were added as fertilizer and only small amounts were lost through leaching [4]. Instead, the systematic discrepancies in the crop growth predictions (e.g. N_{ta} in Fig. 6) to a large extent, can be explained by errors in parameter values (cf. Table 3) and/or the overestimation of N_{Soil} after some fertilizations.

The results would have been different if simulations of crop growth had been done without initializing the mineral-N content with measured values prior to each simulation. The small number of observations made outside the growing season made evaluation of simulated nitrogen transformations in the soil and losses such as denitrification and leaching difficult to evaluate. This was also the case for observations of soil water conditions, with some exceptions. The most interesting observation concerning physical conditions was the water table depth. It was shown that the response of the groundwater depth was a more sensitive indicator of the soil physical properties than the soil water content. Both indicated a bypass flow in a sandy loam soil.

Acknowledgement

The authors wish to thank Dr. R. Ragab, Institute of Hydrology, Wallingford, UK, for his earlier contribution to the development of this CROP-GROWTH submodel whilst working at the Swedish University of Agricultural Sciences, Uppsala. Dr. N. Jarvis at our department is thanked for valuable comments on the manuscript including linguistic revision.

Appendix

List of symbols and parameter values

Parameter values given are those used in the simulations for wheat. Category refers to the CROP-GROWTH submodel except if mentioned otherwise.

Symbol	Explanation	Value	Unit	Category
A_{li}	Leaf area index		–	Output
A_{rel}	Ratio between vertical area of soil compartment and unit horizontal area	0.2–0.4	–	Input[a]
A_{scale}	Scaling coefficient accounting for the geometry of soil aggregates	0.8	–	Input[a]
a_{ls}	Specific leaf area	0.022	$m^2\,g^{-1}$	Input
a_r	Fraction of total root surface found in a soil layer		–	Input
b_g	Fraction of the biomass in plant tissues translocated to grain compartment	0.02	d^{-1}	Input
b_i	Leaf area to total aboveground biomass ratio		$m^2\,g^{-1}$	Auxiliary
b_{io}	b_i at unity W_{ta}	0.048	$m^2\,g^{-1}$	Input
b_{il}	Coefficient relating b_i to W_{ta}	0.0064	$m^2\,g^{-1}$	Input
b_r	Fraction of daily total growth allocated to roots		–	Auxiliary
b_{ro}	b_r minimum	0.15	–	Input
c_u	Fraction of N in the nitrate and ammonia pool possible to be extracted per day	0.16	d^{-1}	Input
c_{um}	Fraction of a deficit in N supply in one layer possible to be replaced by N from the layer just below	1.0	–	Input
c_0	Coefficient used for calculating i_v	0.0252	d^{-1}	Input
c_1	Coefficient used for calculating i_v	−0.153	$°C^{-1}$	Input
c_2	Coefficient used for calculating i_v	3.51	$°C$	Input
c_3	Coefficient used for calculating i_v	−0.301	h^{-1}	Input
c_4	Coefficient used for calculating i_v	9.154	h	Input
D	Day length		h	Input
E, E_p	Evapotranspiration, actual and potential, respectively		$mm\,d^{-1}$	Output[a]
f_N	Response function for nitrogen limitation to growth		–	Auxiliary
f_T	Response function for temperature limitation to growth		–	Auxiliary
f_W	Response function for water limitation to growth		–	Auxiliary
i_v	Index that determines the start of grain development		–	Auxiliary
I	Global radiation, shortwave, 300–3000 nm		$MJ\,m^{-2}\,d^{-1}$	Input
I_i	Shortwave radiation intercepted by canopy		$MJ\,m^{-2}\,d^{-1}$	Auxiliary
k	Light extinction coefficient related to leaf area	0.6	–	Input
k_{mat}	Saturated hydraulic conductivity, excluding the contribution of macropores	Fig. 3b	$mm\,d^{-1}$	Input[a]
k_s	Saturated hydraulic conductivity, including the contribution of macropores	Fig. 3b	$mm\,d^{-1}$	Input[a]
$k(\theta)$	Unsaturated hydraulic conductivity function	Fig. 3b	$mm\,d^{-1}$	Input[a]
L	Characteristic distance in the gradient term in Darcy's equation		m	Input[a]
n_1, n_s, n_r	N concentrations in leaf, straw and root		$g\,N(gDW)^{-1}$	Auxiliary
n_{lMax}	Maximal N concentration in leaf	0.06	$g\,N(gDW)^{-1}$	Input
n_{rMax}	Maximal N concentration in root	0.025	$g\,N(gDW)^{-1}$	Input
n_{sMax}	Maximal N concentration in straw	0.02	$g\,N(gDW)^{-1}$	Input

Symbol	Explanation	Value	Unit	Category
n_{lMin}	Minimum N concentration in leaf	0.01	$g\,N(gDW)^{-1}$	Input
$N_g, N_l, N_r,$				
N_s, N_t, N_{ta}	Amount of N in grain, leaf, root, straw, total and total aboveground		$g\,N\,m^{-2}$	Output
N_{NO_3}, N_{NH_4}	Amount of nitrate-N and ammonium-N in the soil		$g\,N\,m^{-2}$	Output[b]
N_{Soil}	Mineral-N in the soil, $N_{NO_3} + N_{NH_4}$		$g\,N\,m^{-2}$	Output[b]
ψ	Soil water tension		cm water	Output[a]
pF	pF value, defined as $^{10}\log(\psi)$	Fig. 3a	–	Auxiliary[a]
q_{in}	Vertical water flow, positive downwards		$mm\,d^{-1}$	Output[a]
q_{mat}	Vertical water flow, soil matrix		$mm\,d^{-1}$	Output[a]
q_{bypass}	Vertical water flow, soil macropores		$mm\,d^{-1}$	Output[a]
q_{pipe}	Sink flow from one layer to drainage pipes		$mm\,d^{-1}$	Output[a]
S_{mat}	Sorption capacity of aggregates		$mm\,d^{-1}$	Output[a]
t	Time coordinate		d	Auxiliary
t_0	t when plant starts to grow		d	Auxiliary
T	Daily mean of air temperature		°C	Input
T_{Min}	Temperature limit for start of growth	2	°C	Input
T_{Max}	Temperature limit for optimal growth	10	°C	Input
$W_g, W_l, W_s,$				
W_r, W_t, W_{ta}	Accumulated growth of grain, leaf, straw, root, total and total aboveground		$g\,DW\,m^{-2}$	Output
x_{fra}	Fraction nitrate-N of mineral-N		–	Auxiliary
X_{Nu}	Actual daily N uptake by plant		$g\,N\,m^{-2}\,d^{-1}$	Auxiliary
X_{Nd}	Daily N uptake demand by plant		$g\,N\,m^{-2}\,d^{-1}$	Auxiliary
X_{Nm}	Demand of N transferred from one soil layer to the layer below		$g\,N\,m^{-2}\,d^{-1}$	Auxiliary
X_{Np}	Potential daily N uptake from one layer		$g\,N\,m^{-2}\,d^{-1}$	Auxiliary
z	Depth coordinate		m	Auxiliary[a]
z_{gw}	Depth of groundwater table		m	Output[a]
z_{pipe}	Depth of drainage pipes	1.0	m	Input[a]
θ	Liquid water content		vol%	Auxiliary[a]
ε	Total growth per unit of intercepted radiation (300–3000 nm) at optimal temperature, nitrogen and water conditions	2.0	$g\,DW\,MJ^{-1}$	Input

'   : Denotes the daily change of the symbol concerned
(in)   : Denotes a positive change of the symbol concerned
(out) : Denotes a negative change of the symbol concerned
a   : Refers to the SOIL model
b   : Refers to the SOILN model

References

1. Angus JF, Cunningham RB, Moncur MW and Mackenzie DH (1980) Phasic development in field crops. I. Thermal response in the seedling phase. Field Crops Res 3: 365–378

2. Angus JF, Mackenzie DH, Morton R and Schafer CA (1981) Phasic development in field crops. II. Thermal and photoperiodic responses of spring wheat. Field Crops Res 4: 269–283

3. Bergström L and Johnsson H (1988) Simulated nitrogen dynamics and nitrate leaching in a perennial grass ley. Plant and Soil 105: 273–281

4. Bergström L, Johnsson H and Torstensson G (1991) Simulation of soil nitrogen dynamics using the SOILN model. Fert Res 27: 181–188

5. Brooks RH and Corey AT (1964) Hydraulic properties of porous media. Colorado State Univ. Hydrology Paper No. 3 (Fort Collins) 27 p

6. Cannell MGR (1989) Physiological basis of wood production: a review. Scand J For Res 4: 459–490

7. Eckersten H and Ericsson T (1989) Allocation of biomass during growth of willow. In: KL Perttu and PJ Kowalik (eds.): Modelling of energy forestry – growth, water relations and economy. pp 77–85. PUDOC (Wageningen, The Netherlands)

8. Eckersten H and Slapokas T (1990) Biomass production and nitrogen turnover in an irrigated short rotation forest. Agric For Meteor 50: 99–123

9. Ericsson T (1981) Effects of nitrogen stress on growth and nutrition in three *Salix* clones. Physiol Plant 52: 239–244

10. Groot JJR and Verberne ELJ (1991) Response of wheat to nitrogen fertilization, a data set to validate simulation models for nitrogen dynamics in crop and soil. 27: 349–383

11. Hansson A-C (1987) Roots of arable crops: production, growth dynamics and nitrogen content. Department of Ecology and Environmental Research. Report 28. Swedish Univ of Agr Sci (Uppsala) 28 p

12. Ingestad T (1971) A definition of optimum nutrient requirements in birch seedlings. II. Physiol Plant 24: 118–125

13. Larcher W (1980) Physiological plant ecology. Springer Verlag, Berlin, Heidelberg, New York. 303 p

14. Jansson P-E, Borg GC, Lundin L-C and Lindén B (1987) Simulation of soil nitrogen storage and leaching applications to different Swedish agricultural systems. The Swedish National Environmental Protection Board (Solna) Report 3356, 63 p

15. Jansson P-E and Halldin S (1979) Model for annual water and energy flow in layered soil. In: Halldin (ed.) Comparison of Forest Water and Energy Exchange Models, pp 145–163. Int Soc Ecol Modelling (Copenhagen)

16. Jansson P-E and Halldin S (1980) Soil Water and Heat Model. Technical Description. Swedish Coniferous Project. Tech Rep 26. Swedish Univ of Agr Sci (Uppsala) 81 p

17. Jansson P-E, Johnsson H and Alvenäs G (1989) Heat and water processes. In: O Andrén T Lindberg K Paustian and T Rosswall (eds.) Ecology of arable land-Organisms, carbon and nitrogen cycling. Ecol Bull (Copenhagen) 40: 31–40

18. Jansson P-E and Thoms-Hjärpe C (1986) Simulated and measured soil water dynamics of unfertilized and fertilized barley. Acta Agric Scand 36: 162–172

19. Johnsson H (1990) Nitrogen and water dynamics in arable soil – A modelling approach emphasizing nitrogen losses. Department of Soil Sciences. Reports and Dissertations 6. Swedish Univ of Agr Sci (Uppsala) 36 p

20. Johnsson H, Bergström L, Jansson P-E and Paustian K (1987) Simulated nitrogen dynamics and losses in a layered agricultural soil. Agric Ecosystems Environ 18: 333–356.

21. Monteith JL (1977) Climate and the efficiency of crop production in Britain, pp 277–294. Philosophical Transactions of the Royal Society of London B281

22. Mualem Y (1976) A new model for predicting the hydraulic conductivity of unsaturated porous media. Water Resour Res 12: 513–522

23. Pelkonen P (1984) Carbon dioxide exchange in willow clones. In: Perttu K (ed.) Ecology and management of forest biomass production systems. Department of Ecology and Environmental Research. Report 15. Swedish Univ of Agr Sci (Uppsala) pp 187–196

24. Pettersson R (1987) Primary production in arable crops: above-ground growth dynamics, net production and nitrogen uptake. Department of Ecology and Environmental Research. Report 31. Swedish Univ of Agr Sci (Uppsala) 23 p

Fertilizer Research **27**: 331–339, 1991.
© 1991 *Kluwer Academic Publishers.*

WHNSIM – a soil nitrogen simulation model for Southern Germany

B. Huwe & R.R. van der Ploeg
Institute of Soil Science, Hohenheim University, P.O. Box 700562, 700 Stuttgart-70, Germany

Key words: Soil nitrogen, simulation, modelling, nitrate leaching, groundwater contamination

Abstract

A one-dimensional deterministic soil nitrogen simulation model (WHNSIM) is presented. With the model the leaching of soil nitrate, its uptake by plant roots and the mineralization of soil organic nitrogen can be simulated. Basic elements of WHNSIM are differential equations that describe soil water, soil heat and soil solute transport. The equations are solved with a fully implicit finite difference method for a variety of boundary and initial conditions. With WHNSIM the soil nitrogen behavior of arable fields for one or more consecutive years can be described. The model has been calibrated for typical site conditions in Southern Germany. The main features of WHNSIM are discussed, some simulation results are also presented. For site conditions of Southern Germany the model appears to perform adequately.

Notation

Symbols not explained in the text were used as follows:

C = nitrate concentration in the soil solution ($mg\,l^{-1}$)
D = apparent diffusion coefficient ($cm^2\,day^{-1}$)
H = total potential of soil water (cm)
K = soil hydraulic conductivity ($cm\,day^{-1}$)
N = soil or crop nitrogen ($kg\,ha^{-1}$)
S = sink or source for soil nitrate ($mg\,cm^{-3}\,day^{-1}$)
U = soil water extraction rate ($cm^3\,cm^{-3}\,day^{-1}$)
V = soil heat extraction rate ($J\,cm^{-3}\,day^{-1}$)
c_v = soil volumetric heat capacity ($J\,cm^{-3}\,{}^{\circ}C^{-1}$)
c_{vw} = volumetric heat capacity of soil water ($J\,cm^{-3}\,{}^{\circ}C^{-1}$)
h = soil water tension (cm)
t = time (day)
z = soil depth (cm)
γ = psychrometer coefficient ($mbar\,{}^{\circ}C^{-1}$)
Δ = slope of the relation between the temperature and the saturated (water) vapor pressure ($mbar\,{}^{\circ}C^{-1}$)
λ = soil thermal conductivity ($J\,cm^{-1}\,day^{-1}\,{}^{\circ}C^{-1}$)
θ = volumetric water content of soil ($cm^3\,cm^{-3}$)

Introduction

Soil nitrogen is of interest as a major crop nutrient, but also as a potential environmental pollutant. As a consequence, knowledge about the behavior of soil nitrogen is desirable in order to optimize plant growth and crop yield, with environmental side effects that are as small as possible. Many aspects of soil nitrogen behavior are conveniently discussed with use of the soil nitrogen budget equation. For a defined period of time (for example a growing season, a calendar year or a crop rotation) this equation for a flat area for a soil profile of arbitrary depth can be written as:

$$F + A + O + R = E + P + G + B \qquad (1)$$

in which equation the symbols (units $kg\,ha^{-1}$) are used as follows:

F = amount of applied nitrogen fertilizer
A = atmospheric nitrogen deposition
O = nitrogen fixation by soil organisms
R = nitrogen incorporated into the soil with crop residues
E = crop nitrogen uptake that is stored above the soil surface
P = amount of soil nitrogen leached from the soil profile
G = gaseous nitrogen losses into the atmosphere
B = change of amount of nitrogen that is stored in the soil profile under consideration.

In sloped areas it may be necessary to consider also soil nitrogen losses caused by erosion and surface runoff. However, in the present paper the processes of erosion and surface runoff will not be dealt with.

With reference to eq. 1, agricultural management practices should be such that an optimum value of E is obtained for values of F, P and G that are as small as possible. This, however, is not easily achieved since the interrelationships between the various components of the soil nitrogen budget equation are complex and not fully understood. Soil nitrogen simulation models may help to examine such interrelationships

and identify gaps in the knowledge about soil nitrogen behavior. At the same time soil nitrogen simulation models can be used to calculate (or estimate) such components of eq. 1 that cannot, or only with great effort, be measured. To this end WHNSIM (Water, Heat and Nitrogen Simulation) was developed. It combines features of the soil water simulation model of [4], the root nitrogen uptake model of [10], the soil nitrogen mineralization model of [14] and the numerical analysis of [6]. Hence, WHNSIM is primarily constructed for the calculation of deep seepage and nitrate leaching. Therefore, the transport of water and solutes in WHNSIM is treated in great detail, whereas nitrate uptake and crop growth are considered of secondary importance. These quantities, however, are preferably used to evaluate the model performance.

Model description

Transport equations

In the introduction it was mentioned that calibrated soil nitrogen simulation models can be used advantageously to estimate soil nitrate seepage losses. In this respect it is necessary that such models describe the movement of the soil solution. At the same time such models should be able to treat the process of mineralization, which adds nitrate to the soil solution. Since it is well known that soil temperature influences mineralization, it thus is necessary that also soil heat behavior is simulated. For this reason three transport equations build the central part of WHNSIM. They are:

$$\frac{\partial \theta}{\partial t} = \frac{\partial}{\partial z}\left[K(\theta) \cdot \frac{\partial H}{\partial z}\right] - U \qquad (2)$$

for soil water movement,

$$\frac{\partial(c_v \cdot T)}{\partial t} = \frac{\partial}{\partial z}\left(\lambda \cdot \frac{\partial T}{\partial z}\right) - c_{vw} \cdot \frac{\partial(\theta v \cdot T)}{\partial z} - V \qquad (3)$$

for soil heat movement, and

$$\frac{\partial(\theta C)}{\partial t} = \frac{\partial}{\partial z}\left(\theta D \cdot \frac{\partial C}{\partial z}\right) - \frac{\partial(\theta v \cdot C)}{\partial z} - S \qquad (4)$$

for the movement of nitrate with the soil solution. The used symbols are explained at the beginning of this contribution (Notation).

In eqs. 2–4 the symbols U, V and S are used to denote sinks and sources. In eq. 2 U represents root water uptake, in eq. [3] V denotes (for the sake of completeness) the related uptake of heat. In eq. 4 finally, S symbolizes a sink or a source for nitrate due to mineralization of soil organic matter or manure, immobilization or root nitrogen uptake.

Equations 2, 3 and 4 are solved with a fully implicit finite difference procedure. This procedure requires that the soil profile under consideration is subdivided into a finite number of compartments. The height of the compartments in WHNSIM is kept variable in order that soil profiles of arbitrary depth can be dealt with. The tridiagonal system of linear equations that arises for each of the equations 2–4 are solved with the Thomas algorithm, as described in [13]. With use of a Newton-Raphson procedure [6] a proper water balance is maintained.

Soil parameters

The soil parameters, needed for computational work with eqs. 2–4, can be determined experimentally without exception [9]. However, since the laboratory or field determination of some of these parameters is rather cumbersome, it is sometimes more convenient to estimate such parameters from other soil data. In principle WHNSIM requires original, experimentally determined soil characteristics in tabular form. However, WHNSIM also contains options to derive some of the needed characteristics from other soil parameters. The simulations that were carried out for Southern Germany used measured soil water retention curves and hydraulic conductivities at saturation. The unsaturated hydraulic conductivities, however, were calculated with a method of Millington and Quirk in the modification of Jackson [8]. The volumetric heat capacity c_v and the thermal conductivity λ also were calculated from other soil data, according to formulas described in [3]. Finally, the apparent diffusion coefficient D in WHNSIM was calculated as

$$D = D_m + D_h \tag{5}$$

where D_m is the true diffusion coefficient and D_h is the hydrodynamic dispersion. The parameter D_m was calculated as $D_m = \delta\, D_0$ and the parameter D_h as $D_h = \varepsilon v$, where D_0 is the diffusion coefficient for nitrate in water and v is the pore flow velocity. The quantities δ and ε denote the tortuosity factor and the dispersivity, as discussed in [1] or [17]. The dispersivity ε depends strongly on the degree of aggregation of the soil [17] and can be determined from tracer experiments. The tortuosity δ finally, is calculated in WHNSIM according to

$$\delta = (m/\theta)\cdot e^{n\cdot\theta} \tag{6}$$

where θ is the volumetric water content ($cm^3\, cm^{-3}$) and m and n are constants which, according to [11], can be taken as $m = 0.005$ and $n = 10.0$.

Boundary and initial conditions

Before simulations with eqs. 2–4 can be carried out, boundary and initial conditions must be specified. Theoretically, both conditions can be obtained from field measurements. Frequently, however, the required information is incomplete and substitutes are necessary. This is true for the upper boundary of the soil column under consideration (soil surface) as well as for the lower boundary (some arbitrary depth, usually well below the main rooting zone). The simplest boundary at the lower end of the soil column is encountered when this boundary coincides with a groundwater table at constant depth, with constant temperature and constant nitrate concentration. Usually, however, conditions are not as simple and some judgement has to be made. WHNSIM contains a number of options that deal with this lower boundary as it may occur in the field. It can handle Dirichlet (specification of boundary condition in terms of hydraulic head), Neumann (flux) as well as Cauchy (mixed head and flux) conditions.

The boundary conditions at the soil surface (with respect to eqs. 2–4) usually are not simple either. Especially the condition with respect to soil water movement can be rather complex and is neither easily measured nor formulated. This condition will be dealt with in some detail. First, however, the boundary conditions at the soil surface for heat and solute (nitrate) movement will be discussed briefly. So far, they have been treated rather simply in WHNSIM. As far as heat transport is concerned, either the mean daily temperature of the air above the soil has been used, or a cosine curve is constructed from the maximum and minimum daily temperature. To describe the boundary condition at the soil surface for nitrate transport, also a simple approach has been used so far. For periods without precipitation no nitrate flux is crossing the soil surface. For periods with precipitation (rain), measured nitrate concentrations in the influent are used to define the solute flux. In case of mineral fertilizer application, an equivalent nitrate concentration in subsequent rainwater is calculated. Manure is treated as organic matter, which is subject to mineralization after having been incorporated into the soil.

For the simulation of water movement through the soil, a flux condition at the soil surface is formulated. In this respect WHNSIM follows a previous work as described in [4]. Either infiltration or evaporation occurs. The infiltration rate is calculated from the precipitation rate from which the rate of interception is subtracted. The rate of evaporation, in turn, is calculated as described in [2] and depends on the potential evaporation and on the dryness of the soil near the surface. The potential evaporation E_p is calculated from weather and crop data. Its calculation will be discussed in the next section.

Sinks and sources

In eq. 2 the sink term U denotes root water uptake. In WHNSIM this uptake is considered to be a function of the evaporative demand of the atmosphere, the fine root distribution and the soil dryness in the rooting zone. The daily evaporative demand ET_p of the atmosphere in WHNSIM is estimated with a modified Priestley and Taylor equation [12], which can be given as

$$ET_p = \alpha \cdot \frac{\Delta/\gamma}{\Delta/\gamma + 1} \cdot R_n^* \tag{7}$$

where α is a constant (in WHNSIM $\alpha = 0.860$ for grassland and $\alpha = 0.935$ for other crops), Δ/γ is a

temperature-dependent coefficient, see [5], and R_n^* is the equivalent net radiation, in WHNSIM calculated as

$$R_n^* = (1 - \phi) \cdot R_s \tag{8}$$

where ϕ is the albedo of the evaporating surface and R_s is the global radiation. Eq. 7 yields for summer days values for ET_p that are nearly identical with such, calculated with the original Priestley and Taylor equation. For winter days, where some evaporation is observed, the original Priestley and Taylor equation provides frequently negative values for ET_p. With Eq. 7, however, also for such days a positive value for the daily evaporation is calculated. From ET_p of eq. 7 the potential daily transpiration T_p is calculated according to [4] as

$$T_p = ET_p - E_p \tag{9}$$

where E_p is the potential daily evaporation. This potential evaporation E_p in turn is calculated as

$$E_p = (1 - S_C)(ET_p - E_I), \tag{10}$$

with S_C being the soil cover and E_I the eventual evaporation of crop intercepted precipitation. The amount of intercepted precipitation finally is estimated with relations given by [4] as

$$E_I = 0.55 \cdot S_C \cdot P^{0.53 - 0.0085 \cdot (P-5)} \tag{11}$$

for a precipitation rate $P \leq 17$ mm per day, and by

$$E_I = 1.85 \cdot S_C \tag{12}$$

for $P > 17$ mm per day.

After T_p thus is estimated, the potential daily root water uptake rate U_p at depth z is calculated as

$$U_p = \frac{w}{\int_0^L w \cdot dz} \cdot T_p \tag{13}$$

in which expression $w = w(z, t)$ is the root length density at depth z and L is the maximum rooting depth. From U_p of eq. 13, the actual root water uptake rate U at depth z finally is calculated according to [16] as

$$U = f_1 \cdot U_p ,\qquad(14)$$

where f_1 at depth z is a reduction factor, which depends on the soil water tension h at depth z. The general shape of f_1 as a function of h is shown in Fig. 1. Typical values for $lg\,h_1$, $lg\,h_2$, $lg\,h_3$ and $lg\,h_4$ are 0, 1, 3 and 4.2. However, the used values for h_1, h_2, h_3 and h_4 for different sites and different crops showed a considerable variation. The sinks and sources denoted with S in eq. 4 are not as easily calculated. In WHNSIM five sinks and sources are treated (S_1, S_2 and S_3 for mineralization, S_4 for immobilization and S_5 for root nitrogen uptake. Since in soils from Southern Germany usually little ammonium is found, it is assumed that the process of nitrification is much faster than the process of ammonification. Therefore, WHNSIM does not consider ammonium in the soil solution. Thus the sources S_1, S_2 etc. denote nitrate. The sources S_1, S_2 and S_3 are calculated from the different fractions of soil organic material (very easily mineralized, easily mineralized and not easily mineralized) as recognized by [14]. Mineralization as well as immobilization in WHNSIM is described as a first-order reaction. Hence, if the three fractions of organic matter are denoted as N_1, N_2 and N_3, the change of nitrate concentration in the soil solution due to mineralization can be calculated from

$$\frac{dN_i}{dt} = -k_i \cdot N_i , \quad i = 1, 2, 3 \qquad(15)$$

where k_i are coefficients, which depend on soil water tension and soil temperature. Similarly, the immobilization rate dN_2/dt is calculated as

$$\frac{dN_2}{dt} = k_4 \cdot (\theta C) \qquad(16)$$

Hence, the fraction N_2 at depth z increases (immobilization) whenever $k_4 \cdot (\theta C)$ is larger than $k_2 N_2$. This happens, for example, when large amounts of mineral fertilizer are applied.

WHNSIM calculates the coefficients k_i as function of soil water tension h and temperature T according to [14] and [18] as

$$k_i = f_2 \cdot g_i(T), \quad i = 1, 2, \ldots 4 \qquad(17)$$

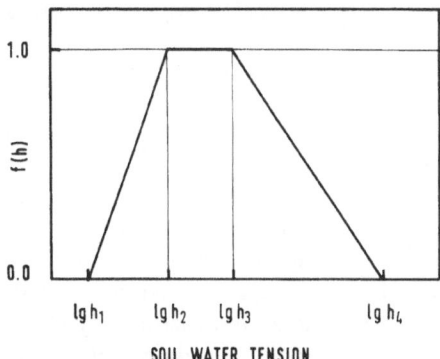

Fig. 1. A schematic representation of the function $f(h)$, used in eqs. 14 and 17.

where f_2, like f_1, is a reduction factor. The general shape of f_2 as a function of h, is also shown in Fig. 1. Typical values for $lg\,h_1$, $lg\,h_2$, $lg\,h_3$ and $lg\,h_4$ in this case are 1, 2, 2.7 and 3.5, but also here the used values for different sites varied considerably. The functions $g_i(T)$ were not considered to be site-specific. In WHNSIM they have the form

$$g_i(T) = K_0^Q \cdot K_{20}^{(1-Q)} \qquad(18)$$

in which expression K_0 and K_{20} are reaction coefficients at 0 and 20°C [14]. Table 1 shows the values of K_0 and K_{20} that WHNSIM uses for Southern Germany. It is remarked that g_1 refers to most easily mineralized organic matter, g_2 and g_3 to easily and not easily mineralized organic matter, and g_4 to immobilization with respect to easily mineralized organic matter (fraction N_2). The values of K_0 and K_{20} for g_2 and g_3 are taken from [14], those for g_4 from [15]. The values for K_0 and K_{20} pertaining to g_1 were set (somewhat arbitrarily) five times as high as those for g_2.

The function Q in eq. 18 is derived from data

Table 1. The reaction coefficents K_0 and K_{20} for the description of mineralization and immobilization (see eq. 18)

Function	Reaction coefficients (day^{-1})	
	K_0	K_{20}
g_1	$7.196 \cdot 10^{-3}$	$8.342 \cdot 10^{-2}$
g_2	$1.439 \cdot 10^{-3}$	$1.668 \cdot 10^{-2}$
g_3	$1.734 \cdot 10^{-4}$	$1.417 \cdot 10^{-3}$
g_4	$3.122 \cdot 10^{-3}$	$2.550 \cdot 10^{-2}$

given by [14]. In WHNSIM Q is expressed as

$$Q = 3999.45/(T + 273) - 13.65 \qquad (19)$$

where T is the soil temperature in degrees Celsius at depth z.

To conclude the section on the sources and the sinks that occur in eqs. 2–4, the sink term S_5 (root nitrogen uptake) must be discussed. In WHNSIM root nitrogen uptake is calculated according to a procedure, described by [10], which we slightly modified. The daily uptake of nitrogen (or the change in the amount of nitrogen stored in the crop, expressed in kg ha^{-1} day^{-1}) is denoted as $(dN/dt)_{act}$. It is the sum of convective and diffusive uptake and hence can be written as

$$\left(\frac{dN}{dt}\right)_{act} = \left(\frac{dN}{dt}\right)_{con} + \left(\frac{dN}{dt}\right)_{dif} \qquad (20)$$

In WHNSIM care is taken that $(dN/dt)_{act}$ is equal to or less than the potential daily nitrogen uptake $(dN/dt)_{pot}$, defined as

$$\left(\frac{dN}{dt}\right)_{pot} = b \cdot W \cdot Z \cdot (Z_{max} - Z) \qquad (21)$$

In eq. 21 the quantity $b \cdot Z \cdot (Z_{max} - Z)$ is a crop-specific time-variable coefficient (units kg nitrogen per kg dry matter of canopy). The quantities Z and Z_{max} (numbers without units) are assigned values that correspond to the instantaneous amount of nitrogen N (kg ha^{-1}) in the canopy and the potential amount of nitrogen N_{max} (kg ha^{-1}), stored in the canopy at the time of harvest. The quantity W in eq. 21 (kg ha^{-1} day^{-1}) denotes the rate of dry matter accumulation and is calculated according to the Bierhuizen and Slatyer procedure as described in [4]. First, the actual amount of daily transpiration T_a is calculated as

$$T_a = \int_0^L U \cdot dz \qquad (22)$$

Next, the quantity W is calculated as

$$W = a \cdot \frac{T_a}{\Delta e} \qquad (23)$$

where Δe is the mean daily pressure deficit of the air above the canopy and a is another crop-specific constant.

After $(dN/dt)_{pot}$ thus has been estimated, the convective nitrogen uptake (eq. 20) is calculated from

$$\left(\frac{dN}{dt}\right)_{con} = \int_0^L U \cdot C \, dz \qquad (24)$$

In case $(dN/dt)_{con}$, as determined with eq. 24, is larger than $(dN/dt)_{pot}$ (eq. 21), $(dN/dt)_{con}$ is reduced and put equal to $(dN/dt)_{pot}$. In this case $(dN/dt)_{dif}$ of eq. 20 is set to zero. In case $(dN/dt)_{con}$ is smaller than $(dN/dt)_{pot}$, a quantity $(dN/dt)_{difpot}$ is calculated as

$$\left(\frac{dN}{dt}\right)_{difpot} = \left(\frac{dN}{dt}\right)_{pot} - \left(\frac{dN}{dt}\right)_{con} \qquad (25)$$

This quantity denotes the potential daily nitrogen uptake by diffusion and is considered to be the upper limit for the actual daily nitrogen uptake $(dN/dt)_{dif}$ of eq. 20. This last quantity is calculated in WHNSIM as

$$\left(\frac{dN}{dt}\right)_{dif} = \int_0^L S_{dif} \cdot dz \qquad (26)$$

in which expression S_{dif} is the nitrogen uptake per unit volume of soil. This quantity in turn is calculated as

$$S_{dif} = M(z) \cdot D_m \cdot \frac{C - C_r}{d} \qquad (27)$$

with C being the nitrogen (nitrate) concentration in the bulk solution and C_r the concentration at the root surface. The symbol d ($d = 0.1$ mm) represents a characteristic length, which is described by [10]. The quantity $M(z)$ in eq. 27 denotes the total root surface density at depth z, and D_m is the molecular diffusion coefficient for nitrate, that was discussed previously (eq. 5). For $C_r = 0$ a maximum value for S_{dif}, and correspondingly for $(dN/dt)_{dif}$, is calculated. In case $(dN/dt)_{dif}$ is smaller than or equal to $(dN/dt)_{difpot}$ of eq. 25, $(dN/dt)_{dif}$ represents the actual daily nitrogen uptake by diffusion. If, however, $(dN/dt)_{dif}$ is larger than $(dN/dt)_{difpot}$, a new value for C_r is calculated according to

$$C_r = C \cdot (1 - \tau) \qquad (28)$$

with τ given as

$$\tau = \frac{\left(\dfrac{dN}{dt}\right)_{difpot}}{\displaystyle\int_0^L S_{difm} \cdot dz} \qquad (29)$$

in which expression S_{difm} represents S_{dif} from eq. 26, calculated for $C_r = 0$. In this respect it is noted that WHNSIM calculates L, w (eq. 13) and M (eq. 27) as functions of the cumulative dry matter production. However, during the growing season these quantities should be measured once or twice so that a calibration is possible. These and other measurements are required in order that meaningful simulations with WHNSIM can be carried out. They will be discussed briefly in the next section.

Some results and data requirements

The model WHNSIM was tested and calibrated with field data collected at Hohenheim and Renningen between 1984 and 1987. At Renningen the soil nitrogen dynamics of a crop rotation (winter wheat, sugar beet, winter wheat, corn) was investigated. At Hohenheim the soil nitrogen behavior of extensively used grassland was studied. Materials and methods are described in detail in Huwe and Van der Ploeg [7]. Among others, soil matric potential, soil temperature and soil nitrate in the soil profile were determined weekly. During the growing season dry matter accumulation, root distribution and nitrogen uptake were determined a number of times. Dates of fertilization and amounts of N-fertilizer

were recorded and also the atmospheric input of nitrate (wet deposition) was measured. At Hohenheim as well as at Renningen, an agroclimatic weather station collected regularly a variety of parameters needed in WHNSIM.

Besides the field work at Hohenheim and Renningen, additional soil nitrogen studies were conducted at Öhringen and at Bruchsal. Here, however, the measuring programs were less comprehensive. For example, neither at Öhringen nor at Bruchsal a weather station was located at the research site. Nevertheless, the data collected at Öhringen (1984–1987) and Bruchsal (1988–1989) were also used to evaluate the performance of WHNSIM.

As far as weather data are concerned, WHNSIM is able to process data from the German Weather Bureau, like mean daily temperature, mean daily air humidity, daily hours with bright sunshine and daily sums for global radiation. In case the Weather Bureau cannot provide radiation data, these data can be estimated from other weather data with a method described in [19].

Primary crop data required by WHNSIM are day of emergence and day of harvest. Also crop specific constants, like b and Z_{max} (eq. 21) or a (eq. 23), have to be known. Table 2 shows values for these variables as used by WHNSIM for Southern Germany. Other data, like soil cover and root distribution, are considered as functions of the total amount of dry matter, see also [4]. It is recommended, though, that occassionally during the growing season for which soil nitrogen simulation studies are planned, relevant crop parameters are measured directly.

Substitutes for not directly determined soil parameters were discussed previously. For all sites in Southern Germany for which simulations with WHNSIM were carried out, at least the moisture retention curves and the hydraulic con-

Table 2. Crop-specific constants for Southern Germany, as used in WHNSIM

Crop	Crop constants		
	a ($kg \cdot mbar \cdot mm^{-1} \cdot ha^{-1}$)	b –	N_{max} ($kg \cdot ha^{-1}$)
Winter wheat	$0.28 \cdot 10^3$	$0.16 \cdot 10^{-5}$	350
Sugar beet	$0.50 \cdot 10^3$	$0.96 \cdot 10^{-6}$	290
Corn	$0.49 \cdot 10^3$	$0.18 \cdot 10^{-5}$	225
Grass	$0.26 \cdot 10^3$	$0.54 \cdot 10^{-5}$	200

338

ductivities (at saturation) were determined experimentally. Also the organic matter and the nonorganic matter content (especially the quartz content) were determined without exception. It was tried occasionally to determine the various soil organic matter fractions (N_1, N_2 and N_3) from incubation experiments. Usually, however, they were estimated with use of preliminary test runs. To give an impression about the performance of the model, a few comparisons of measured and calculated data will be made. They are shown in Fig. 2 (above-ground dry matter production), Fig. 3 (the corresponding cumulative nitrogen uptake), Fig. 4 (soil temperature) and Fig. 5 (nitrate concentration in the soil solution). To save space, no further comparisons will be presented here. An interested reader may find additional information in [7]. That bulletin, as well as a documented copy of WHNSIM, can be obtained from the senior author on request.

Conclusions

A soil nitrogen model was developed with which the nitrate seepage, the crop nitrogen uptake as

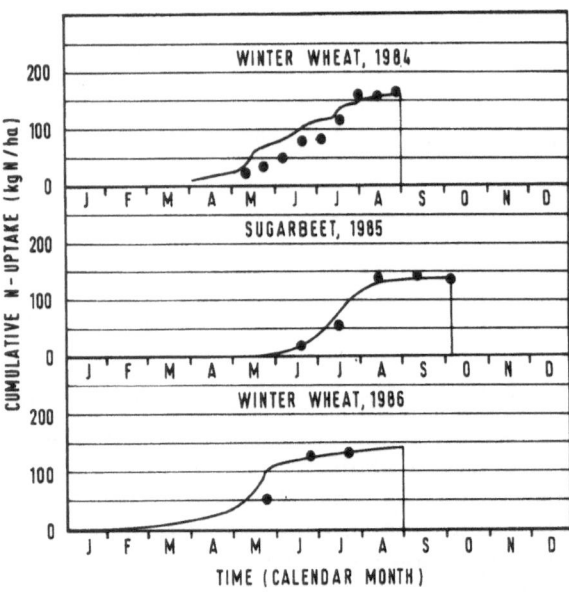

Fig. 3. A comparison of calculated (solid lines) and measured amounts of cumulative nitrogen uptake for the crops dealt with in Fig. 2.

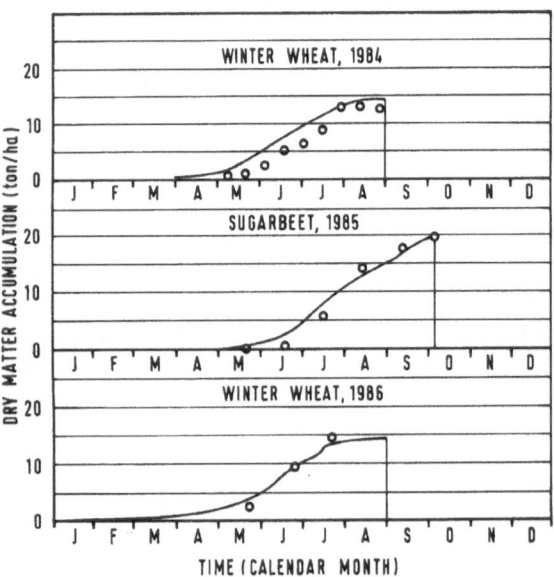

Fig. 2. A comparison of calculated (solid lines) and measured dry matter values for crops in Southern Germany (Renningen): winter wheat (1984), sugar beet (1985) and winter wheat (1986).

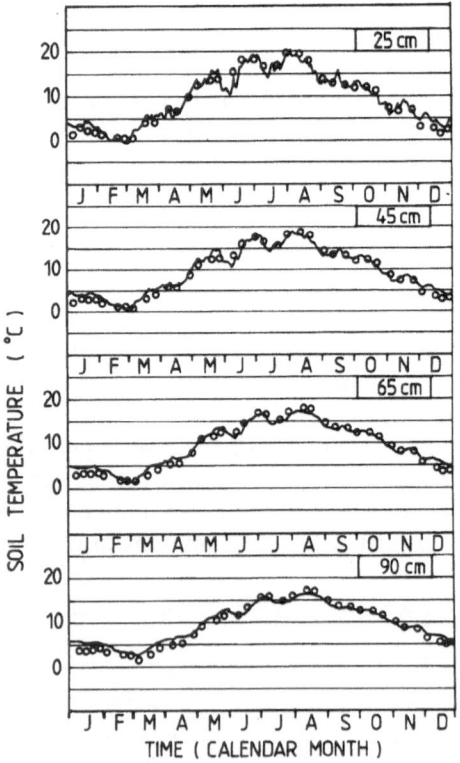

Fig. 4. A comparison of calculated (solid lines) and measured soil temperatures under grassland (Hohenheim) in 1986.

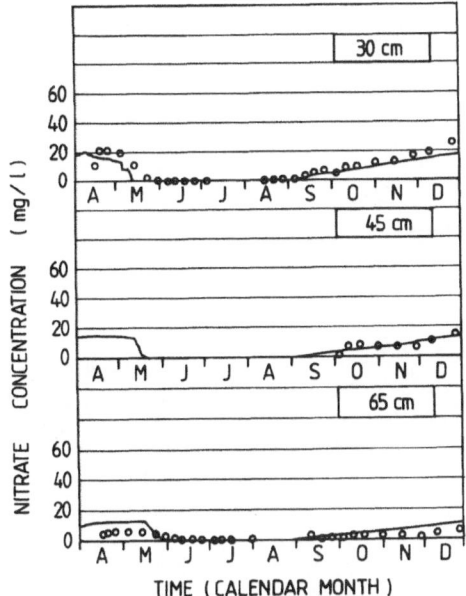

Fig. 5. A comparison of calculated (solid lines) and measured nitrate concentrations in the soil solution under winter wheat in 1984 (Renningen).

well as the mineralization of soil organic matter under field conditions can be simulated. For a number of crops and years, model results were compared with measured data for various sites in Southern Germany. After calibration, which involved mainly the reduction factors for water uptake and mineralization, a fair agreement between measured and simulated data was obtained. Therefore, it is concluded that the model is a valuable tool to describe soil nitrogen behavior of agricultural fields.

Acknowledgement

The research reported on was financially supported by the German Research Foundation (DFG).

References

1. Bear J (1972) Dynamics of fluids in porous media. American Elsevier Publishing Co, New York
2. Beese F, Van der Ploeg RR and Richter W (1977) Test of a soil water model under field conditions. Soil Sci Soc Am J 41: 979–984
3. Blom TJM and Troelstra SR (1972) Simulation model of the combined transport of water and heat produced by a thermal gradient in porous media. Report no. 6, Department Theoretical Production Ecology. Agricultural University, Wageningen, The Netherlands
4. Feddes RA, Kowalik PJ and Zaradny H (1978) Simulation of field water use and crop yield. PUDOC, Wageningen, The Netherlands
5. Hanks RJ and Ashcroft GL (1980) Applied soil physics. Springer-Verlag, Berlin-Heidelberg-New York
6. Hornung U and Messing W (1984) Poröse Medien – Methoden und Simulation. Verlag Beiträge zur Hydrologie Ilse Nippes, Kirchzarten, Fed Rep Germany
7. Huwe B and Van der Ploeg RR (1988) Modelle zur Simulation des Stickstoffhaushaltes von Standorten mit unterschiedlicher landwirtschaftlicher Nutzung. Mitteilungen Institut für Wasserbau, Universität Stuttgart (FRG). Volume 69, 213 p
8. Jackson RD (1972) On the calculation of the hydraulic conductivity. Soil Sci Soc Am Proc 36: 380–382
9. Klute A (1986) Methods of soil analysis. Agronomy 9. ASA, Madison (Wisconsin), USA
10. McIsaac G, Martin D and Watts D (1985) Users guide to NITWAT – a nitrogen and water management model. Agric Eng Dept, University of Nebraska, Lincoln (Nebraska), USA
11. Olsen SR and Kemper WD (1968) Movement of nutrients to plant roots. Adv Agron 30: 91–151
12. Priestley CHB and Taylor RJ (1972) On the assessment of surface heat flux and evaporation using large-scale parameters. Month Weath Rev 100: 81–92
13. Remson I, Hornberger GM and Molz FJ (1971) Numerical methods in subsurface hydrology. Wiley-Interscience, New York
14. Richter J, Nuske A, Habenicht W and Bauer J (1982) Optimized N-mineralization parameters of loess soils from incubation experiments. Plant and Soil 68: 378–388
15. Syring HM and Sauerbeck D (1985) Ein Modell zur quantitativen Beschreibung des Stickstoffumsatzes im System Boden-Pflanze. Z dt geol Gesellschaft 136: 461–472
16. Van der Ploeg RR and Benecke P (1981) Evaluation of one and two-dimensional water flow models and field validation of unsaturated water flow. *In*: Iskandar IK (ed) Modeling Wastewater Renovation, pp 92–114. John Wiley and Sons, New York-Chichester-Brisbane-Toronto
17. Van Genuchten MT and Wierenga PJ (1986) Solute dispersion coefficients and retardation factors. *In*: Klute A (ed) Methods of soil analysis, pp 1025–1054. Agronomy 9. ASA, Madison (Wisconsin), USA
18. Van Veen JA and Frissel MJ (1981) Simulation model of the behaviour of N in soil. *In*: Frissel MJ and Van Veen JA (eds) Simulation of nitrogen behaviour of soil-plant systems, pp 126–144. PUDOC, Wageningen, The Netherlands
19. Zaradny H and Van der Ploeg RR (1982) Calculation of the shortwave radiation flux from weather station data in evapotranspiration studies. Z Pflanzenernähr Bodenkd 145: 611–622

Fertilizer Research **27**: 341–347, 1991.
© 1991 *Kluwer Academic Publishers.*

A comparison of the performance of N simulation models in the prediction of N_{min} on farmers' fields in the spring

S. Otter-Nacke & H. Kuhlmann
Centre for Plant Nutrition and Environmental Research Hanninghof, Hanninghof 35, DW-4408 Dülmen, Germany

Key words: N dynamics, spring mineral N, simulation models, cereals

Abstract

The performance of three different models, which simulate changes in the inorganic N content of the soil, was evaluated in respect of their ability to predict N_{min} content in the spring under cereal crops. The models of British, Dutch and German origin, were tested using data from farmers' fields supplied by 70 farmers over two growing seasons in FRG. The models were run between harvest of the previous crop and spring of the following year, and predictions of N_{min} in the spring compared to soil measurements. The performance of the models was assessed by counting the number of cases in which predictions agreed within 10 or 20 kg (N) ha^{-1} of the measurements. Predictions were less than ± 10 kg (N) ha^{-1} of measured values in only 30–44% and 28–55% of cases in 1988 and 1989, respectively. Predictions were less than ± 20 kg (N) ha^{-1} of measured values in 62–70% and 68–82% of cases in 1988 and 1989, respectively. Predictions in 1989 were better because the initial N_{min} content in the autumn was included in the model input. None of the models tested had been designed to use input data of the type available to farmers. It is concluded that, at present, the results are too variable for any of the models to be used with confidence as tools to aid in N fertilizer recommendations.

Introduction

Contamination of drinking water by nitrate is of increasing concern to the population of Europe. Furthermore, for farmers, nitrogen leached from the profile is a waste which increases the cost of production and reduces their revenues. For environmental and economic reasons, the correct amount and timing of N fertilizer application to crops is of great importance.

Improvement in recommendations may be possible by using computer models which simulate nitrogen leaching, net mineralization, and N uptake by crops, and therefore the supply of nitrogen in the soil. Using easily available soil and meteorological data, farmers with the aid of personal computers may be able to predict the nitrogen amount (N_{min} content) in the soil in the spring down to a depth of 90 cm, and thereby replace the need for soil measurements of N_{min}

in the spring, upon which recommendations for the spring fertilizer application are commonly based [7].

Information is therefore required to assess the value of different models as predictive tools to aid in N fertilizer recommendations. The objectives of this study were to adapt the models to run using easily accessible soil and meteorological data available to farmers, to predict the mineral nitrogen content in the soil (0–90 cm) in early spring, and to compare the performance of different models.

Materials and methods

Three different models were available for testing:

– Model A, originally developed for Great Britain by Addiscott and Whitmore [1, 9]

– Model B, originally developed for Germany by Kersebaum and Richter [5, 6]

– Model C, originally developed for The Netherlands by Groot and De Willigen [3, 4]

The farmers who initiated this investigation are located in the Eastern part of Westphalia, Germany. It is common practice within this extension group to test the nitrogen content of fields in early spring. Soil samples are taken by farmers and sent to laboratories for analysis. The amounts of mineral nitrogen in the three top layers (of a thickness of 30 cm each) are the basis for determining the spring fertilizer application following the 'N_{min}-method' [7].

Model input

Essential input data to simulation models are weather data. High regional variability of precipitation required local measurements. Thus, data of the German Meteorological Office as published in weekly reports as well as from an automatic weather station at one site and rainfall measurements from 24 farm locations were used. Meteorological data included minimum, maximum, and mean air temperatures, mean soil temperatures at 10 and 20 cm depth, air saturation deficit, wind velocity, duration of daily sunshine, and global radiation (only available from very few stations).

Mineralization rates and amount of leached nitrogen are dependent on weather conditions during winter. Both winters considered were relatively mild compared to the 30-year average.

While the winter of 1987/88 was relatively wet, precipitation in the following winter season was below average until late February.

Input data were collected via a specially developed questionnaire from 70 farmers as well as from experimental fields at the University of Hannover, covering an area adjacent to Westphalia. Data included soil characteristics and site specific data concerning the previous crop and fertilizer history (Table 1). Two years' field data were collected: 1987/88 and 1988/89. The initial mineral nitrogen content after harvest of the previous crop was known only in the second season. Simulation runs were executed without the knowledge of the measured spring mineral nitrogen values to avoid altering of input data in order to achieve a better agreement between simulated and measured values.

Model output

Model output was compiled and used to calculate statistics such as the mean absolute and bias errors. The intercept (a) and slope (b) of a simple regression of the form $y = a + b * x$ are interpreted as quality measures. The closer a is to 0 and b is to 1 the higher the predictive value of the simulations. The coefficient of determination of this regression (r^2) is a measure of the contribution of simulations to the variation of measured values [2].

In 1988 a total of 126 sites was used for comparison between measurements and simulations by model B. The number of comparisons

Table 1. Data available as input to simulation models

Site characteristics:	– altitude, groundwater level
	– soil texture, bulk density (per layer)
	– organic C and N contents
	– grades of slope, waterlogging, stoniness
	– underground material, type and depth
	– depth and spacing of drainage
	– pasture in site's history
Actual crop:	– variety
	– sowing date and density
	– type, date, and amount of organic or mineral fertilizer
Previous crop:	– yield, harvest date
	– harvest method, date and depth of residue incorporation
	– mineral nitrogen content in 3 separate layers after harvest of previous crop (1989 only)
Catch crop:	– sowing date, height
	– harvest method, date and depth of residue incorporation

with the two other models was smaller because of restrictions to crops and/or types of organic fertilizer. The number of comparisons between measurements and simulations for the third layer (60–90 cm) is sometimes smaller than for the top two layers because soil samples were taken from these two top layers only for certain soil profiles because of insufficient rooting depth. In these cases total amounts of nitrogen for the profile include only two layers. In 1989, 68 and 69 sites were used for simulations with models A and B respectively. Due to restrictions mentioned above there were only 53 suitable cases for model C. In 1989, mineral nitrogen contents after harvest of the previous crop were available to initialize the models. The additional question to be answered was whether the knowledge of this initial value resulted in improved model performance.

The soil testing method used involves errors due to soil variability which Severin, Kersebaum and Richter [8] on the basis of their experience assume to be 20 kg (N) ha^{-1}. This value was taken as a maximum permitted divergence of the simulations compared to measurements of mineral nitrogen. A limit of practical importance to users such as farmers would be in the order of 10 kg (N) ha^{-1}.

Results and discussion

Figures 1–3 show scattergrams of measured versus simulated values of mineral nitrogen in early spring of 1988, in which the dotted lines indicate a ± 20 kg (N) ha^{-1} divergence from the 1:1 line. This limit was chosen as a maximum permitted divergence between the simulated and measured nitrogen contents.

Figures 4–6 is a corresponding series of graphs for the following year (1989). Unlike 1988 there were true outliers. In one particular case two of the models (model A and model B) suggested values of about tenfold the measured nitrogen content whereas model B gave a perfectly matching simulation. These extreme values were marked in the scattergrams and excluded from further statistical evaluation (Figs. 4 and 5).

Table 2 is a compilation of all the linear regressions calculated including those depicted in

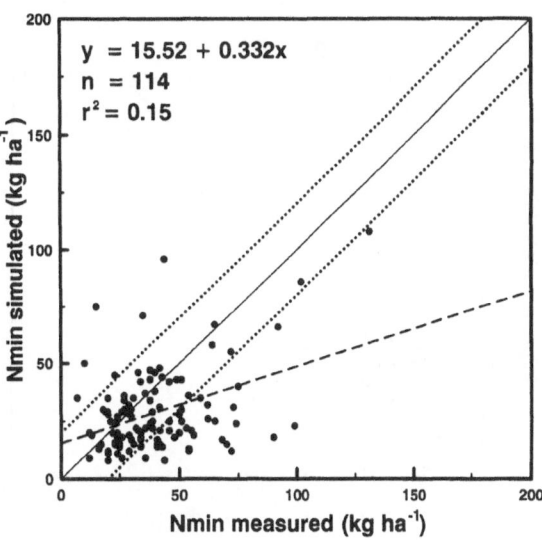

Fig. 1. Comparison between measured and simulated mineral nitrogen in the soil profile (0–90 cm) in the spring of 1988. Test runs with the model of Whitmore (A). Dotted lines represent a 20 kg (N) ha^{-1} deviation from the solid 1:1 line; the regression function is depicted in a broken line.

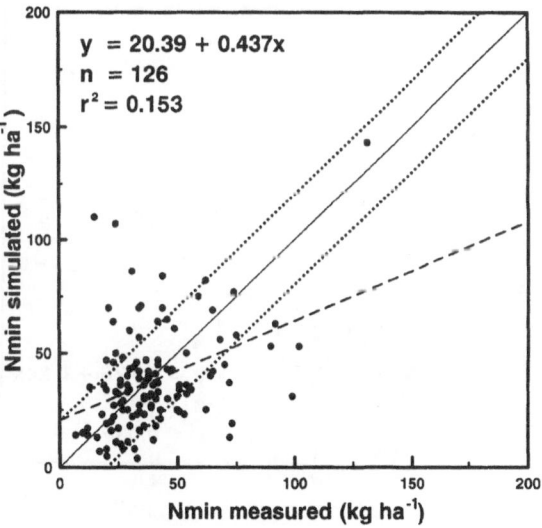

Fig. 2. Comparison between measured and simulated mineral nitrogen in the soil profile (0–90 cm) in the spring of 1988. Test runs with the model of Kersebaum (B). Dotted lines represent a 20 kg (N) ha^{-1} deviation from the solid 1:1 line; the regression function is depicted in a broken line.

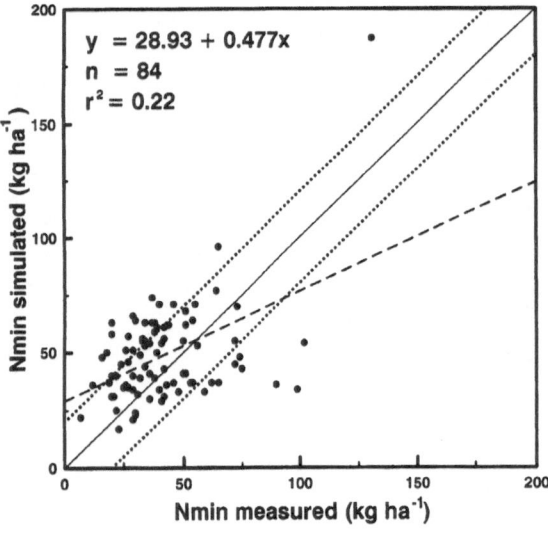

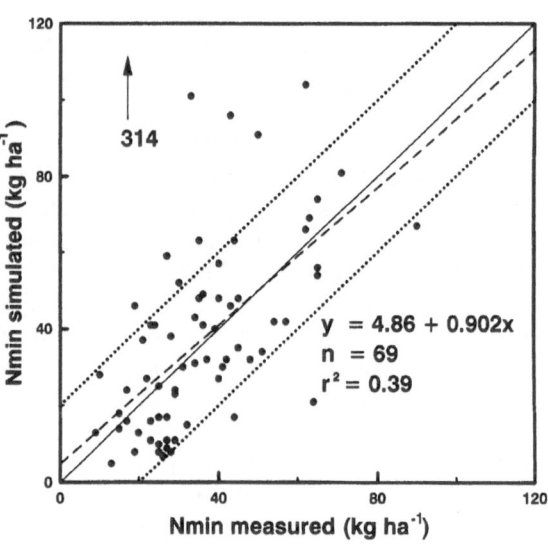

Fig. 3. Comparison between measured and simulated mineral nitrogen in the soil profile (0–90 cm) in the spring of 1988. Test runs with the model of Groot (C). Dotted lines represent a 20 kg (N) ha^{-1} deviation from the solid 1:1 line; the regression function is depicted in a broken line.

Fig. 5. Comparison between measured and simulated mineral nitrogen in the soil profile (0–90 cm) in the spring of 1989. Test runs with the model of Kersebaum (B). Dotted lines represent a 20 kg (N) ha^{-1} deviation from the solid 1:1 line; the regression function is depicted in a broken line.

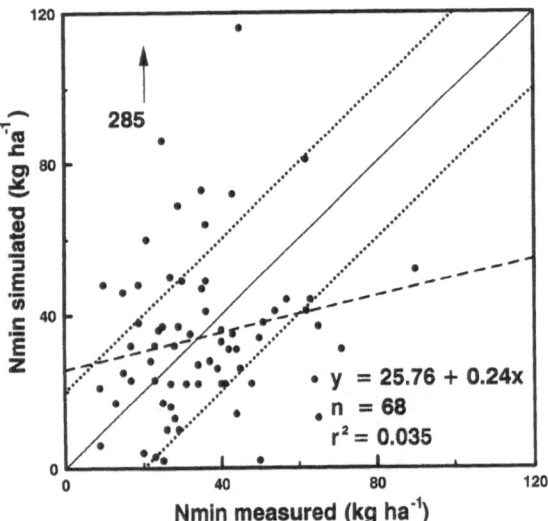

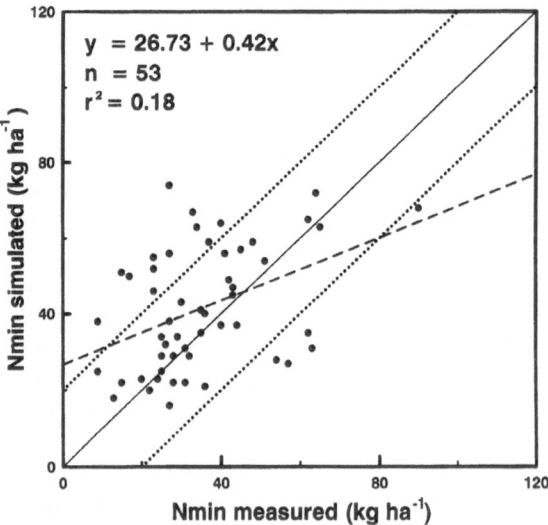

Fig. 4. Comparison between measured and simulated mineral nitrogen in the soil profile (0–90 cm) in the spring of 1989. Test runs with the model of Whitmore (A). Dotted lines represent a 20 kg (N) ha^{-1} deviation from the solid 1:1 line; the regression function is depicted in a broken line.

Fig. 6. Comparison between measured and simulated mineral nitrogen in the soil profile (0–90 cm) in the spring of 1989. Test runs with the model of Groot (C). Dotted lines represent a 20 kg (N) ha^{-1} deviation from the solid 1:1 line; the regression function is depicted in a broken line.

Table 2. Regression equations of linear regressions for all model tested

Year	Model	Layer depth (cm)	n	Regression equation $y = a + b * x$	Regression coefficient (r^2)
1988	A	0–30	106	$y = 3.83 + 0.141x$	0.050
	Whitmore	30–60	106	$y = 8.09 + 0.213x$	0.050
		60–90	89	$y = 8.30 + 0.077x$	0.017
		0–90	114	$y = 15.52 + 0.332x$	0.150
	B	0–30	126	$y = 1.85 + 0.141x$	0.223
	Kersebaum	30–60	126	$y = 4.49 + 0.213x$	0.223
		60–90	126	$y = 9.15 + 0.077x$	0.178
		0–90	126	$y = 20.39 + 0.332x$	0.153
	C	0–30	84	$y = 8.58 + 0.672x$	0.174
	Groot	30–60	84	$y = 10.12 + 0.575x$	0.229
		60–90	72	$y = 9.72 + 0.162x$	0.099
		0–90	84	$y = 28.93 + 0.447x$	0.221
1989	A	0–30	68	$y = 8.30 + 0.052x$	0.004
	Whitmore	30–60	68	$y = 10.34 + 0.620x$	0.057
		60–90	42	$y = 11.91 + 0.164x$	0.150
		0–90	68	$y = 25.76 + 0.240x$	0.035
	B	0–30	69	$y = 1.14 + 0.903x$	0.442
	Kersebaum	30–60	69	$y = 4.59 + 0.547x$	0.115
		60–90	43	$y = 9.05 + 0.905x$	0.227
		0–90	69	$y = 4.86 + 0.902x$	0.387
	C	0–30	53	$y = 14.1 + 0.179x$	0.050
	Groot	30–60	53	$y = 13.0 + 0.247x$	0.040
		60–90	37	$y = 12.1 + 0.052x$	0.070
		0–90	53	$y = 26.7 + 0.420x$	0.178

Figs. 1–6. In 1988 relatively small coefficients of determination (r^2) were found using model A. However, r^2 was considerably higher for the total profile compared to separate layers. Some of the errors seem to have been compensated for by others. In the case of model B the contrary is true: r^2 calculated for the total profile considered is smaller than for single layers.

Model C produces similarly good predictions as indicated by r^2 for the whole profile. However, the top and bottom layers considered separately provide much lower coefficients and the number of cases tested was the smallest of the three models.

In 1989, the models A and C performed very similarly to the previous year. Only model B shows an improved overall performance as indicated by the statistical parameters considered in Table 3. The percentage of cases with a difference of more than 20 kg (N) ha^{-1} is 29 to 38% in 1988, but only 18 to 32% in 1989. This suggests predictions in the second year to be more reli-

able. Also, the number of cases with differences less than 10 kg (N) ha^{-1} was larger in 1989 than in 1988 except for model A. While producing 44% good simulations in 1988 there were only 28% of the cases with differences being 10 kg (N) ha^{-1} or less.

Model C provides the contrasting situation in that small differences occur in nearly 55% of the cases in 1989 compared to 30% in 1988. Model B was able to predict 40 and 47% of the cases in 1988 and 1989, respectively, with an error of 10 kg (N) ha^{-1} or less. Knowledge of the initial nitrogen content seems to increase the goodness of fit.

Ignoring the outliers in 1989 mentioned above, the maximum errors were between 47 and 71 kg (N) ha^{-1}. These extreme values are not surprising in a statistical sense because they are characteristic of normal distributions. For the practical purpose of nitrogen predictions the implication that the actual error could be none or a multiple of the real value is unacceptable for a serious

Table 3. Statistical parameters of test runs. ME = Maximal Error, MBE = Mean Bias Error, MAE = Mean Absolute Error

Year	Model	n	Differences					ME	MBE	MAE
			<10 (kg(N) ha^{-1})		>20 (kg(N) ha^{-1})					
			no.	%	no.	%				
1988	A	114	50	43.9	39	34.2	76	−11.03	18.0
	B	126	50	39.7	37	29.4	95	−1.60	17.0
	C	84	25	29.8	32	38.1	65	0.02	18.9
1989	A	68	19	27.9	21	30.9	71	−1.17	19.2
	B	69	32	47.1	12	17.6	68	1.37	14.3
	C	53	29	54.7	17	32.1	47	6.25	13.7

user who may be obliged in the future not to exceed legal limits of nitrate in the percolating water.

Table 3 also gives the mean absolute errors (MAE, kg (N) ha^{-1}) and mean bias errors (MBE, kg (N) ha^{-1}), the latter taking into account signs of the differences between measurements and simulations. In 1988 all models had mean absolute errors of 17 to 19 kg (N) ha^{-1}, being close to the limit of 20 kg (N) ha^{-1}. While model A also shows a similar MAE of 19.2 kg (N) ha^{-1} in 1989, the other two models produced errors of 14 kg (N) ha^{-1} on average in 1989.

The mean bias error can be used to indicate if a model has a general tendency to overpredict. This is true in the case of model C which over-predicted in general by 6 to 7 kg (N) ha^{-1} in both years. Model B appears to be balanced at least in the two years under investigation. This applies to model A too, except for the first season when the tendency to produce too low nitrogen values was rather obvious.

All three of the models tested were originally designed for use of input data of a scientific level of accuracy. It was a severe restriction to all models to be run on the type of information which is available to well informed farmers. Knowledge of the initial nitrogen content of the profile considerably increased the quality of simulation. Occurrence of extreme values, the relatively high percentage of cases with errors of more than 20 kg (N) ha^{-1}, and the mean absolute error suggest that modified model approaches are needed to increase the reliability of predictions.

A simple calculation was performed to assess the practical use of the models. A mean N_{min} value was calculated from the data for both 1988 and 1989. It was found that a similar number of measurements fell within ± 20 kg (N) ha^{-1} of the mean N_{min} value, as had fallen within ± 20 kg (N) ha^{-1} of the model predictions and measured values. These results show that model performance can be improved by inclusion of the measured mineral N content in the soil after harvest of the previous crop in the model input. Soil sampling and analysis at this time has obvious practical advantages over similar measurements in spring. However, it is concluded that, at present, model predictions of N_{min} content in spring deviate at too great a regularity from measurements, for the current models to be used with any confidence to aid in N fertilizer recommendations.

Acknowledgements

Thanks are due to Messrs. Whitmore, Addiscott, Groot, and Kersebaum for permitting the use of their models and for performing most of the test runs, to A. Kläser for assistance with the input of the meteorological data, and to P.D. Seward for his valuable suggestions on the manuscript.

References

1. Addiscott TM and Whitmore AP (1987) Computer simulation of changes in soil mineral nitrogen and crop nitrogen during autumn, winter and spring. J Agric Sci Camb 109: 141–157
2. Gomez KA and Gomez AA (1984) Statistical procedures for agricultural research. J Wiley & Sons, New York
3. Groot JJR (1987) Simulation of the nitrogen balance in a system of winter wheat and soil. Simulation Report CABO-TT 13, Agricultural University Wageningen

4. Groot JJR and De Willigen P (1991) Simulation of nitrogen balance in the soil and a winter wheat crop. Fert Res 27: 261–272

5. Kersebaum KC (1989) Die Simulation der Stickstoff-Dynamik von Ackerböden. Ph.D. Thesis University Hannover

6. Kersebaum KC and Richter J (1991) Modelling nitrogen dynamics in a plant-soil system with a simple model for advisory purposes. Fert Res 27: 273–281

7. Scharpf HC and Wehrmann J (1975) Die Bedeutung des Mineralstickstoffvorrates des Bodens zu Vegetationsbeginn für die Bemessung der N-Düngung zu Winterweizen. Landwintsch Forsch 32/I: 100–114

8. Severin K, Kersebaum KC and Richter J (1985) Die Simulation der Stickstoff-Dynamik im Winterhalbjahr zur Berechnung des anorganischen N-Vorrats zu Vegetationsbeginn im Vergleich mit unterschiedlichen Meßverfahren. VDLUFA-Schriftenreihe, 16, Kongreßband 1985: 129–135

9. Whitmore AP, Coleman KW, Bradbury NJ and Addiscott TM (1991) Simulation of nitrogen in soil and winter wheat crops: Modelling nitrogen turnover through organic matter. Fert Res 27: 283–291

Fertilizer Research **27**: 349–383, 1991.

Response of wheat to nitrogen fertilization, a data set to validate simulation models for nitrogen dynamics in crop and soil

J.J.R. Groot & E.L.J. Verberne
Institute for Soil Fertility Research, P.O. Box 30003, 9750 RA Haren (Gn), The Netherlands

Key words: Winter wheat, nitrogen uptake, mineral nitrogen, soil water content, dry matter production

Abstract

A data set originating from winter wheat experiments at three locations during two years is described. The purpose is to provide sufficient data for testing simulation models for soil nitrogen dynamics, crop growth and nitrogen uptake. Each experiment comprised three different nitrogen treatments, and observations were made at intervals of two or three weeks. The observations included measurements of soil mineral nitrogen content, soil water content, groundwater table, dry matter production and dry matter distribution, nitrogen uptake, nitrogen distribution and root length density.

Introduction

On 5 and 6 June 1990, the Institute for Soil Fertility Research in Haren (the Netherlands), organized a workshop on the occasion of its centennial celebration. The topic of the workshop was modelling of the nitrogen cycle in the soil and uptake of nitrogen by the crop, and the major purpose was to compare various models which are currently in use. The comparison concerned the principles of the models (e.g., their basic assumptions, the processes which are taken into account), their actual performance and the possibility of using the models for extension purposes, e.g., improvement of nitrogen fertilizer recommendations. The extent to which nitrogen dynamics could be simulated was discussed during the workshop, but the major concern was how simulations could be improved, i.e., which processes are not yet fully understood.

To compare the different models, all participants ran their model with the same data set prior to the workshop. It was up to the participants to decide to what degree of detail they wished to perform their simulations and which

part of the data they used to parameterize their model. The data set originated from a series of winter wheat field experiments [4]. In this paper, a description of the experimental sites and the experimental techniques is given. The results of the experiments are presented in the *appendix*. In the data set several parameters were measured in a rather crude manner or derived from general soil data bases. Still, much more detail is given than will be available for future application of models in farmers' fields.

Experiments

During the growing seasons 1982–1983 and 1983–1984, field experiments with winter wheat were conducted to obtain data on crop growth and development, nitrogen uptake and soil nitrogen dynamics in different nitrogen treatments. Originally the experiments were intended to provide sufficient data to develop a simulation model for crop growth and nitrogen uptake [4]. The observations included both crop and soil samplings, and we considered the data to be sufficiently accurate for a comparison of different simulation models.

In each year, one experiment was located near Wageningen on the experimental farm 'the Bouwing' (silty clay loam) and two experiments were located in the newly reclaimed polders on the experimental farms 'the Eest' (silty loam) and 'PAGV' (silty loam). Each experiment contained three different nitrogen treatments; all other nutrients were considered to be nonlimiting. In Table 1 a summary of information concerning the experimental sites is given. For the nitrogen treatments the reader is referred to the *appendix* (Tables 1).

Soil data

Soil physical characteristics

The Bouwing. The soil water retention curves and hydraulic conductivity curves for the Bouwing (Fig. 1) were taken from a series of standar-

dized curves [7] on the basis of soil classification. The fields on which the experimental plots were laid out at the Bouwing were naturally drained. On the whole farm, only two fields are tile-drained, and a comparison between tile-drained and naturally drained plots did not show any difference in soil water content. This is due to the presence of layers of gravel at a depth of about 100–120 cm, which act as drains.

The Eest, PAGV. Soil water retention and conductivity for the Eest and PAGV were measured in the field on a site considered representative of both experimental locations in the Polders (Fig. 2) [2]. The relation between pressure head and volumetric water content was determined in the immediate vicinity of tensiometers in the soil. Water content was measured gravimetrically using cylinders having a volume of $100\,cm^3$. Conductivity was measured by the 'hot air' method [1].

At the Eest and at PAGV, ditches are 300 m apart. At the ditch the depth of the drains is 110–120 cm, in the middle of the field 90 cm.

Table 1. Description of the experimental sites

	the Eest	the Bouwing	PAGV
Location	Nagele	Randwijk	Lelystad
	52°37'N, 5°45'E	51°57'N, 5°45'E	52°30'N, 5°30'E
Soil type	Silty loam	Silty clay loam	Silty loam
Maximum rooting depth	120 cm	100 cm	100 cm
Wheat variety	Arminda	Arminda	Arminda
1982:			
Sowing date	19 October	21 October	25 October
Previous crop	potatoes	potatoes	sugar beets
Harvest prev.crop	3rd week September	24–27 September	4 September
Yield prev.crop	?	50 t ha^{-1}	?
Fertilizer prev.crop	270 kg(N) ha^{-1}	220 kg(N) ha^{-1}	150 kg (N) ha^{-1}
1983:			
Sowing date	21 October	27 October	21 October
Previous crop	potatoes	potatoes	sugar beets
Harvest prev.crop	1st week October	3,4 October	3 October
Yield prev.crop	55 t ha^{-1}	40 t ha^{-1}	?
Fertilizer prev.crop	270 kg(N) ha^{-1}	220 kg(N) ha^{-1}	195 kg(N) ha^{-1}

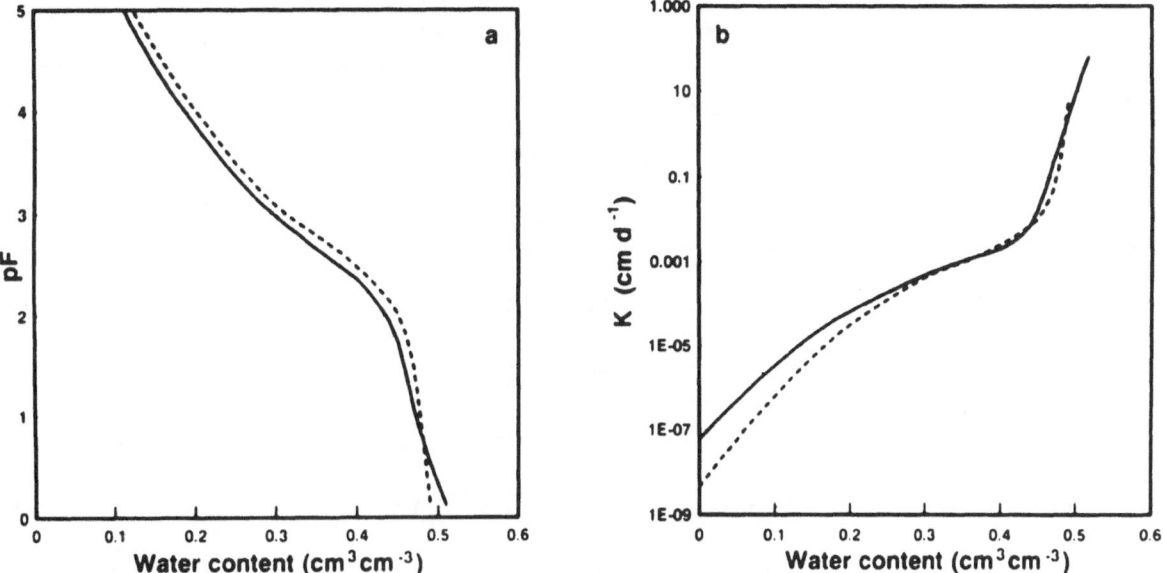

Fig. 1. Soil water retention (a) and conductivity (b) curves for the layers 0–40 cm (——) and 40–100 cm (– – –) on the experimental farm the Bouwing. From [7].

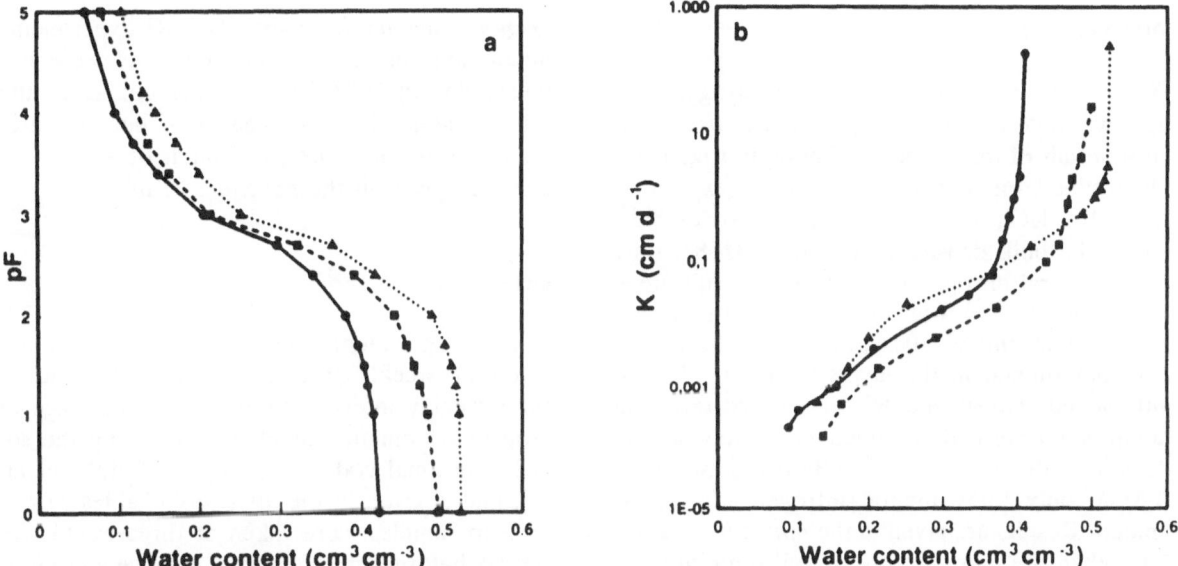

Fig. 2. Soil water retention (a) and conductivity (b) curves for the layers 0–25 cm (——), 25–40 cm (– – –) and 40–100 cm (· · · ·) on the experimental farms the Eest and PAGV (after [2]).

The distance between drains is 12 m at the Eest and 10 m at PAGV.

Particle size distribution, percentage organic matter, and bulk density

The soil texture is classified on the basis of the relative proportions of the various soil separates in the soil, according to the SSSA soil texture classification system [5]. Particle size distribution for the upper 100 cm of the soil profile, the percentage organic matter in different layers and the bulk density are given in Table 2, for each of the three experimental locations.

Table 2. Particle size distribution (expressed in % (v/v) of mineral material), bulk density (g cm^{-3}) and soil organic matter content (%, w/w) for the 0–30, 30–60 and 60–100 cm layers at the experimental locations

	the Bouwing			the Eest			PAGV		
	0–30	30–60	60–100[a]	0–30	30–60	60–100	0–30	30–60	60–100
< 2 μm	35	35	35	20	10	–	20	10	–
2–50 μm	55	55	55	46	65	–	46	65	–
> 50 μm	10	10	10	34	25	–	34	25	–
bulk density	1.35	1.40	1.35	1.3	1.1	1.0	1.3	1.1	1.0
organic matter	2.8	1.4	1.2	3.0	2.0	1.8	2.3	1.8	1.8

[a] layers of gravel at approximately 100–120 cm

Soil sampling

Soil mineral nitrogen. Soil mineral nitrogen content was determined in 8 replicates per treatment at intervals of three weeks. Mineral nitrogen was determined on a weight basis after extraction with 1M KCl, and was converted to kg ha^{-1} using the bulk density values. In 1982–83, only total mineral nitrogen was determined (nitrate + ammonium), and was measured in the layers 0–30, 30–60 and 60–100 cm. In 1983–84 nitrogen was determined in the layers 0–20, 20–30, 30–40, 40–60, 60–80 and 80–100 cm. Nitrate and ammonium were determined separately at the locations the Eest and the Bouwing, while at PAGV only total mineral nitrogen was determined. Results are given in the *appendix* (Tables 2). When the dates shown for soil sampling and nitrogen application are the same, the soil was sampled before the nitrogen was applied.

Soil water content. Soil cores of 100 cm^3 were taken in 8 replicates per treatment at intervals of three weeks; soil water content was determined gravimetrically. In 1982–83, the soil water content was determined in the layers 0–30, 30–60 and 60–100 cm, in 1983–84 in the layers 0–20, 20–30, 30–40, 40–60, 60–80 and 80–100 cm. Results are given in the *appendix* (Tables 3).

Groundwater level. In the 1982–83 experiments fluctuations in the groundwater level were not measured. In 1983–84, the time course of the groundwater table was measured, but measurements started late, i.e., in March or April. Results are given in the *appendix* (Tables 7).

Crop data

In each experiment crop development was monitored at weekly intervals in 1982–83 and at three-weekly intervals in 1983–84. The stage of crop development was characterized by the so-called decimal code according to Zadoks et al. [8], and is given in the *appendix* (Tables 3).

Crop samples were taken at three-weekly intervals before anthesis and at two-weekly intervals after anthesis. At each sampling date 0.5 m^2 crop was harvested in eight replicates. The last sampling date always coincided with the final harvest. From each sample, the number of plants (before tillering) or the number of tillers (after tillering) was determined. From each of the 0.5 m^2 samples a subsample of 25 plants (before tillering) or 25 tillers (after tillering) was taken. These subsamples were divided into green leaves, dead leaves (yellow or brown leaves), stems + sheaths, chaff and grain. Of each component, the contents of dry matter and nitrogen were determined. Dry matter (expressed in $kg\,ha^{-1}$), Leaf Area Index ($m^2\,m^{-2}$), amounts and contents of nitrogen in each of the plant organs ($kg\,ha^{-1}$ and $g\,kg^{-1}$ respectively) are given as averages of eight replications in the *appendix* (Tables 5 and Tables 6).

In the 1983–84 experiments root length density was measured three times during the growing season, using the core break method [3]; the results are presented graphically in the *appendix* (Figs. 1). Cylindrical auger samples of 7 cm diameter and 10 cm length were divided into two parts, approximately in the middle, and the numbers of roots counted on the two sides were added and recorded. Data on the number of root intersections per unit surface area, N, were converted to estimates of root length density, L_{rv}. The conversion formula used was: $L_{rv} = 5\,N$, based on the calibration of the core break method for winter wheat by Drew and Saker [3]. The sampling scheme for the core samples was chosen according to Van Noordwijk et al. [6]. The data in the *appendix* are averages of 8 replications.

The original data set [4] is more detailed than the data set given in the *appendix*. Besides the crop measurements mentioned, all leaf samples were subdivided into 6 leaf layers, and the dry matter content and area of each layer were measured separately. In the experiment at the Bouwing in 1982–83, leaf length and leaf width of all leaves on main stem, first tiller and second tiller were measured at weekly intervals.

Weather data

For 1982, 1983 and 1984, daily weather data (maximum and minimum temperature (°C), daily total global radiation ($MJ\,m^{-2}\,d^{-1}$), rainfall ($mm\,d^{-1}$), vapour pressure at 9.00 a.m. (mbar), average wind speed ($m\,s^{-1}$)) were obtained from the meteorological station Wageningen, at a distance of 7 km from the experimental farm the Bouwing, and from the meteorological Station Swifterbant, at a distance of 15 km from experimental farms the Eest and PAGV. The annual time courses of minimum temperature, maximum temperature, total global radiation and rainfall for 1982, 1983 and 1984 are given in the *appendix* (Figs. 2).

Note

A more detailed description of the data [4] can be obtained from the Centre for Agrobiological Research, Wageningen. Weather data on floppy disk are available from the junior author upon request.

Acknowledgements

The workshop was financially supported by the Dutch Fertilizer Institute, Norsk Hydro's Hanninghof Research Station, FRG, the European Community, and the British Fertiliser Manufacturers Association.

References

1. Arya LM, Farrel DA and Blake GR (1975) A field study of soil water depletion patterns in presence of growing soybean roots. I. Determination of hydraulic properties of the soil. Soil Sci Soc Am Proc 39: 424–430
2. De Vos JA, Inckel MS and Raats PAC (1990) The water balance of the unsaturated zone, considered within the context of sustainable agriculture. In: Van Genuchten R (ed) Proc Workshop 'Indirect methods for estimating the hydrolic properties of unsaturated soils', October 11–13 1989, Riverside, California (in press)
3. Drew MC and Saker LR (1980) Assessment of a rapid method using soil cores, for estimating the amount and distribution of roots in the field. Plant and Soil 55: 297–305
4. Groot JJR (1987) Simulation of nitrogen balance in a system of winter wheat and soil. Simulation Report CABO-TT no. 13, Centre for Agrobiological Research, and Department of Theoretical Production Ecology, Agricultural University, Wageningen, 69 p
5. Soil Science Society of America (1987) Glossary of soil science terms. Soil Science Society of America, Madison, Wisconsin, 44 p
6. Van Noordwijk M, Floris J and De Jager A (1985) Sampling schemes for estimating root density distribution in cropped fields. Neth J Agric Sci 33: 241–262
7. Wösten JHM, Bannink MH and Beuving J (1986) Waterretentie- en doorlatendheidskarakteristieken van boven- en ondergrond in Nederland: de Staringreeks. ICW-rapport 18, Stiboka-rapport 1932, Wageningen, p 24 and p 60
8. Zadoks JC, Chang TT and Konzak CF (1974) A decimal code for the growth stages of cereals. Weed Res 14: 415–421

354

Appendix

Index

I. The Bouwing, 1982–1983

Table I-1. Nitrogen treatments: time (yymmdd) and rate of application (kg ha^{-1}). Nitrogen was applied as Calcium Ammonium Nitrate

Treatment	830214	830513	830622
N1	0	0	0
N2	0	60	0
N3	0	120	40

Table I-2. Mineral nitrogen (ammonium + nitrate) content (kg ha^{-1}) of soil layers 0–30, 30–60 and 60–100 cm

Date	Treatment N1			Treatment N2			Treatment N3		
yymmdd	0–30	30–60	60–100	0–30	30–60	60–100	0–30	30–60	60–100
830207	19.7	23.1	70.7						
830228	18.9	33.1	78.1						
830328	8.1	17.3	60.2						
830418	8.5	12.9	73.0						
830509	8.7	8.9	40.4						
830531	9.9	5.2	7.7	10.3	4.9	6.3	22.0	6.0	9.3
830613	6.7	3.0	5.6	5.5	4.5	9.2	10.9	4.2	4.9
830704	8.1	5.1	7.2	5.5	5.1	8.8	24.6	6.9	10.2
830718	8.5	5.6	7.5	9.1	5.4	8.7	25.0	15.4	20.7
830801	14.2	12.6	10.7	17.4	10.1	13.8	56.3	21.7	27.7

Table I-3. Volumetric water content (cm^3 cm^{-3}) of soil layers 0–30, 30–60 and 60–100 cm

Date	Treatment N1			Treatment N2			Treatment N3		
yymmdd	0–30	30–60	60–100	0–30	30–60	60–100	0–30	30–60	60–100
830207	.428	.377	.339						
830228	.390	.407	.365						
830328	.412	.389	.376						
830418	.403	.401	.401						
830509	.426	.387	.376						
830531	.412	.390	.375	.414	.386	.383	.399	.383	.375
830613	.354	.339	.364	.349	.353	.373	.347	.358	.371
830704	.286	.301	.349	.219	.305	.330	.270	.296	.332
830718	.202	.248	.266	.203	.246	.264	.197	.231	.254
830801	.211	.269	.304	.356	.272	.106	.228	.264	.290

Table I-4. Decimal codes (DC) for crop development according to Zadoks et al. (1974) Weed Res 14: 415–421

Date	DC	Date	DC	Date	DC	Date	DC
830207	21	830321	22	830502	31	830613	57
830214	22	830328	23	830509	32	830620	61
830221	22	830405	23	830516	33	830627	70
830228	22	830411	24	830524	37	830704	77
830307	22	830418	24	830531	38	830718	87
830314	23	830425	30	830607	45	830725	91

Table I-5. Dry matter distribution (kg ha^{-1}) and Leaf Area Index (LAI)

Date	Total kg ha^{-1}	Green leaf kg ha^{-1}	Dead leaf kg ha^{-1}	Stem+ sheaths kg ha^{-1}	Chaff kg ha^{-1}	Grain kg ha^{-1}	LAI m^2 m^{-2}
Treatment N1							
830207	64	55		10			0.1
830228	72	57		15			0.1
830328	132	106		26			0.3
830418	452	369		83			0.8
830509	1861	945	0	916	0	0	2.2
830524	3661	1418	7	2237	0	0	3.3
830613	7750	1297	265	5011	1177	0	3.0
830704	12128	1173	503	6631	1796	2025	2.5
830718	14672	190	1271	4938	1735	6538	0.3
830801	14182	0	1317	4594	1541	6730	0.0
Treatment N2							
830524	4034	1636	2	2395	0	0	3.8
830613	8475	1827	156	5178	1315	0	4.2
830704	13879	1514	637	7179	2174	2375	3.3
830718	15995	343	1348	5498	2007	6799	0.5
830801	17685	0	1731	5739	2113	8103	0.0
Treatment N3							
830524	4013	1726	7	2280	0	0	4.1
830613	9230	2193	93	5517	1428	0	5.4
830704	13420	1672	523	6829	2285	2111	3.4
830718	17134	371	1583	5904	2205	7071	0.5
830801	17183	0	1864	5786	2080	7454	0.0

Table I-6. Nitrogen content of dry matter and nitrogen distribution

Date	g/kg	Total kg/ha	Green leaf g/kg	kg/ha	Dead leaf g/kg	kg/ha	Stem+sheaths g/kg	kg/ha	g/kg	Grain kg/ha	g/kg	Chaff kg/ha
Treatment N1												
830207	50.3	3.2										
830228	53.6	3.8										
830328	57.0	7.5										
830418	54.2	24.5										
830509	34.1	63.4	43.5	41.1	0.0	0.0	24.4	22.3	0.0	0.0	0.0	0.0
830524	20.7	75.8	33.5	47.5	0.0	0.0	12.7	28.3	0.0	0.0	0.0	0.0
830613	13.6	105.2	31.4	40.8	19.0	5.0	7.5	37.6	0.0	0.0	18.5	21.8
830704	10.3	124.3	23.5	27.6	12.3	6.2	5.2	34.7	19.0	38.5	9.6	17.3
830718	7.7	112.4	9.5	1.8	6.4	8.2	2.2	10.6	13.0	84.8	4.0	7.0
830801	10.2	143.9	12.6	0.0	8.0	10.6	2.7	12.2	17.2	115.5	3.7	5.7
Treatment N2												
830524	27.2	109.8	41.3	67.6	0.0	0.0	17.6	42.2	0.0	0.0	0.0	0.0
830613	18.2	154.4	37.1	67.7	20.7	3.2	10.8	55.7	0.0	0.0	21.1	27.8
830704	12.4	171.5	24.7	37.4	12.9	8.2	7.6	54.2	20.5	48.8	10.5	22.9
830718	11.8	188.0	14.9	5.1	9.9	13.4	3.8	20.9	20.0	136.1	6.2	12.5
830801	12.0	213.1	12.6	0.0	10.1	17.4	2.6	15.0	21.2	171.3	4.4	9.4
Treatment N3												
830524	30.8	123.5	44.2	76.3	0.0	0.0	20.7	47.2	0.0	0.0	0.0	0.0
830613	20.4	188.3	39.5	86.7	23.9	2.2	12.2	67.2	0.0	0.0	22.6	32.2
830704	14.2	190.4	27.7	46.3	16.9	8.8	8.9	60.5	22.5	47.6	11.9	27.1
830718	11.8	202.9	12.6	4.7	9.1	14.4	4.4	26.1	20.3	143.4	6.5	14.3
830801	14.8	254.4	12.6	0.0	11.2	20.8	3.6	20.8	27.4	204.0	4.2	8.8

II. The Bouwing, 1983–1984

Table II-1. Nitrogen treatments: time (yymmdd) and rate of application (kg ha^{-1}). Nitrogen was applied as Calcium Ammonium Nitrate

Treatment	840217	840509	840606
N1	70	0	0
N2	70	60	40
N3	70	120	40

Table II-2. Mineral nitrogen (ammonium, nitrate, ammonium + nitrate) content ($kg \cdot ha^{-1}$) of soil layers 0–20, 20–30, 30–40, 40–60, 60–80 and 80–100 cm

ammonium

Date yymmdd	Treatment N1						Treatment N2						Treatment N3					
	0–20	20–30	30–40	40–60	60–80	80–100	0–20	20–30	30–40	40–60	60–80	80–100	0–20	20–30	30–40	40–60	60–80	80–100
840214	0.0	0.7	0.0	0.0	0.0	0.0												
840313	8.1	0.7	0.0	0.0	0.0	0.0												
840403	2.7	1.4	0.0	1.4	0.0	0.0												
840424	2.7	1.4	0.0	0.0	0.0	0.0												
840508	4.1	1.4	0.7	0.0	0.0	0.0												
840528	2.7	1.4	0.7	0.0	0.0	0.0	2.7	1.4	0.7	0.0	0.0	0.0	5.4	2.7	0.7	0.0	0.0	0.0
840619	0.0	0.7	0.7	1.4	1.4	1.4	2.7	1.4	1.4	2.8	2.7	2.7	2.7	2.7	0.7	1.4	0.0	0.0
840703	2.7	1.4	1.4	1.4	0.0	0.0	2.7	1.4	0.7	2.8	0.0	0.0	4.1	0.7	0.7	0.0	0.0	0.0
840717	2.7	1.4	0.7	1.4	1.4	0.0	2.7	0.0	0.0	0.0	0.0	0.0	6.8	4.1	1.4	2.8	2.7	2.7
840807	4.1	1.4	1.4	2.8	2.7	2.7	2.7	1.4	1.4	2.8	2.7	2.7	5.4	2.7	1.4	2.8	4.1	2.7

nitrate

Date yymmdd	Treatment N1						Treatment N2						Treatment N3					
	0–20	20–30	30–40	40–60	60–80	80–100	0–20	20–30	30–40	40–60	60–80	80–100	0–20	20–30	30–40	40–60	60–80	80–100
840214	0.0	0.0	0.7	4.2	8.1	14.9												
840313	40.5	6.1	4.2	7.0	12.2	18.9												
840403	41.9	14.2	5.6	8.4	12.2	21.6												
840424	20.3	14.9	4.9	12.6	17.6	20.3												
840508	16.2	7.4	1.4	7.0	8.1	18.9												
840528	4.1	2.0	0.0	1.4	2.7	4.1	14.9	6.1	1.4	2.8	2.7	6.8	32.4	12.2	2.8	2.8	1.4	5.4
840619	4.1	2.7	1.4	2.8	0.0	1.4	24.3	21.6	4.9	12.6	6.8	6.8	40.5	36.5	9.1	21.0	12.2	13.5
840703	1.4	0.7	0.0	0.0	0.0	0.0	8.1	4.7	1.4	4.2	2.7	2.7	18.9	10.1	2.8	7.0	4.1	8.1
840717	2.7	1.4	0.7	1.4	1.4	0.0	6.8	2.7	1.4	2.8	1.4	1.4	18.9	5.4	2.8	5.6	2.7	5.4
840807	2.7	2.0	1.4	2.8	2.7	1.4	8.1	4.7	2.1	5.6	2.7	2.7	17.6	8.8	4.2	8.4	5.4	5.4

ammonium+nitrate

Date yymmdd	Treatment N1						Treatment N2						Treatment N3					
	0–20	20–30	30–40	40–60	60–80	80–100	0–20	20–30	30–40	40–60	60–80	80–100	0–20	20–30	30–40	40–60	60–80	80–100
840214	0.0	0.7	0.7	4.2	8.1	14.9												
840313	48.6	6.8	4.2	7.0	12.2	18.9												
840403	44.6	15.5	5.6	9.8	12.2	21.6												
840424	23.0	16.2	4.9	12.6	17.6	20.3												
840508	20.3	8.8	2.1	7.0	8.1	18.9												
840528	6.8	3.4	0.7	1.4	2.7	4.1	17.6	7.4	2.1	2.8	2.7	6.8	37.8	14.9	3.5	2.8	1.4	5.4
840619	4.1	3.4	2.1	4.2	1.4	2.7	27.0	23.0	6.3	15.4	9.5	9.5	43.2	39.2	9.8	22.4	12.2	13.5
840703	4.1	2.0	1.4	1.4	0.0	0.0	10.8	6.1	2.1	7.0	2.7	2.7	23.0	10.8	3.5	7.0	4.1	8.1
840717	5.4	2.7	1.4	2.8	1.4	0.0	9.5	2.7	1.4	2.8	1.4	1.4	25.7	9.5	4.2	8.4	5.4	8.1
840807	6.8	3.4	2.8	5.6	5.4	4.1	10.8	6.1	3.5	8.4	5.4	5.4	23.0	11.5	5.6	11.2	9.5	8.1

Table II-3. Volumetric water content (cm³ cm⁻³) of soil layers 0–20, 20–30, 30–40, 40–60, 60–80 and 80–100 cm

Date ymmdd	Treatment N1						Treatment N2						Treatment N3					
	0–20	20–30	30–40	40–60	60–80	80–100	0–20	20–30	30–40	40–60	60–80	80–100	0–20	20–30	30–40	40–60	60–80	80–100
840214	.334	.344	.347	.363	.355	.344												
840313	.330	.348	.353	.352	.356	.341												
840403	.306	.328	.336	.347	.346	.339												
840424	.267	.277	.338	.332	.331	.322												
840508	.239	.259	.289	.319	.331	.328												
840528	—	.318	.313	.314	.313	.318	.315	.311	.312	.317	.318	.320	.313	.311	.306	.315	.316	.323
840619	.256	.270	.302	.309	.323	.326	.252	.256	.294	.300	.323	.319	.256	.259	.294	.313	.329	.332
840703	.256	.252	.266	.295	.317	.318	.247	.250	.260	.296	.321	.322	.242	.241	.257	.290	.308	.322
840717	.299	.296	.293	.272	.273	.282	.295	.289	.270	.261	.278	.290	.298	.295	.276	.261	.282	.285
840807	.233	.234	.258	.270	.285	.284	.223	.223	.239	.260	.278	.278	.221	.224	.239	.264	.282	.283

Table II-4. Decimal codes (DC) for crop development according to Zadoks et al. (1974) Weed Res 14: 415–421

Date	DC	Date	DC	Date	DC	Date	DC
840213	12	840424	23	840618	59	840716	80
840312	22	840507	32	840702	70	840807	90
840402	21	840528	39				

Table II-5. Dry matter distribution (kg ha^{-1}) and Leaf Area Index (LAI)

Date	Total kg ha^{-1}	Green leaf kg ha^{-1}	Dead leaf kg ha^{-1}	Stem+ sheaths kg ha^{-1}	Chaff kg ha^{-1}	Grain kg ha^{-1}	LAI m^2 m^{-2}
Treatment N1							
840213	33	33	0	0			0.1
840312	44	44	0	0			0.1
840402	86	82	0	4			0.1
840424	477	367	0	110			0.8
840507	1523	925	32	566			2.2
840528	3643	1295	9	2339	0	0	3.3
840618	7957	1465	44	5234	1215	0	3.5
840702	10635	1511	209	6883	1556	478	2.5
840716	12336	1022	456	6431	1712	2715	2.2
840806	14486	604	676	4720	1713	6774	1.3
840821	15162	0	1134	4740	1709	7581	0
Treatment N2							
840528	3562	1367	5	2190	0	0	3.4
840618	7792	1605	29	4949	1209	0	3.6
840702	9966	1543	205	6210	1589	419	3.0
840716	11903	1129	434	6153	1696	2492	2.5
840806	15018	693	620	4680	2967	6058	1.5
840821	16231	0	1309	5538	1830	7554	0
Treatment N3							
840528	3628	1372	7	2249	0	0	3.5
840618	7443	1677	29	4662	1074	0	4.0
840702	10292	1637	219	6404	1596	437	3.2
840716	12245	1275	404	6322	1863	2381	3.0
840806	14720	741	693	5121	1870	6295	1.7
840821	17111	0	1380	5644	1920	8168	0

Table II-6. Nitrogen content of dry matter and nitrogen distribution

Date	g/kg	Total kg/ha	Green leaf g/kg	kg/ha	Dead leaf g/kg	kg/ha	Stem+sheaths g/kg	kg/ha	g/kg	Grain kg/ha	g/kg	Chaff kg/ha
Treatment N1												
840213	48.5	1.6										
840312	52.8	2.3										
840402	54.3	4.7										
840424	51.7	24.6										
840507	33.2	50.5	44.2	40.9	26.8	0.9	15.4	8.7				
840528	23.6	86.0	54.0	69.9	0	0	6.9	16.2				
840618	15.5	123.1	32.5	47.6	0	0	9.9	51.8	0	0	19.6	23.8
840702	11.9	126.6	27.1	40.9	18.4	3.8	7.3	50.5	25.3	12.1	12.4	19.3
840716	12.5	153.8	28.0	28.6	16.2	7.4	6.8	44.0	20.0	54.2	11.5	19.7
840806	11.3	164.1	18.5	11.2	11.6	7.8	4.5	21.2	16.9	114.7	5.4	9.3
840821	12.8	194.7	0	0	10.8	12.3	3.5	16.6	20.3	153.5	7.2	12.3
Treatment N2												
840528	27.5	97.8	63.2	86.4	0	0	5.2	11.4				
840618	21.4	166.3	55.1	88.5	0	0	8.2	40.3	0	0	31.1	37.6
840702	16.0	159.7	55.1	85.0	32.1	6.6	1.7	10.5	42.8	17.9	25.0	40.0
840716	14.0	166.0	29.8	33.6	15.6	6.8	8.6	53.0	20.8	51.8	12.3	20.9
840806	13.2	198.3	22.1	15.3	14.3	8.9	6.8	31.7	19.9	120.5	7.4	21.9
840821	16.2	262.9	0	0	15.1	19.8	6.0	33.4	25.9	195.6	7.7	14.1
Treatment N3												
840528	29.0	105.1	74.6	102.3	0	0	1.2	2.7				
840618	25.2	187.5	69.2	116.1	0	0	6.0	28.2	0	0	40.3	43.3
840702	18.2	187.3	57.3	93.8	32.5	7.1	3.9	25.0	45.1	19.7	26.1	41.7
840716	16.2	198.3	45.6	58.2	24.9	10.1	1.7	11.0	34.0	81.0	20.4	38.0
840806	13.6	199.4	22.2	16.4	14.4	10.0	7.4	38.0	19.2	121.0	7.5	13.9
840821	16.1	275.5	0	0	15.4	21.2	6.7	37.7	24.9	203.0	7.1	13.6

Table II-7. Time course of groundwater table

Date	Depth (cm)	Date	Depth (cm)	Date	Depth (cm)	Date	Depth (cm)
840402	138	840507	120	840614	127	840713	136
840409	153	840514	123	840618	125	840716	131
840416	140	840521	132	840624	134	840723	135
840424	120	840528	124	840702	136	840730	139
840501	131	840604	116	840709	140	840806	148

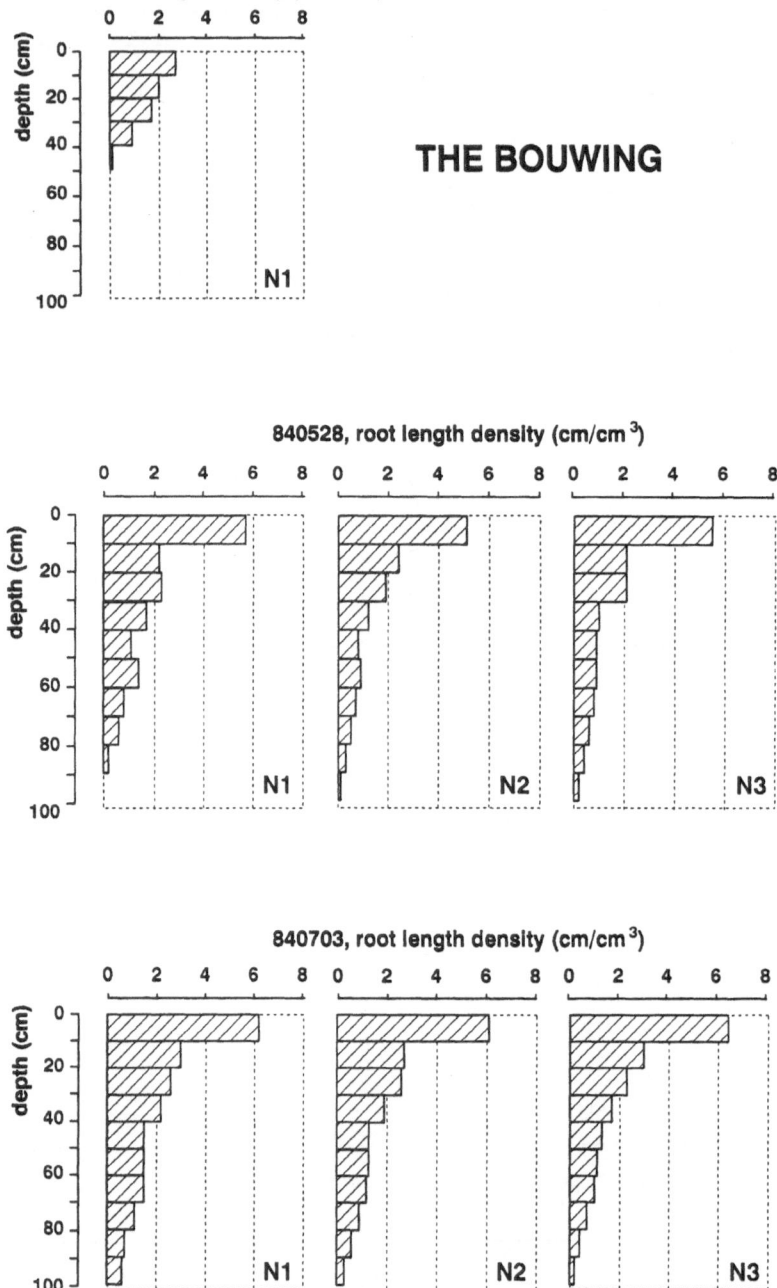

Fig. IV-1. Root length density as a function of depth in treatments N1, N2 and N3 at the experimental site the Bouwing, 1984.

III. The Eest, 1982–1983

Table III-1. Nitrogen treatments: time (yymmdd) and rate of application (kg ha^{-1}). Nitrogen was applied as Calcium Ammonium Nitrate

Treatment	830214	830511	830621
N1	0	0	0
N2	0	60	0
N3	0	120	40

Table III-2. Soil mineral nitrogen (ammonium + nitrate) content (kg ha^{-1}) of soil layers 0–30, 30–60 and 60–100 cm

Date yymmdd	Treatment N1			Treatment N2			Treatment N3		
	0–30	30–60	60–100	0–30	30–60	60–100	0–30	30–60	60–100
830209	6.7	24.5	62.6						
830302	51.8	18.2	46.1						
830330	11.1	10.6	39.0						
830420	3.2	3.5	6.3						
830511	5.8	4.3	6.0						
830525	4.4	2.8	8.1	14.4	2.9	6.2	60.8	10.4	8.6
830615	2.5	2.0	2.6	6.9	3.4	3.4	18.5	8.7	4.9
830706	2.7	1.9	3.0	3.4	2.6	3.0	26.5	8.2	14.3
830720	7.3	5.6	10.2	7.8	6.3	9.6	23.0	14.3	23.3
830803	7.0	6.3	9.5	7.3	5.5	8.7	30.2	10.1	12.4

Table III-3. Volumetric water content (cm^3 cm^{-3}) of soil layers 0–30, 30–60 and 60–100 cm

Date yymmdd	Treatment N1			Treatment N2			Treatment N3		
	0–30	30–60	60–100	0–30	30–60	60–100	0–30	30–60	60–100
830209	.411	.425	.481						
830302	.357	.341	.397						
830330	.385	.412	.567						
830420	.351	.353	.396						
830511	.378	.378	.473						
830525	.392	.405	.514	.380	.408	.509	.393	.417	.490
830615	.298	.337	.452	.286	.326	.444	.271	.298	.429
830706	.271	.262	.398	.278	.242	.389	.232	.216	.356
830720	.231	.265	.444	.201	.246	.426	.196	.203	.374
830803	.255	.307	.469	.242	.240	.421	.237	.213	.419

Table III-4. Decimal Codes (DC) for crop development according to Zadoks et al. (1974) Weed Res 14: 415–421

Date	DC	Date	DC	Date	DC	Date	DC
830216	22	830322	24	830426	30	830530	39
830223	23	830328	24	830502	30	830606	44
830301	24	830405	24	830509	31	830613	59
830308	23	830412	24	830516	32	830620	61
830316	23	830418	24	830524	38	830627	72

Table III-5. Dry matter distribution ($kg\,ha^{-1}$) and Leaf Area Index (LAI)

Date	Total $kg\ ha^{-1}$	Green leaf $kg\ ha^{-1}$	Dead leaf $kg\ ha^{-1}$	Stem+ sheaths $kg\ ha^{-1}$	Chaff $kg\ ha^{-1}$	Grain $kg\ ha^{-1}$	LAI $m^2\ m^{-2}$
Treatment N1							
830209	127						0.3
830302	134						0.3
830330	361						0.9
830420	627						1.3
830511	2720	1101	32	1587	0	0	2.1
830525	2783	813	25	1945	0	0	1.7
830615	6294	817	209	4362	906	0	1.7
830706	8510	499	479	4531	1064	1937	1.0
830720	9417	2	852	3462	1029	4072	0
830803	10095	0	942	3577	1080	4496	0
Treatment N2							
830525	3641	1260	12	2369	0	0	2.7
830615	7941	1532	82	5012	1315	0	3.5
830706	12458	1092	497	6629	1718	2523	2.3
830720	14608	200	1259	5413	1804	5932	0.4
830803	17423	0	1753	6125	2049	7495	0
Treatment N3							
830525	2839	1095	2	1742	0	0	2.2
830615	8352	1978	79	5030	1266	0	4.6
830706	13627	1673	465	7120	2158	2211	3.8
830720	14802	588	1143	5593	1845	5633	1.2
830803	17964	0	1900	6258	2174	7632	0

Table III-6. Nitrogen content of dry matter and nitrogen distribution

Date	g/kg	Total kg/ha	Green leaf g/kg	kg/ha	Dead leaf g/kg	kg/ha	Stem+sheaths g/kg	kg/ha	Grain g/kg	kg/ha	Chaff g/kg	kg/ha
Treatment N1												
830209	52.4	8.7										
830302	52.0	8.8										
830330	51.5	24.0										
830420	38.2	31.1										
830511	18.3	49.8	28.7	31.6	0	0	11.5	18.2	0	0	0	0
830525	13.8	38.5	26.5	21.6	0	0	8.7	16.9	0	0	0	0
830615	9.8	61.8	22.6	18.5	11.7	2.5	5.8	25.4	0	0	17.1	15.5
830706	7.0	59.2	11.4	5.7	7.1	3.4	3.6	16.2	13.7	26.6	6.9	7.3
830720	7.2	67.9	0	0	5.1	4.4	2.3	7.9	12.9	52.4	3.1	3.2
830803	7.6	76.4	13.7	0	6.1	5.7	3.0	10.9	12.4	55.9	3.7	4.0
Treatment N2												
830525	23.1	84.2	38.6	48.6	0	0	15.0	35.6	0	0	0	0
830615	17.1	135.8	34.5	52.9	22.6	1.8	10.6	53.1	0	0	21.2	27.9
830706	9.3	116.2	19.2	20.9	10.3	5.1	4.9	32.3	16.9	42.7	8.9	15.2
830720	9.2	133.7	12.5	2.5	7.1	8.9	3.0	16.5	16.4	97.5	4.6	8.3
830803	10.7	187.1	13.7	0	8.9	15.6	2.5	15.1	19.8	148.6	3.9	7.9
Treatment N3												
830525	31.8	90.4	46.4	50.8	0	0	22.8	39.6	0	0	0	0
830615	19.6	163.5	35.0	69.3	14.5	1.1	13.8	69.5	0	0	18.7	23.6
830706	16.6	225.7	32.2	53.8	16.2	7.5	10.7	76.1	26.0	57.5	14.3	30.8
830720	9.8	145.7	13.7	8.0	8.6	9.9	4.5	25.4	16.5	92.7	5.3	9.7
830803	13.1	235.4	13.7	0	11.3	21.5	3.5	21.7	23.7	180.9	5.2	11.3

IV. The Eest, 1983–1984

Table IV-1. Nitrogen treatments: time (yymmdd) and rate of application (kg ha^{-1}). Nitrogen was applied as Calcium Ammonium Nitrate

Treatment	840217	840511	840621
N1	50	60	0
N2	50	60	40
N3*	50	60	40

*) Due to an error, treatments N2 and N3 received the same amount of N

Table IV-2. Mineral nitrogen (ammonium, nitrate, ammonium + nitrate) content (kg ha^{-1}) of soil layers 0–20, 20–30, 30–40, 40–60, 60–80 and 80–100 cm

Date	Treatment N1						Treatment N2						Treatment N3					
yymmdd	0–20	20–30	30–40	40–60	60–80	80–100	0–20	20–30	30–40	40–60	60–80	80–100	0–20	20–30	30–40	40–60	60–80	80–100
ammonium																		
840216	1.3	1.3	0.6	0.0	0.0	3.0												
840315	9.1	1.3	1.2	1.0	2.0	3.0												
840405	2.6	1.3	0.6	1.0	2.0	2.0												
840426	3.9	1.9	1.2	1.0	2.0	2.0												
840510	2.6	1.3	0.6	0	2.0	3.0												
840530	2.6	1.3	0.6	0	0	2.0	2.6	0.6	0.6	2.1	2.0	2.0	3.9	1.3	1.2	1.0	1.0	2.0
840621	2.6	0.6	1.2	1.0	0	2.0	1.3	0.6	0	0	1.0	0	1.3	0	0	0	0	2.0
840705	0	0	0	0	1.0	1.0	1.3	0.6	0	0	0	1.0	2.6	0.6	0	0	1.0	2.0
840719	1.3	0.6	0	0	0	0	1.3	0.6	0	0	0	2.0	2.6	1.3	0	1.0	2.0	2.0
840809	2.6	1.9	1.2	2.1	2.0	3.0	3.9	2.6	1.2	2.1	2.0	4.0	2.6	1.3	1.2	2.1	3.0	4.0
nitrate																		
840216	1.3	0.6	0	2.1	15.0	11.0												
840315	52.0	11.1	6.6	8.4	18.0	19.0												
840405	40.3	9.8	6.0	8.4	20.0	16.0												
840426	11.7	14.3	2.4	4.2	16.0	12.0												
840510	2.6	3.9	0.6	1.0	10.0	8.0												
840530	11.7	0	0	0	0	0	7.8	0	0	0	0	1.0	16.9	1.3	0	0	0	1.0
840621	2.6	1.3	0	0	0	0	5.2	1.9	0	0	0	1.0	7.8	2.6	1.2	0	0	2.0
840705	0	0	0	0	0		0	0	0	0	0	0	0	0	0	0	0	0
840719	3.9	1.3	0	0	0	1.0	2.6	0.6	0	0	0	1.0	2.6	1.3	0	0	0	2.0
840809	1.3	1.3	0	1.0	1.0	1.0	1.3	1.3	0	0	1.0	0	2.6	1.3	0	1.0	1.0	0
ammonium+nitrate																		
840216	2.6	1.9	0.6	2.1	15.0	14.0												
840315	61.1	12.4	7.8	9.4	20.0	22.0												
840405	42.9	11.1	6.6	9.4	22.0	18.0												
840426	15.6	16.3	3.6	5.3	18.0	14.0												
840510	5.2	5.2	1.2	1.0	12.0	11.0												
840530	14.3	1.3	0.6	0	0	2.0	10.4	0.6	0.6	2.1	2.0	3.0	20.8	2.6	1.2	1.0	1.0	3.0
840621	5.2	1.9	1.2	1.0	0	2.0	6.5	2.6	0	0	1.0	1.0	9.1	2.6	1.2	0	0	4.0
840705	0	0	0	0	1.0	1.0	1.3	0.6	0	0	0	1.0	2.6	0.6	0	0	1.0	2.0
840719	5.2	1.9	0	0	0	1.0	3.9	1.3	0	0	0	3.0	5.2	2.6	0	1.0	2.0	4.0
840809	3.9	3.3	1.2	3.1	3.0	4.0	5.2	3.9	1.2	2.1	3.0	4.0	5.2	2.6	1.2	3.1	3.0	4.0

Table IV-3. Volumetric water content (cm³ cm⁻³) of soil layers 0–20, 20–30, 30–40, 40–60, 60–80 and 80–100 cm

Date yymmdd	Treatment N1						Treatment N2						Treatment N3					
	0–20	20–30	30–40	40–60	60–80	80–100	0–20	20–30	30–40	40–60	60–80	80–100	0–20	20–30	30–40	40–60	60–80	80–100
840216	.374	.399	.422	.428	.502	.535												
840315	.352	.371	.359	.397	.479	.511												
840405	.381	.381	.394	.425	.490	.497												
840426	.286	.247	.358	.406	.503	.513												
840510	.263	.257	.345	.381	.471	.495												
840530	.362	.372	.382	.439	.500	.514	.367	.383	.398	.430	.478	.497	.358	.359	.405	.437	.495	.512
840621	.302	.315	.341	.395	.459	.496	.292	.333	.347	.416	.480	.491	.292	.319	.358	.426	.488	.472
840705	.345	.322	.343	.405	.500	.522	.345	.334	.334	.389	.494	.499	.341	.328	.367	.390	.494	.489
840719	.325	.334	.372	.407	.497	.524	.329	.339	.370	.418	.503	.514	.324	.341	.380	.444	.489	.494
840809	.276	.307	.307	.388	.475	.496	.267	.301	.309	.371	.465	.477	.274	.288	.311	.399	.438	.472

Table IV-4. Decimal Codes (DC) for crop development according to Zadoks et al. (1974) Weed Res 14: 415–421

Date	DC	Date	DC	Date	DC	Date	DC
840215	21	840404	23	840509	32	840620	59
840314	22	840425	25	840530	38	840704	65

Table IV-5. Dry matter distribution (kg ha^{-1}) and Leaf Area Index (LAI)

Date	Total kg ha^{-1}	Green leaf kg ha^{-1}	Dead leaf kg ha^{-1}	Stem+ sheaths kg ha^{-1}	Chaff kg ha^{-1}	Grain kg ha^{-1}	LAI m^2 m^{-2}
Treatment N1							
840215	71	71					0.1
840314	104	104					0.2
840404	221	179		45		0.4	
840425	1027	737		291			1.6
840509	2548	1311	114	1124			3.1
840529	3860	1507	0	2353	0	0	3.6
840620	9823	1776	67	6429	1551	0	4.6
840704	11940	1675	89	7923	1671	582	3.7
840718	13790	1325	361	7287	2090	2727	2.8
840808	16145	526	987	5584	1916	7132	1.0
840822	16309	0		6934*	1940	7435	–
Treatment N2							
840529	3974	1513	0	2461	0	0	3.6
840620	10015	1954	48	6466	1547	0	4.8
840704	11992	1835	46	7619	1952	540	4.0
840718	13749	1527	314	7074	2131	2704	3.2
840808	18426	868	813	6372	2193	8181	1.9
840822	16926	0		7328*	1982	7617	–
Treatment N3							
840529	4108	1604	0	2504	0	0	3.8
840620	10005	1981	25	6416	1583	0	5.2
840704	11439	1704	103	7306	1772	553	3.7
840718	14330	1585	298	7355	2128	2964	3.4
840808	17290	910	738	5920	2078	7644	1.9
840822	16623	0		7139*	1963	7521	–

*) dead leaf + stem + sheaths

Table IV-6. Nitrogen content of dry matter and nitrogen distribution

Date	Total g/kg	Total kg/ha	Green leaf g/kg	Green leaf kg/ha	Dead leaf g/kg	Dead leaf kg/ha	Stem+sheaths g/kg	Stem+sheaths kg/ha	Grain g/kg	Grain kg/ha	Chaff g/kg	Chaff kg/ha
Treatment N1												
840215	49.3	3.5										
840314	54.2	5.6										
840404	57.5	12.7	57.8	10.4			51.7	2.3				
840425	43.1	44.3	46.5	34.3			34.5	10.0				
840509	26.2	66.6	33.5	43.9	23.5	2.7	17.8	20.0				
840529	25.9	99.9	38.5	58.0	0	0	17.8	58.0	0	0	0	0
840620	13.7	134.3	28.7	51.0	0	0	8.8	56.2	0	0	17.5	27.1
840704	11.1	132.4	25.8	43.2	10.8	1.0	7.5	59.6	19.0	11.0	10.6	17.7
840718	11.3	155.1	24.4	32.3	11.5	4.1	6.2	45.2	18.9	51.5	10.6	22.0
840808	9.4	152.4	12.8	6.7	7.3	7.2	1.9*	10.3*	16.6	118.2	5.2	10.0
840822	10.4	168.7	16.1	0.0			3.5*	24.3*	18.3	136.2	4.2	8.2
Treatment N2												
840529	25.7	102.1										
840620	16.0	160.2	54.0	105.4	0	0	1.3	8.4	0	0	30.0	46.3
840704	14.0	167.8	33.0	60.6	15.6	0.7	9.2	70.0	22.7	12.3	12.5	24.4
840718	13.3	182.2	26.9	41.1	13.4	4.2	7.7	54.2	21.3	57.7	11.7	25.0
840808	12.5	230.3	17.8	15.4	9.7	7.9	5.4*	34.4*	19.4	158.3	6.5	14.2
840822	12.1	205.0	16.1	0			4.7*	34.5*	21.1	160.5	5.1	10.0
Treatment N3												
840529	27.9	114.6										
840620	15.6	156.3	50.9	100.8	0	0	2.0	12.7	0	0	27.1	42.8
840704	13.6	155.1	32.5	55.3	13.2	1.4	9.2	66.9	21.8	12.0	11.0	19.5
840718	13.2	188.9	27.3	43.3	12.3	3.7	7.4	54.1	21.7	64.4	11.0	23.4
840808	11.3	164.7	16.1	14.7	8.8	6.5	4.9*	29.0*	17.3	132.1	6.0	12.4
840822	13.5	224.6	16.1	0			5.5*	39.0*	23.1	173.7	6.1	11.9

*) dead leaf + stem + sheaths

Table IV-7. Time course of groundwater table

Date	Depth (cm)	Date	Depth (cm)	Date	Depth (cm)	Date	Depth (cm)
840404	84	840509	137	840613	103	840718	98
840411	86	840516	137	840620	117	840725	106
840418	92	840523	136	840627	125	840808	135
840425	94	840530	108	840704	135	840815	143
840502	108	840606	80	840711	59	840822	150

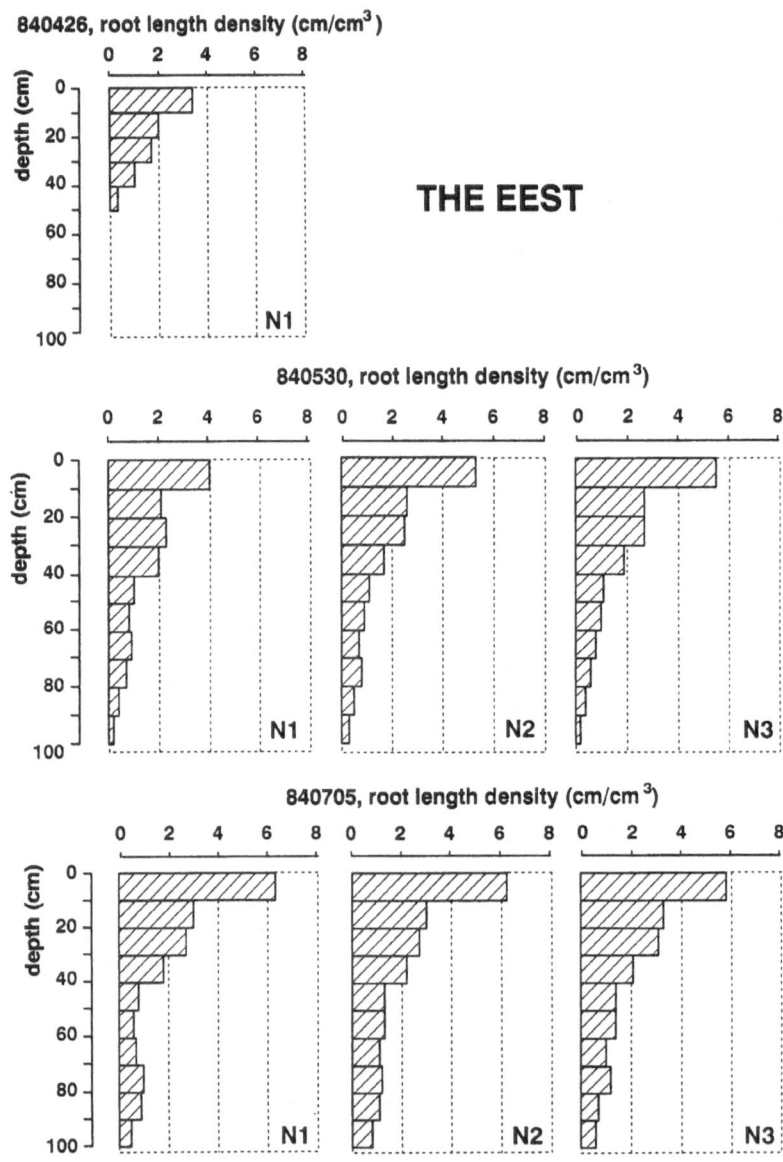

THE EEST

Fig. IV-1. Root length density as a function of depth in treatments N1, N2 and N3 at the experimental site the Eest, 1984.

V. PAGV, 1982–1983

Table V-1. Nitrogen treatments: time (yymmdd) and rate of application (kg ha^{-1}). Nitrogen was applied as Calcium Ammonium Nitrate

Treatment	830216	830510	830610
N1	80	0	0
N2	60	80	0
N3	60	140	40

Table V-2. Mineral nitrogen (ammonium + nitrate) content (kg ha^{-1}) of soil layers 0–30, 30–60 and 60–100 cm

Date	Treatment N1			Treatment N2			Treatment N3		
yymmdd	0–30	30–60	60–100	0–30	30–60	60–100	0–30	30–60	60–100
830208	9.0	12.3	25.0						
830301	10.9	15.1	20.9						
830329	9.9	12.3	31.4						
830419	7.4	15.5	21.0						
830510	7.3	5.9	17.0						
830524	6.1	2.8	10.2	8.9	2.4	6.8	24.4	2.4	10.8
830614	2.5	1.3	2.9	2.3	1.3	4.8	3.4	1.8	4.0
830705	1.6	0.9	2.3	1.6	1.1	2.7	9.8	3.0	4.2
830719	5.1	4.3	7.2	4.9	4.3	7.2	7.2	5.3	8.1
830802	7.7	3.9	5.9	5.7	3.7	5.0	6.7	4.1	5.6

Table V-3. Volumetric water content (cm^3 cm^{-3}) of soil layers 0–30, 30–60 and 60–100 cm

Date	Treatment N1			Treatment N2			Treatment N3		
yymmdd	0–30	30–60	60–100	0–30	30–60	60–100	0–30	30–60	60–100
830208	.356	.302	.397						
830301	.362	.399	.520						
830329	.334	.346	.422						
830419	.309	.321	.381						
830510	.344	.364	.394						
830524	.353	.325	.407	.355	.340	.415	.356	.319	.407
830614	.257	.286	.369	.243	.272	.378	.237	.272	.387
830705	.195	.190	.368	.186	.186	.368	.169	.185	.350
830719	.136	.179	.308	.126	.181	.316	.126	.182	.291
830802	.194	.198	.327	.191	.200	.330	.231	.187	.299

Table V-4. Decimal Codes (DC) for crop development according to Zadoks et al. (1974) Weed Res 14: 415–421

Date	DC	Date	DC	Date	DC	Date	DC
830208	22	830323	23	830426	25	830530	38
830215	22	830329	23	830503	30	830606	43
830222	22	830406	23	830509	31	830613	56
830301	22	830413	24	830516	31	830620	60
830308	22	830419	25	830524	38	830628	71
830316	22						

Table V-5. Dry matter distribution (kg ha^{-1}) and Leaf Area Index (LAI)

Date	Total kg ha^{-1}	Green leaf kg ha^{-1}	Dead leaf kg ha^{-1}	Stem+ sheaths kg ha^{-1}	Chaff kg ha^{-1}	Grain kg ha^{-1}	LAI m^2 m^{-2}
Treatment N1							
830208	101	83		18			0.2
830301	100	81		18			0.2
830329	272	212		60			0.5
830419	740	603		138			1.2
830510	2491	1062	51	1378			2.2
830524	4661	1543	69	3049			3.2
830614	8212	1857	202	6152			3.2
830705	12317	1016	486	7095	1715	2005	2.1
830719	13220	10	1168	4826	1505	5710	0.0
830802	13586	0	1164	4648	1518	6255	0.0
Treatment N2							
830524	4504	1768	53	2683			3.6
830614	8962	2120	182	6661			4.6
830705	13866	1293	513	7775	2191	2094	2.7
830719	15268	234	1193	5528	1946	6367	0.4
830802	16345	0	1454	5407	2041	7442	0.0
Treatment N3							
830524	4746	1959	74	2713			4.3
830614	9267	2468	176	6623			5.5
830705	15180	1721	560	8066	2435	2398	3.5
830719	16842	735	1055	5968	2140	6944	1.3
830802	17776	0	1593	5761	2143	8279	0.0

Table V-6. Nitrogen content of dry matter and nitrogen distribution

Date	g/kg	Total kg/ha	Green leaf g/kg	kg/ha	Dead leaf g/kg	kg/ha	Stem+sheaths g/kg	kg/ha	g/kg	Grain kg/ha	g/kg	Chaff kg/ha
Treatment N1												
830208	47.4	4.8										
830301	49.7	5.0										
830329	49.3	13.4										
830419	47.9	35.5										
830510	23.4	58.2	33.4	35.4	0	0	16.5	22.7	0	0	0	0
830524	16.4	76.4	29.0	44.7	0	0	10.4	31.7	0	0	0	0
830614	12.6	103.3	30.0	55.7	31.2	6.3	6.7	41.3	0	0	0	0
830705	6.6	81.3	15.4	15.6	6.2	3.0	2.9	20.7	13.6	27.3	8.5	14.6
830719	4.7	62.0	0	0	5.4	6.3	1.2	5.6	8.1	46.0	2.7	4.1
830802	7.8	105.4	12.8	0	5.5	6.4	1.8	8.2	13.6	84.9	3.9	5.9
Treatment N2												
830524	24.2	109.1	38.3	67.8	0	0	15.4	41.4	0	0	0	0
830614	15.5	139.3	35.8	75.8	16.4	3.0	9.1	60.5	0	0	0	0
830705	10.3	142.3	22.9	29.6	10.7	5.5	5.0	38.7	21.5	45.0	10.7	23.5
830719	5.8	89.1	6.2	1.5	4.7	5.6	1.8	9.8	10.3	65.7	3.3	6.5
830802	10.0	163.5	12.8	0	6.5	9.4	2.0	10.6	18.2	135.3	4.0	8.1
Treatment N3												
830524	29.2	138.8	42.2	82.6	0	0	20.7	56.1	0	0	0	0
830614	19.0	176.1	39.6	97.6	21.1	3.7	11.3	74.7	0	0	0	0
830705	11.8	179.6	27.0	46.5	12.6	7.1	5.8	46.8	22.6	54.1	10.3	25.0
830719	10.5	176.5	12.8	9.4	8.2	8.6	3.6	21.7	17.3	120.4	7.6	16.3
830802	13.0	231.1	12.8	0	8.5	13.5	2.4	14.1	23.5	194.8	4.1	8.8

VI. PAGV, 1983–1984

Table VI-1. Nitrogen treatments: time (yymmdd) and rate of application (kg ha^{-1}). Nitrogen was applied as Calcium Ammonium Nitrate

Treatment	840217	840514	840608
N1	80	0	0
N2	80	60	40
N3	80	120	40

Table VI-2. Mineral nitrogen (ammonium + nitrate) content (kg ha^{-1}) of soil layers 0–20, 20–30, 30–40, 40–60, 60–80 and 80–100 cm

Date	Treatment N1						Treatment N2						Treatment N3					
yymmdd	0–20	20–30	30–40	40–60	60–80	80–100	0–20	20–30	30–40	40–60	60–80	80–100	0–20	20–30	30–40	40–60	60–80	80–100
ammonium+nitrate																		
831103	9	5	8	15	7	7												
840118	8	4	7	14	19	19												
840215	0	1	0	0	7	8												
840314	74	14	4	7	9	16												
840403	63	19	5	11	12	14												
840425	31	18	4	11	15	22												
840508	12	7	0	4	8	9												
840529	1	0	0	0	0	4	5	1	0	0	1	4	18	9	1	1	3	7
840620	8	4	1	6	5	7	12	5	3	4	7	3	22	7	1	4	4	4
840704	1	0	0	0	0	0	1	1	0	1	3	3	5	2	0	3	1	4
840719	4	3	0	0	0	1	5	3	0	0	0	1	8	3	1	0	3	3
840808	4	2	1	3	3	3	5	3	1	4	3	3	7	3	2	3	3	4

Table VI-3. Volumetric water content (cm^3 cm^{-3}) of soil layers 0–20, 20–30, 30–40, 40–60, 60–80 and 80–100 cm

Date	Treatment N1						Treatment N2						Treatment N3					
yymmdd	0–20	20–30	30–40	40–60	60–80	80–100	0–20	20–30	30–40	40–60	60–80	80–100	0–20	20–30	30–40	40–60	60–80	80–100
840215	.322	.321	.331	.354	.415	.440												
840314	.329	.327	.318	.357	.390	.435												
840403	.306	.317	.315	.372	.400	.412												
840425	.240	.242	.307	.343	.385	.417												
840508	.207	.248	.279	.307	.374	.417												
840529	.325	.300	.285	.331	.399	.429	.317	.298	.307	.365	.399	.431	.321	.304	.307	.348	.396	.430
840620	.273	.267	.279	.333	.385	.422	.263	.277	.281	.333	.379	.399	.259	.280	.304	.318	.359	.394
840704	.311	.284	.267	.285	.342	.408	.288	.257	.265	.295	.352	.403	.284	.251	.243	.299	.342	.394
840719	.325	.304	.301	.314	.359	.417	.319	.317	.287	.309	.370	.399	.306	.296	.267	.309	.381	.408
840808	.261	.248	.271	.289	.342	.412	.248	.240	.243	.291	.351	.383	.240	.242	.217	.259	.339	.390

Table VI-4. Decimal Codes (DC) for crop development according to Zadoks et al. (1974) Weed Res 14: 415–421

Date	DC	Date	DC	Date	DC	Date	DC
840207	16	840326	21	840507	31	840618	55
840213	18	840402	21	840514	32	840625	59
840220	14	840410	22	840521	33	840702	63
840227	18	840419	23	840528	37	840709	72
840305	18	840424	24	840604	42	840716	74
840312	19	840501	28	840612	45	840723	77
840319	22	840503	27				

Table VI-5. Dry matter distribution (kg ha^{-1}) and Leaf Area Index (LAI)

Date	Total kg ha^{-1}	Green leaf kg ha^{-1}	Dead leaf kg ha^{-1}	Stem+ sheaths kg ha^{-1}	Chaff kg ha^{-1}	Grain kg ha^{-1}	LAI m^2 m^{-2}
Treatment N1							
840214	50						0.1
840313	79						0.1
840403	164	126	38				0.3
840424	850	625	226				0.4
840508	2574	1432	1065				0.5
840528	4648	1589	15	3044	0	0	3.7
840619	9376	1490	146	6528	1213	0	3.4
840703	11197	1189	373	7803	1415	419	1.9
840717	13461	956	525	7808	1593	2579	1.9
840808	17549	184	1284	6567	1904	7610	0.3
840820	15902	0		6883*	1656	7363	0.0
Treatment N2							
840528	4859	1756	10	3093	0	0	4.1
840619	9143	1701	141	6144	1156	0	3.6
840703	12249	1603	374	8220	1631	421	2.9
840717	14474	1304	537	8094	1911	2628	2.7
840808	18558	652	843	6820	2102	8141	1.0
840820	17159	0		7288*	1842	8028	0
Treatment N3							
840528	4875	1848	6	3020	0	0	4.4
840619	9516	1895	166	6286	1170	0	4.3
840703	12876	1753	469	8608	1675	372	3.0
840717	14051	1366	581	7883	1934	2287	3.0
840808	19024	958	814	7281	2250	7722	1.6
840820	17133	0		7522*	1919	7692	0

*) dead leaf + stem + sheaths

Table VI-6. Nitrogen content of dry matter and nitrogen distribution

Date	g/kg	Total kg/ha	Green leaf g/kg	kg/ha	Dead leaf g/kg	kg/ha	Stem+sheaths g/kg	kg/ha	Grain g/kg	kg/ha	Chaff g/kg	kg/ha
Treatment N1												
840313	53.4	4.2										
840403	56.2	9.2	58.9	7.4			51.2	1.9				
840424	49.3	41.9	55.2	34.5			41.8	9.4				
840508	27.3	70.3	35.8	51.3			20.1	21.4				
840528	21.6	100.2	36.0	57.2	0	0	14.1	43.0	0	0	0	0
840619	11.4	107.1	26.6	39.6	10.0	1.5	7.3	47.7	0	0	15.2	18.5
840703	8.5	95.4	21.9	26.0	12.6	4.7	4.9	38.3	23.9	10.0	11.6	16.4
840717	9.6	129.7	23.3	22.2	9.5	5.0	5.0	39.0	18.8	48.4	9.4	15.0
840808	7.6	133.4	9.3	1.7	5.9	7.6	1.6[*]	10.2[*]	14.0	106.8	3.8	7.2
840820	11.5	182.1	0	0			2.7[*]	18.6[*]	21.1	155.3	5.0	8.2
Treatment N2												
840528	25.4	123.4	39.9	70.2	0	0	17.2	53.2	0	0	0	0
840619	14.4	131.6	31.0	52.7	12.0	1.7	9.6	58.8	0	0	16.0	18.5
840703	12.4	151.2	30.5	48.9	12.1	4.5	8.0	65.3	27.0	11.4	13.0	21.1
840717	11.2	162.1	25.2	32.9	9.4	5.1	6.6	53.7	20.0	52.6	9.4	17.9
840808	10.1	187.8	13.6	8.9	10.4	8.7	3.6[*]	24.8[*]	16.6	135.5	4.7	9.9
840820	11.2	191.9	0	0			3.5[*]	25.4[*]	19.7	158.1	4.6	8.4
Treatment N3												
840528	28.4	138.3	41.5	76.8	0	0	20.4	61.5	0	0	0	0
840619	17.2	163.4	31.7	60.1	34.1	5.7	12.2	76.7	0	0	18.0	21.0
840703	13.7	177.0	33.1	57.9	12.2	5.7	9.3	80.0	27.7	10.3	13.8	23.0
840717	13.2	185.3	28.7	39.2	11.5	6.7	8.3	65.1	22.8	52.2	11.5	22.2
840808	12.3	234.0	19.4	18.5	8.8	7.2	5.7[*]	41.5[*]	19.7	152.4	6.4	14.4
840820	12.9	220.1	0	0			4.4[*]	32.8[*]	23.0	176.7	5.5	10.6

[*]) dead leaf + stem +sheaths

Table VI-7. Time course of groundwater table

Date	Depth (cm)	Date	Depth (cm)	Date	Depth (cm)	Date	Depth (cm)
840312	86	840418	102	840521	160	840625	150
840319	96	840424	111	840528	133	840702	161
840326	99	840501	126	840604	116	840709	173
840402	95	840507	136	840612	117	840716	134
840410	95	840514	150	840618	134	840723	135

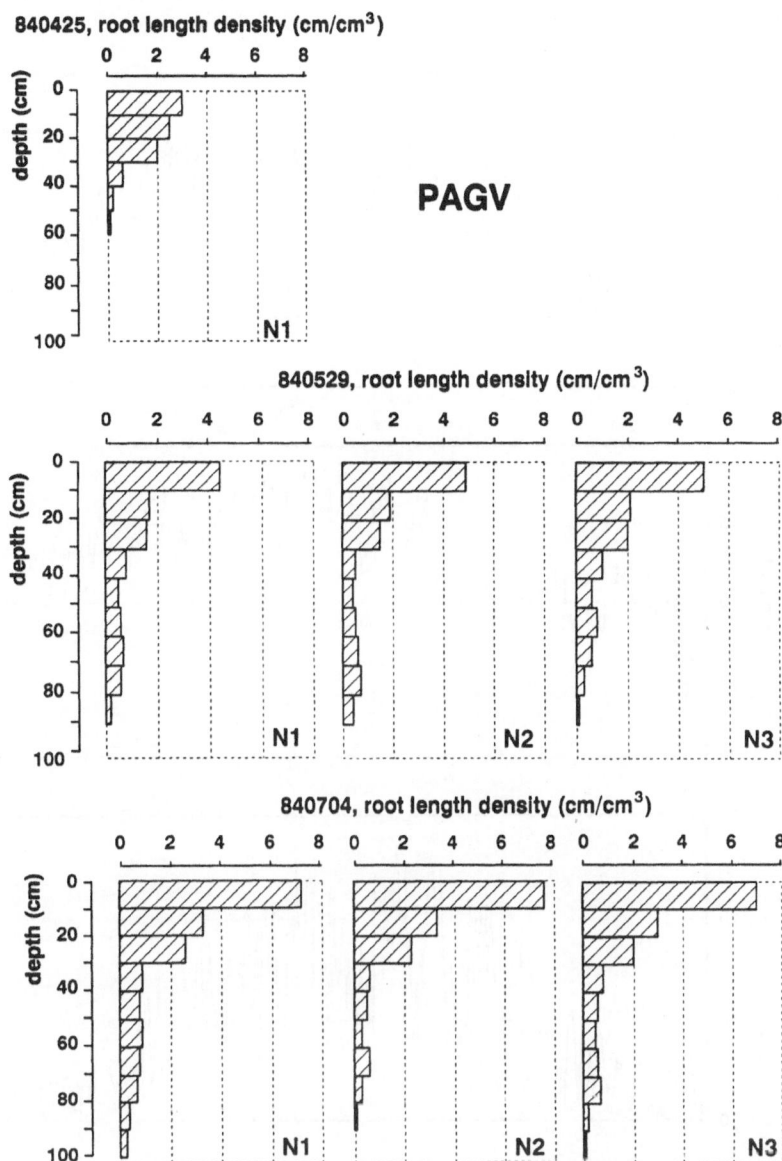

Fig. VI-1. Root length density as a function of depth in treatments N1, N2 and N3 at the experimental site PAGV, 1984.

378

VII. Weather Data

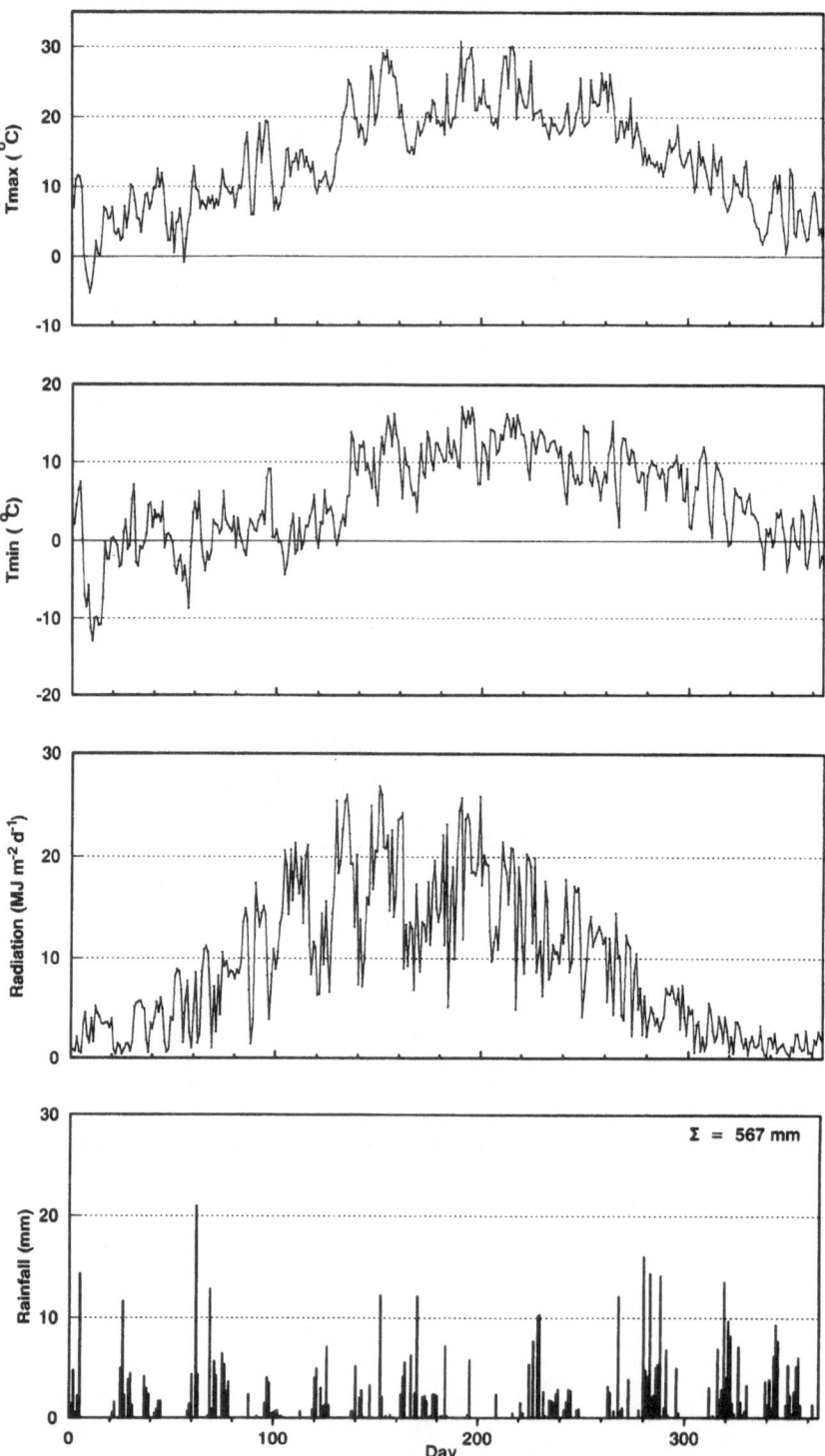

Fig. VII-1. Annual course of maximum temperature, minimum temperature, total global radiation and cumulative rainfall for meteo-station Wageningen, 1982.

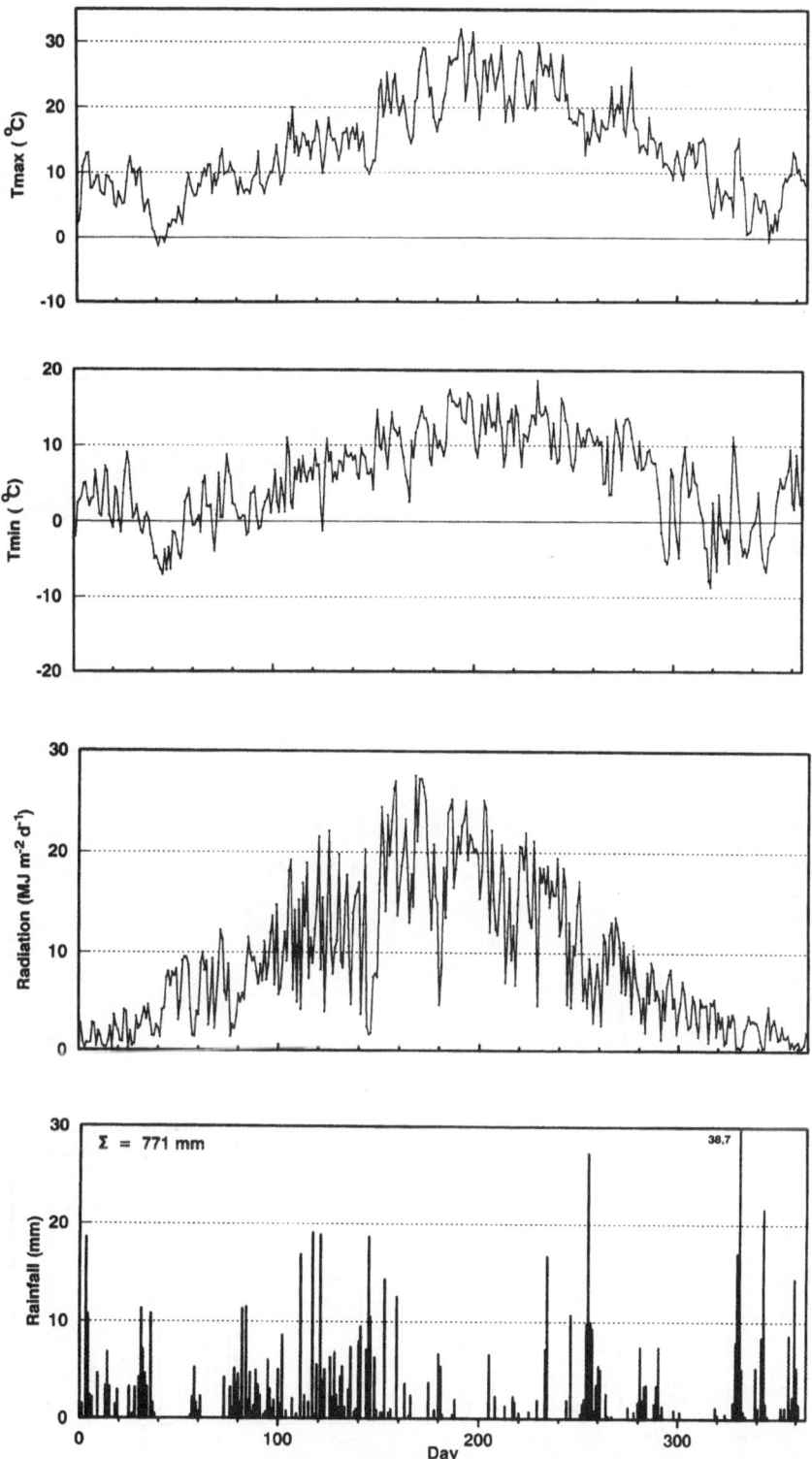

Fig. VII-2. Annual course of maximum temperature, minimum temperature, total global radiation and cumulative rainfall for meteo-station Swifterbant, 1982.

380

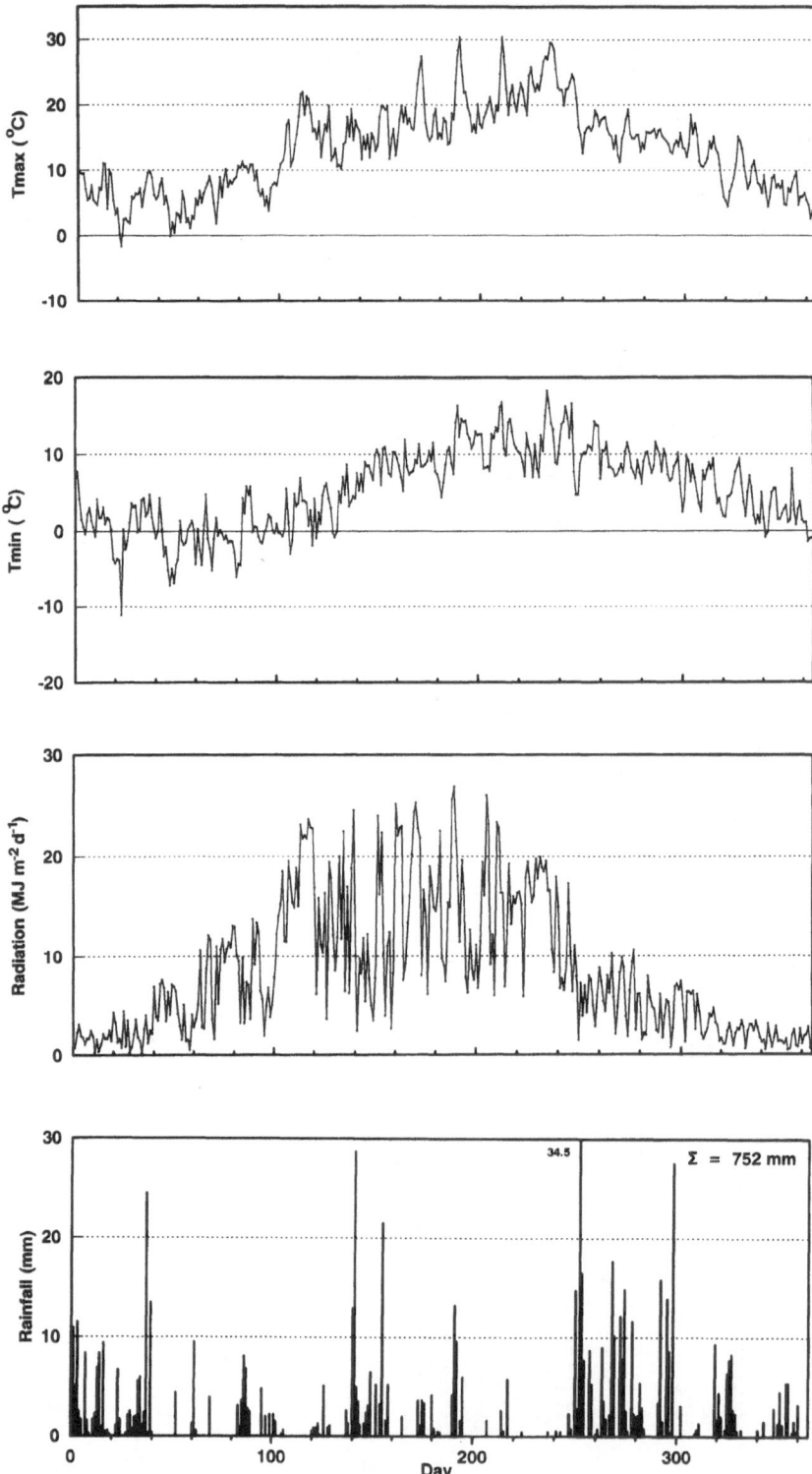

Fig. VII-3. Annual course of maximum temperature, minimum temperature, total global radiation and cumulative rainfall for meteo-station Wageningen, 1983.

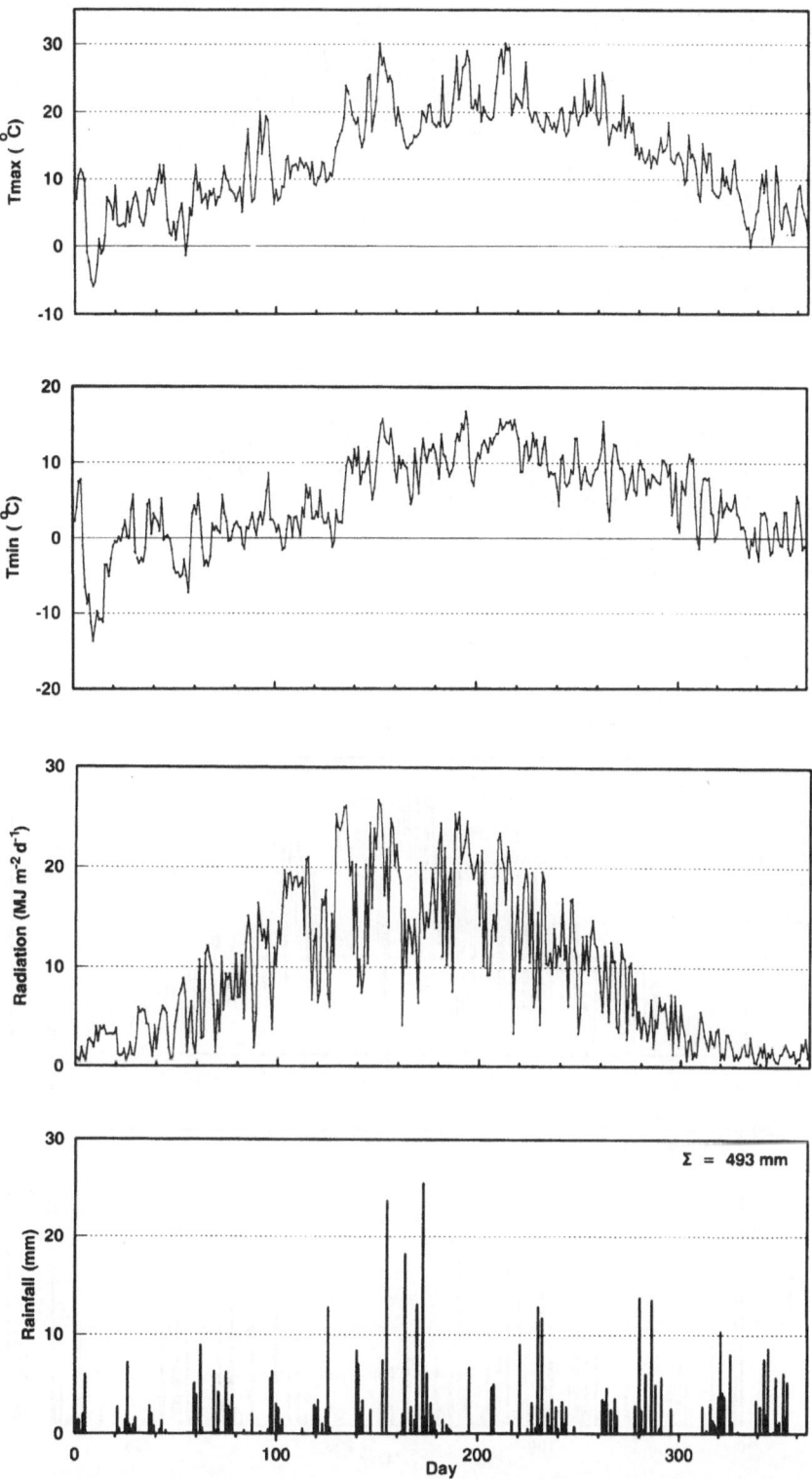

Fig. VII-4. Annual course of maximum temperature, minimum temperature, total global radiation and cumulative rainfall for meteo-station Swifterbant, 1983.

382

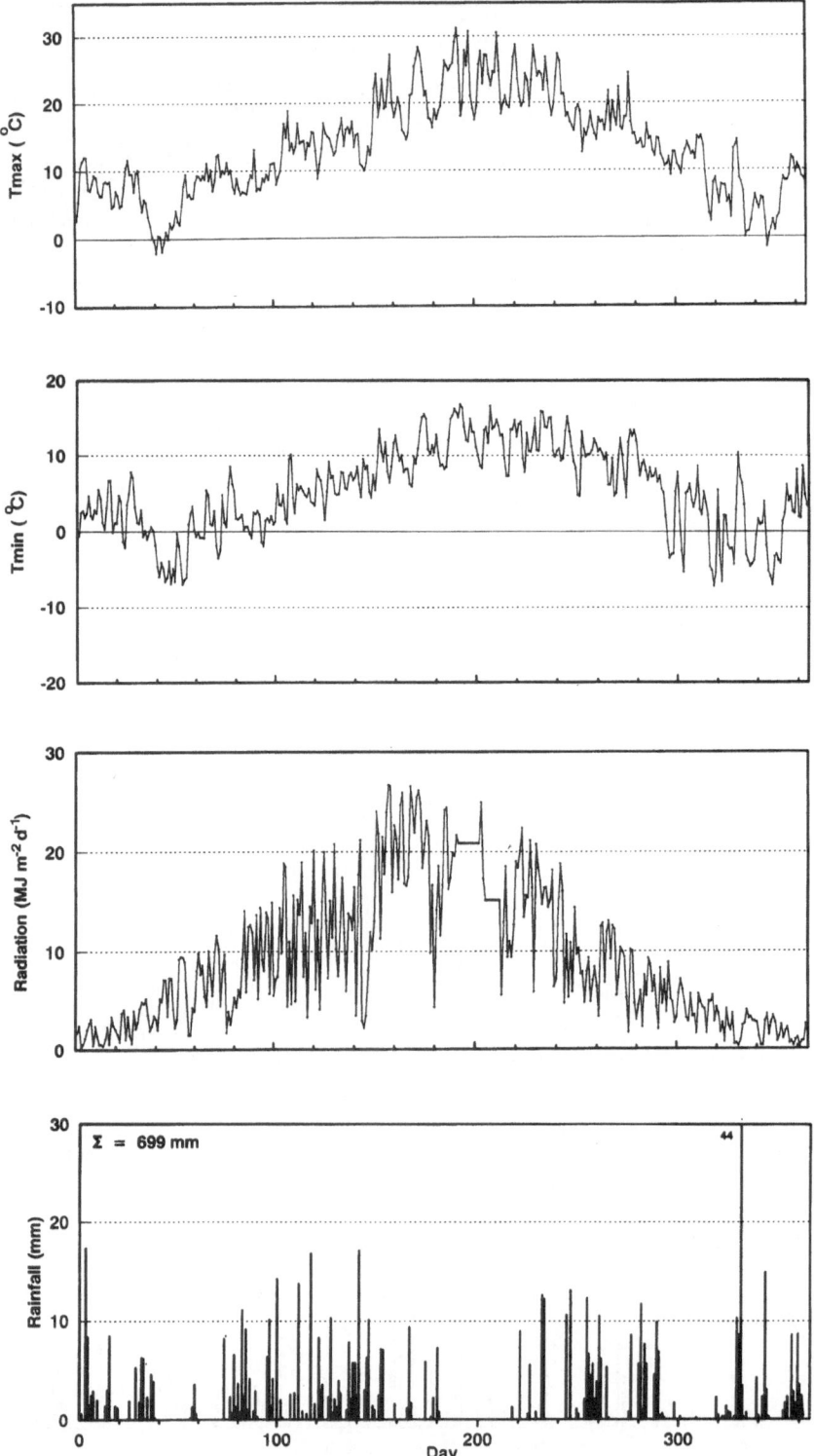

Fig. VII-5. Annual course of maximum temperature, minimum temperature, total global radiation and cumulative rainfall for meteo-station Wageningen, 1984.

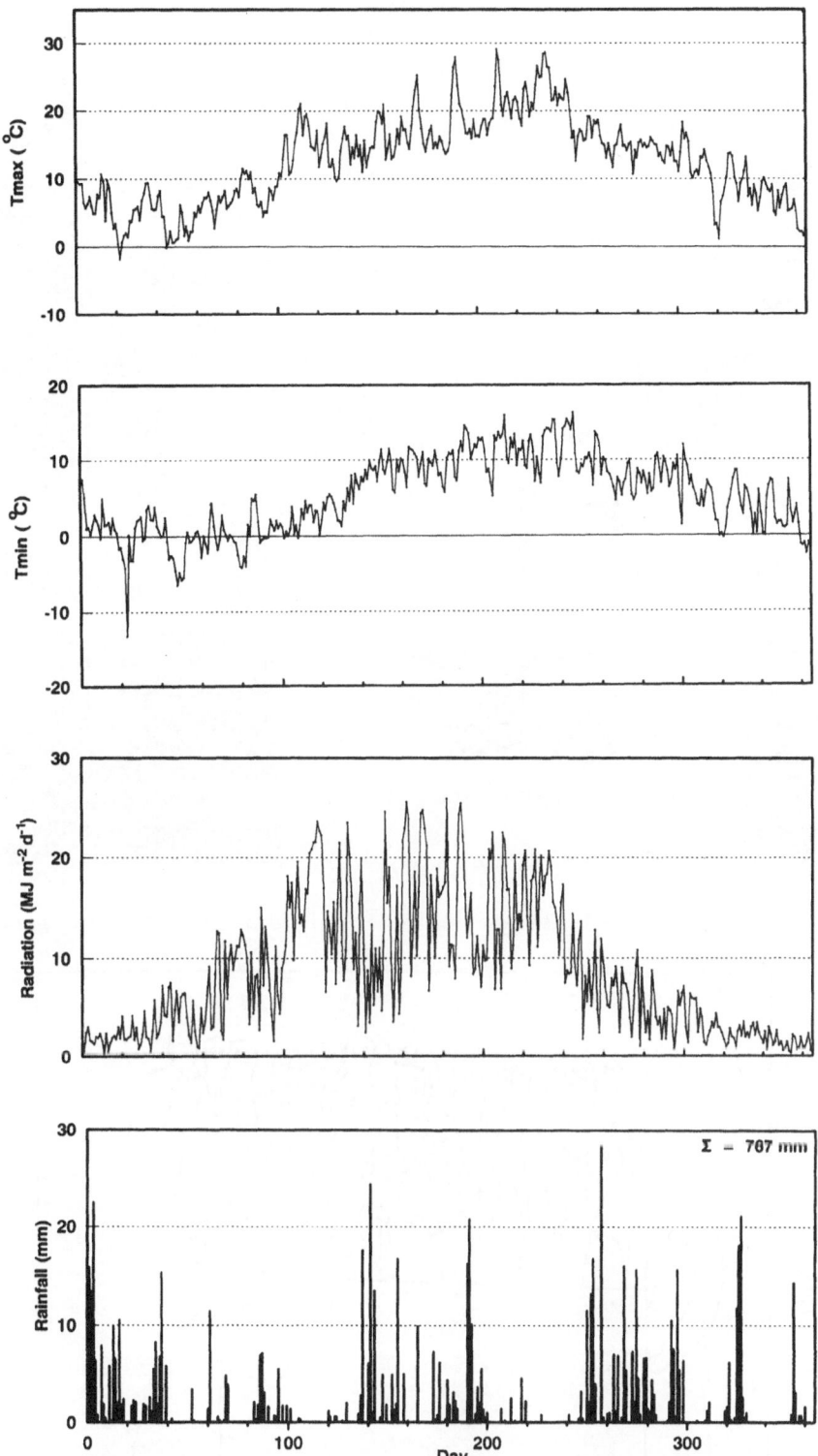

Fig. VII-6. Annual course of maximum temperature, minimum temperature, total global radiation and cumulative rainfall for meteo-station Swifterbant, 1984.

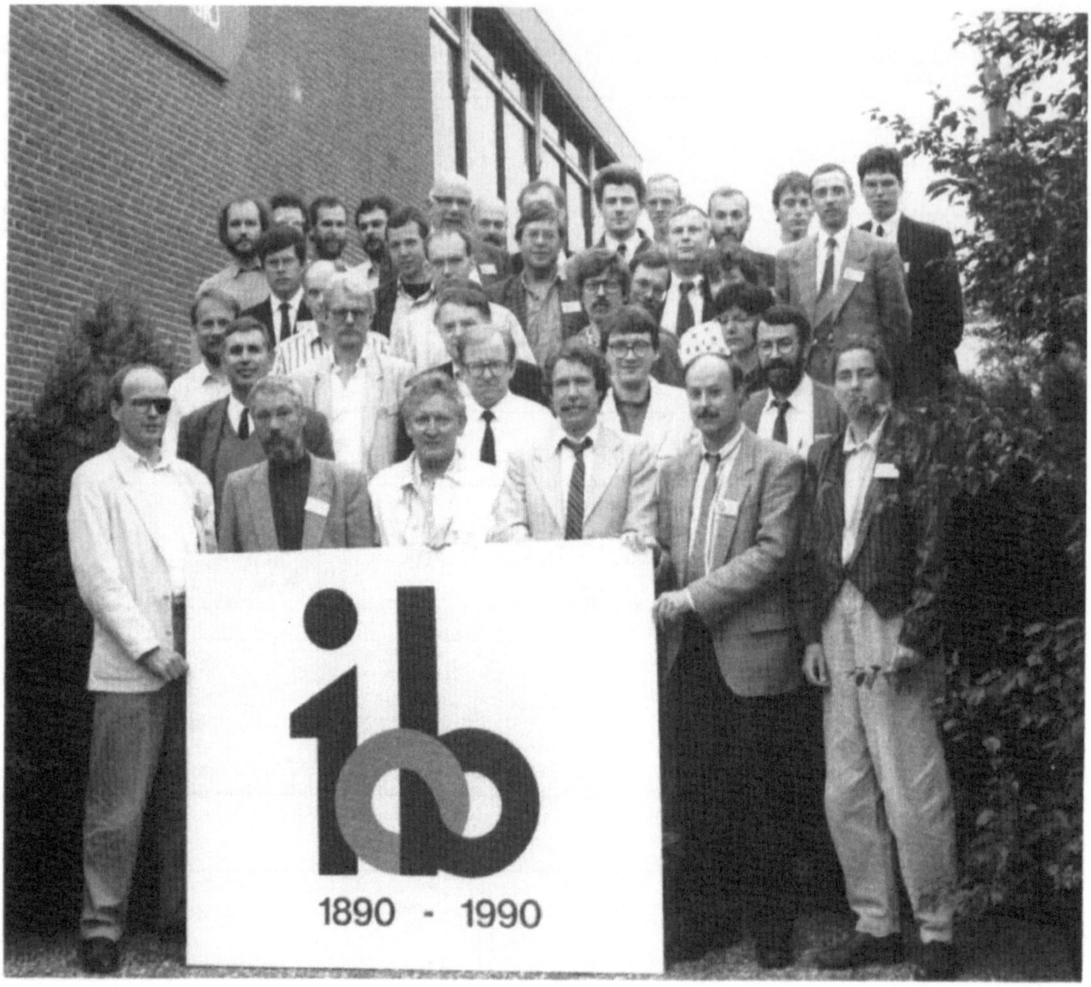
1890 - 1990

1
2
3
4
5
6
7
8
9
10
11
12
13
14
15
16
17
18
19
20
21
22
23
24
25
26
27
28
29
30
31
32
33
34
35
36

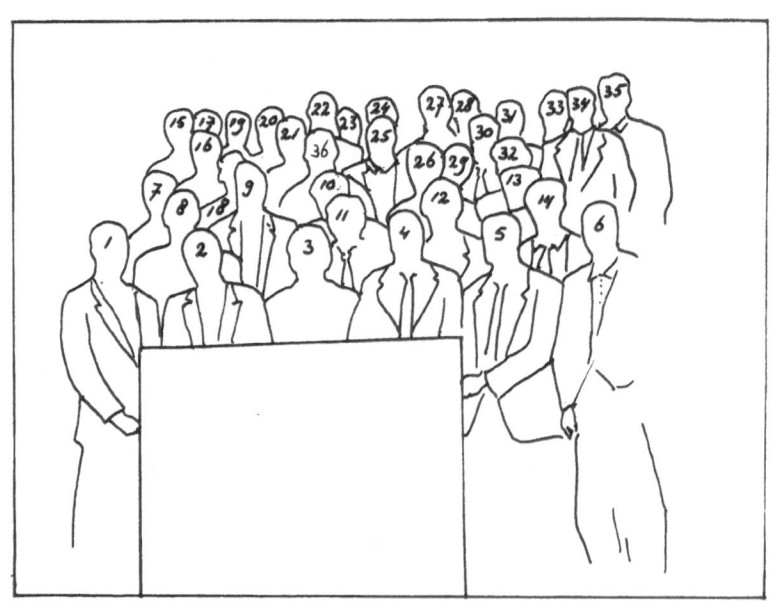

Fertilizer Research **27**: 385–386, 1991.

List of participants

The numbers correspond to numbers on the picture

1. J.J.R. Groot, Institute for Soil Fertility Research, P.O. Box 30003, 9750 RA Haren (Gn.), The Netherlands
2. H. van Keulen, Centre for Agrobiological Research (CABO), P.O. Box 14, 6700 AA Wageningen, The Netherlands
3. N.E. Nielsen, Section of Soil and Water and Plant Nutrition, Department of Agricultural Sciences, The Royal Veterinary and Agricultural University, Thorvaldsensvej 40, DK-1871 Frederiksberg C, Copenhagen, Denmark
4. R.F. Grant, Department of Soil Science, University of Alberta, Edmonton, Alberta, Canada T6G 2E3
5. J.J. Neeteson, Institute for Soil Fertility Research, P.O. Box 30003, 9750 RA Haren (Gn.), The Netherlands
6. E.L.J. Verberne, Institute for Soil Fertility Research, P.O. Box 30003, 9750 RA Haren (Gn.), The Netherlands
7. S. Hansen, Section of Soil and Water and Plant Nutrition, Department of Agricultural Sciences, The Royal Veterinary and Agricultural University, Thorvaldsensvej 40, DK-1871 Frederiksberg C, Copenhagen, Denmark
8. R.R. van der Ploeg, Institute of Soil Science, Hohenheim University, P.C. Box 700562, DW-7000 Stuttgart-70, Germany
9. L. Bergström, Swedish University of Agricultural Sciences, Division of Water Management, P.O. Box 7072, S-75007 Uppsala, Sweden
10. P. Döring, Akademie der Landwirtschaftswissenschaften der DDR, Postfach 1295, DO-1086 Berlin, Germany
11. H.E. Jensen, Section of Soil and Water and Plant Nutrition, Department of Agricultural Sciences, The Royal Veterinary and Agricultural University, Thorvaldsensvej 40, DK-1871 Frederiksberg C, Copenhagen, Denmark
12. T. Kartschall, Akademie der Landwirtschaftswissenschaften der DDR, Postfach 1295, DO-1086, Berlin, Germany
13. M. Swerts, Laboratory of Land Management, Katholieke Universiteit Leuven, Kardinaal Mercierlaan 92, B3030 Belgium
14. C. Ramos, Instituto Valenciano de Investigaciones Agrarias, Apdo. Oficial, 46113 Moncada, Spain
15. K.C. Kersebaum, Institute of Geography & Geoecology, Technical University of Braunschweig, Langer Kamp 19c, DW-3300 Braunschweig, Germany
16. F. Lafolie, Institut National de la Recherche Agronomique, Station de Science du Sol, Domaine St. Paul, B.P. 91, 84143 Montfavet, France
17. J. Richter, Institute of Geography & Geoecology, Technical University of Braunschweig, Langer Kamp 19c, DW-3300 Braunschweig, Germany
18. G. Torstensson, Swedish University of Agricultural Sciences, Division of Water Management, P.O. Box 7072, S-75007 Uppsala, Sweden
19. P.-E. Jansson, Department of Soil Sciences, Swedish University of Agricultural Sciences, P.O. Box 7014, S-75007 Uppsala, Sweden
20. B. Huwe, Institute of Soil Science, Hohenheim University, P.O. Box 700562, DW-7000 Stuttgart-70, Germany
21. J.G. Kroes, The Winand Staring Centre for Integrated Land, Soil and Water Research, P.O. Box 125, 6700 AC Wageningen, The Netherlands
22. P.E. Rijtema, The Winand Staring Centre for Integrated Land, Soil and Water Research, P.O. Box 125, 6700 AC Wageningen, The Netherlands
23. T.M. Addiscott, AFRC Institute of Arable Crops Research, Rothamsted Experimental Station, Harpenden, Hertfordshire AL5 2JQ, United Kingdom
24. P. van Erp, Dutch Fertilizer Research Institute, P.O. Box 30003, 9750 RA Haren (Gn.), The Netherlands
25. P. de Willigen, Institute for Soil Fertility Research, P.O. Box 30003, 9750 RA Haren (Gn.), The Netherlands
26. H. Svendsen, Section of Soil and Water and Plant Nutrition, Department of Agricultural Sciences, The Royal Veterinary and Agricultural University, Thorvaldsensvej 40, DK-1871

Frederiksberg C, Copenhagen, Denmark

27. F. Cabon, Centre d'Informatique Géologique, Ecole Nationale Supérieure des Mines de Paris, 35, Rue Saint-Honoré, 77305 Fontainebleau, France

28. W. Stol, Centre for Agrobiological Research (CABO), P.O. Box 14, 6700 AA Wageningen, The Netherlands

29. A.P. Whitmore, AFRC Institute of Arable Crops Research, Rothamsted Experimental Station, Harpenden, Hertfordshire AL5 2JQ, United Kingdom

30. E. Ledoux, Centre d'Informatique Géologique, Ecole Nationale Supérieure des Mines de Paris, 35, Rue Saint-Honoré, 77305 Fontainebleau, France

31. W. Mirschel, Research Centre of Soil Fertility Müncheberg, Academy of Agricultural Sciences of the GDR, DO-1278 Müncheberg, Wilhelm-Pieck-Strasse 72, Germany

32. N.J. Bradbury, AFRC Institute of Arable Crops Research, Rothamsted Experimental Station, Harpenden, Hertfordshire AL5 2JQ, United Kingdom

33. M. VanClooster, Laboratory of Land Manage-

ment, Katholieke Universiteit Leuven, Kardinaal Mercierlaan 92, B3030 Belgium

34. H. Kretschmer, Research Centre of Soil Fertility Müncheberg, Academy of Agricultural Sciences of the GDR, DO-1278 Müncheberg, Wilhelm-Pieck-Strasse 72, Germany

35. H. Vereecken, Laboratory of Land Management, Katholieke Universiteit Leuven, Kardinaal Mercierlaan 92, B3030 Belgium

36. H. Johnson, Swedish University of Agricultural Sciences, Division of Water Management, P.O. Box 7072, S-75007 Uppsala, Sweden

Not on the picture:

37. S. Otter-Nacke, Norsk Hydro Agrar, Landwirtschaftliche Forschung Hanninghof, Hanninghof 35, DW-4408 Dülmen, Germany

38. P.F.J. van Burg†, late editor-in-chief Fertilizer Research

39. D.C. Hardwick, Fertiliser Manufacturers Association, Greenhill House, Thorpe Wood, Peterborough, PE3 6GF, United Kingdom

40. H. Eckersten, Department of Soil Sciences, Swedish University of Agricultural Sciences, P.O. Box 7014, S-75007 Uppsala, Sweden

Developments in Plant and Soil Sciences

1. J. Monteith and C. Webb (eds.): *Soil Water and Nitrogen in Mediterranean-type Environments*. 1981
   ISBN 90-247-2406-6
2. J. C. Brogan (ed.): *Nitrogen losses and Surface Run-off from Landspreading of Manures*. 1981
   ISBN 90-247-2471-6
3. J. D. Bewley (ed.): *Nitrogen and Carbon Metabolism*. 1981 ISBN 90-247-2472-4
4. R. Brouwer, I. Gašparíková, J. Kolek and B. C. Loughman (eds.): *Structure and Function of Plant Roots*. 1981 ISBN 90-247-2510-0
5. Y. R. Dommergues and H. G. Diem (eds.): *Microbiology of Tropical Soils and Plant Productivity*. 1982
   ISBN 90-247-2624-7
6. G. P. Robertson, R. Herrara and T. Rosswall (eds.): *Nitrogen Cycling in Ecosystems of Latin America and the Caribbean*. 1982 ISBN 90-247-2719-7
7. D. Atkinson, K. K. S. Bhat, M. P. Coutts, P. A. Mason and D. J. Read (eds.): *Tree Root Systems and Their Mycorrhizas*. 1983 ISBN 90-247-2821-5
8. M. R. Sarić and B. C. Loughman (eds.): *Genetic Aspects of Plant Nutrition*. 1983 ISBN 90-247-2822-3
9. J. R. Freney and J. R. Simpson (eds.): *Gaseous Loss of Nitrogen From Plant-Soil Systems*. 1983
   ISBN 90-247-2820-7
10. United Nations Economic Commission for Europe (ed.): *Efficient Use of Fertilizers in Agriculture*. 1983
    ISBN 90-247-2866-5
11. J. Tinsley and J. F. Darbyshire (eds.): *Biological Processes and Soil Fertility*. 1984 ISBN 90-247-2902-5
12. A. D. L. Akkermans, D. Baker, K. Huss-Danell and J. D. Tjepkema (eds.): Frankia *Symbioses*. 1984
    ISBN 90-247-2967-X
13. W. S. Silver and E. C. Schröder (eds.): *Practical Application of* Azolla *for Rice Production*. 1984
    ISBN 90-247-3068-6
14. P. G. L. Vlek (ed.): *Micronutrients in Tropical Food Crop Production*. 1985 ISBN 90-247-3085-6
15. T. P. Hignett (ed.): *Fertilizer Manual*. 1985 ISBN 90-247-3122-4
16. D. Vaughan and R. E. Malcolm (eds.): *Soil Organic Matter and Biological Activity*. 1985
    ISBN 90-247-3154-2
17. D. Pasternak and A. San Pietro (eds.): *Biosalinity in Action*. Bioproduction with Saline Water. 1985
    ISBN 90-247-3159-3
18. M. Lalonde, C. Camiré and J. O. Dawson (eds.): Frankia *and Actinorhizal Plants*. 1985
    ISBN 90-247-3214-X
19. H. Lambers, J. J. Neeteson and I. Stulen (eds.): *Fundamental, Ecological and Agricultural Aspects of Nitrogen Metabolism in Higher Plants*. 1986 ISBN 90-247-3258-1
20. M. B. Jackson (ed.): *New Root Formation in Plants and Cuttings*. 1986 ISBN 90-247-3260-3
21. F. A. Skinner and P. Uomala (eds.): *Nitrogen Fixation with Non-Legumes* (Proceedings of the 3rd Symposium, Helsinki, 1984). 1986 ISBN 90-247-3283-2
22. A. Alexander (ed.): *Foliar Fertilization*. 1986 ISBN 90-247-3288-3
23. H. G. v.d. Meer, J. C. Ryden and G. C. Ennik (eds.): *Nitrogen Fluxes in Intensive Grassland Systems*. 1986
    ISBN 90-247-3309-X
24. A. U. Mokwunye and P. L. G. Vlek (eds.): *Management of Nitrogen and Phosphorus Fertilizers in Sub-Saharan Africa*. 1986 ISBN 90-247-3312-X
25. Y. Chen and Y. Avnimelech (eds.): *The Role of Organic Matter in Modern Agriculture*. 1986
    ISBN 90-247-3360-X
26. S. K. De Datta and W. H. Patrick Jr. (eds.): *Nitrogen Economy of Flooded Rice Soils*. 1986
    ISBN 90-247-3361-8
27. W. H. Gabelman and B. C. Loughman (eds.): *Genetic Aspects of Plant Mineral Nutrition*. 1987
    ISBN 90-247-3494-0
28. A. van Diest (ed.): *Plant and Soil: Interfaces and Interactions*. 1987 ISBN 90-247-3535-1

Developments in Plant and Soil Sciences

29. United Nations Economic Commission for Europe and FAO (eds.): *The Utilization of Secondary and Trace Elements in Agriculture*. 1987   ISBN 90-247-3546-7
30. H. G. v.d. Meer, R. J. Unwin, T. A. van Dijk and G. C. Ennik (eds.): *Animal Manure on Grassland and Fodder Crops*. Fertilizer or Waste? 1987   ISBN 90-247-3568-8
31. N. J. Barrow: *Reactions with Variable-Charge Soils*. 1987   ISBN 90-247-3589-0
32. D. P. Beck and L. A. Materon (eds.): *Nitrogen Fixation by Legumes in Mediterranean Agriculture*. 1988
                                                                                      ISBN 90-247-3624-2
33. R. D. Graham, R. J. Hannam and N. C. Uren (eds.): *Manganese in Soils and Plants*. 1988
                                                                                      ISBN 90-247-3758-3
34. J. G. Torrey and J. L. Winship (eds.): *Applications of Continuous and Steady-State Methods to Root Biology*. 1989   ISBN 0-7923-0024-6
35. F. A. Skinner, R. M. Boddey and I. Fendrik (eds.): *Nitrogen Fixation with Non-Legumes* (Proceedings of the 4th Symposium, Rio de Janeiro, 1987). 1989   ISBN 0-7923-0059-9
36. B. C. Loughman, O. Gašparíková and J. Kolek (eds.): *Structural and Functional Aspects of Transport in Roots*. 1989   ISBN 0-7923-0060-2; Pb 0-7923-0061-0
37. P. Planquaert and R. Haggar (eds.): *Legumes in Farming Systems*. 1990   ISBN 0-7923-0134-X
38. A. E. Osman, M. M. Ibrahim and M. A. Jones (eds.): *The Role of Legumes in the Farming Systems of the Mediterranean Areas*. 1990   ISBN 0-7923-0419-5
39. M. Clarholm and L. Bergström (eds.): *Ecology of Arable Land – Perspectives and Challenges*. 1989
                                                                                      ISBN 0-7923-0424-1
40. J. Vos, C. D. van Loon and G. J. Bollen (eds.): *Effects of Crop Rotation on Potato Production in the Temperate Zones*. 1989   ISBN 0-7923-0495-0
41. M. L. van Beusichem (ed.): *Plant Nutrition – Physiology and Applications*. 1990   ISBN 0-7923-0740-2
42. N. El Bassam, M. Dambroth and B.C. Loughman (eds.): *Genetic Aspects of Plant Mineral Nutrition*. 1990
                                                                                      ISBN 0-7923-0785-2
43. Y. Chen and Y. Hadar (eds.): *Iron Nutrition and Interactions in Plants*. 1991   ISBN 0-7923-1095-0
44. J. J. R. Groot, P. de Willigen and E. L. J. Verberne (eds.): *Nitrogen Turnover in the Soil-Crop System*. 1991
                                                                                      ISBN 0-7923-1107-8

Kluwer Academic Publishers – Dordrecht / Boston / London